Participatory Land Planning: Theory, Methods, and Case Studies

Participatory Land Planning: Theory, Methods, and Case Studies

Guest Editors

Oswaldo Ledesma González
Rafael Merinero Rodríguez

MDPI

Basel • Beijing • Wuhan • Barcelona • Belgrade • Novi Sad • Cluj • Manchester

Guest Editors

Oswaldo Ledesma González
Department of Geography
University of Valladolid
Valladolid
Spain

Rafael Merinero Rodríguez
Department of Sociology
Pablo de Olavide University
Seville
Spain

Editorial Office
MDPI AG
Grosspeteranlage 5
4052 Basel, Switzerland

This is a reprint of the Special Issue, published open access by the journal *Land* (ISSN 2073-445X), freely accessible at: https://www.mdpi.com/journal/land/special_issues/493OHF41B2.

For citation purposes, cite each article independently as indicated on the article page online and as indicated below:

Lastname, A.A.; Lastname, B.B. Article Title. *Journal Name* **Year**, *Volume Number*, Page Range.

ISBN 978-3-7258-7466-8 (Hbk)
ISBN 978-3-7258-7467-5 (PDF)
https://doi.org/10.3390/books978-3-7258-7467-5

Cover image courtesy of Rafael Merinero Rodríguez

Contents

About the Editors

Oswaldo Ledesma González

Oswaldo Ledesma González is a Lecturer in the Department of Geography at the University of Valladolid, Spain. His research has focused on territorial and urban planning, strategic territorial planning processes, public policy analysis, governance networks, and the territorial analysis of tourism. He holds a Bachelor's degree in Geography from the University of La Laguna and a Master's degree in Management and Promotion of Local Development from the University of Valencia, where he received the Extraordinary Award. He subsequently earned his PhD in Geography from the University of La Laguna through the Law, Society, and Tourism Program, graduating with the highest distinction and receiving the Extraordinary Award. He also holds a University Expert Certificate in University Teaching from the University of La Laguna. He has participated in more than a dozen European, national, and regional research projects, as well as knowledge transfer agreements for the public and private sectors. He has undertaken international research stays at the Autonomous University of Aguascalientes (Mexico), the University of Algarve (Portugal), and the University of Costa Rica (Costa Rica). Throughout her career, he has combined university teaching and research with consulting work for the public sector, developing territorial diagnoses, tourism sustainability plans, local action plans for urban agendas, and other related projects.

Rafael Merinero Rodríguez

Rafael Merinero Rodríguez is a Full Professor in the Department of Sociology at Pablo de Olavide University, where he teaches Political Sociology and Public Policy Analysis. He holds a PhD with European Distinction from the University of Seville, a Master's degree in European Union Studies from the National University of Distance Education (UNED), and has been a visiting researcher in the Department of Architecture and Urban Studies at the Polytechnic University of Milan, the Institute of Social Sciences at the University of Lisbon, and the Department of Economic Policy at the University of Siena. His research focuses on governance processes in the field of urban and metropolitan development public policies. He is the Principal Investigator of the research projects "The Governance of Metropolitan Areas in Andalusia: Challenges for the Development of the Strategic Model" (ERDF Funds), "The Metropolitan Dimension of the Dos Hermanas Urban Agenda," and "Urban Agendas and the Seville Metropolitan Area." The results of his research have been published in prominent national and international journals such as *Local Government Studies*, *Tourist Management*, *Revista Española de Investigaciones Sociológicas*, *Boletín de la Asociación de Geógrafos Españoles*, and *Ciudad y Territorio*, as well as being published by prestigious publishing houses such as Springer, Tirant Lo Blanc, Catarata, and Síntesis.

land

MDPI

Article

Use of the Sociogram in Participatory Planning in Contexts of Social Exclusion: A Comparative Case Study in Cordoba Neighbourhoods, Spain

Antonio Delgado-Baena and Antonio Sianes *

Research Institute on Policies for Social Transformation, Universidad Loyola Andalucía, 41704 Dos Hermanas, Sevilla, Spain; adelgadob@uloyola.es
* Correspondence: asianes@uloyola.es; Tel.: +34-957-222-100 (ext. 2220)

Abstract: Urban regeneration and spatial planning have adopted a new participatory approach in recent decades, highlighting the importance of integrating the community in urban decision-making processes, especially in disadvantaged and socially excluded areas. In this context, the sociogram emerges as an essential tool for collaborative governance, allowing the visualization and analysis of the dynamics between the different actors involved. This study employs a comparative case study approach in three disadvantaged neighbourhoods in Córdoba, Spain, to examine how the sociogram can facilitate more effective and democratic participation in urban planning. Using heat maps, scatter plots and average analysis, relationships between actors are identified and characterized, providing a solid basis for more inclusive and equitable planning decisions. This analysis not only reveals the practical utility of the sociogram in participatory research but also underscores its theoretical relevance in building resilient and cohesive communities. Findings confirm the sociogram's effectiveness in mapping stakeholder dynamics and enhancing participatory governance, ultimately fostering more informed and inclusive urban planning processes.

Keywords: participatory planning; urban regeneration; collaborative governance; stakeholders; sociogram; comparative case studies

1. Introduction

The study of stakeholder networks is prevalent in socio-community interventions and participatory research processes, significantly influencing what is known as collaborative governance [1]. This is especially relevant in urban contexts, where collective action is required to transform a city and impact its inhabitants [2]. However, to achieve multistakeholder partnerships, the identification of stakeholders, their relationships and their initial positions is compulsory.

To achieve these goals, recent research situates social cartography as a suitable theoretical and empiric approach [3–6]. Social cartography is able to visually capture the social, economic, cultural and political dimensions that characterize a given community by representing the perceptions, experiences and knowledge of communities about their own territory [7]. To contribute to this body of literature, our analysis relies on the sociogram technique, one tool of social cartography applied for the identification of relevant stakeholders in any given communitarian practice.

In this study, we focus our analysis on a comparative case study of three disadvantaged neighbourhoods in the city of Córdoba, in southern Spain. Our study is pertinent as it contributes to this body of literature but also sheds light on such social processes. It is thus also relevant as it contributes to the comprehension of patterns of how stakeholders behave in contexts of social exclusion, something strongly needed in socio-communitarian interventions [8–11].

1.1. Theoretical Framework: Urban Regeneration as a Response to Territorial and Social Exclusion

Spatial planning has configured places as infrastructure for social planning [12]. The urban model is mediated by historical, cultural, political and economic factors in such a way that cities reflect and reproduce social stratification [13]. Together with neoliberalizing policies and the growth of the urban population, it has multiplied the spaces of social segregation [2].

As a response, hand in hand with rights approaches [14–16], social inclusion has been incorporated into urban development agendas [17]. Urban planning determines how cities are inhabited, producing and dialectically reproducing the urban culture of social inclusion or exclusion [18]. However, in current urban models, social inequality persists in territories characterized by poverty, physical and social isolation and a lack of access to resources. Viewed as ghettos or slums, exclusion and poverty in these spaces have only increased and maximized through what has been called urbanism of exception [19,20].

The most recent literature shows how these processes of territorial and social exclusion go hand in hand. While the former imposes physical limitations and hinders access to habitability resources, the latter stigmatizes its inhabitants and keeps them in a situation of poverty. To address these issues, governments and urban planners have moved from processes of city renewal, revitalization or rehabilitation to urban regeneration, which has governance implications [21]. Urban regeneration relies on the government actions with which the common interest of a place is managed [22,23].

Urban regeneration stands thus as a comprehensive intervention process for multidimensional improvements in urban spaces. Its aim is keeping territories alive and socially balanced, promoting social cohesion through the development of public policies that regenerate urban spaces from the detrimental dynamics of deterioration, segregation, social imbalance and conflicts of interest [24].

Human beings coexist in spaces where they interact, building and rebuilding such spaces through social praxis [13,25]. Thus, neighbourhoods arise from the interactions of inhabitants in their daily lives through processes of recognition between actants [26], linked to very specific times and spaces within territories [27]. In deprived territories, such relationship networks are woven to articulate the power of the neighbours to survive with dignity in the midst of sometimes unworthy living conditions.

Studying these networks involves understanding the neighbourhood as a semantic field where actant networks interact instead of understanding the neighbourhood as infrastructure [12]. To do this, Latour [26,28] raises the need to approach the reality of places from the mapping of relationships that involve people, ideas and technologies analysed as a whole. To better understand a process, its interconnections with other phenomena must be studied [29–31]. In this sense, the cartography of a territory becomes social cartography.

Social mapping integrates dimensions and processes [32] as a research process in the social context [5]. However, because cartography serves a political function, i.e., spatial and societal planning [33], the need arises for critical cartography to reveal the complexity of what is not part of the official representation of reality [34]. In this regard, the sociogram has stood out as an adequate means to reveal the intricate network of processes and phenomena that occur in a territory [35,36].

1.2. Conceptual Framework: The Sociogram as Tool to Promote Collaborative Governance

Rhodes [37] defines governance as the process of governing with and through networks, which implies interactions among strategic actors [38]. Given the difficulty of quantifying the nature of networks [12], sociopraxis or participatory methodologies, as termed by Villasante [39], arise. This participatory approach proposes that urban regeneration processes require a collaborative governance process [40]. This activates a complex network of stakeholders with different participation roles, representing a paradigm shift [6,41].

In this sense, collaborative governance is deemed a democratic and pluralistic mode of governance [40] that requires the collaboration of the different stakeholders in the management of power at the local level. Thus, authors such as Ran and Qi [42] have valued the sharing of power in collaborative governance, always subject to contingency factors, as a network of stakeholders is a group with different coexistent interests. Hence, the identification of all stakeholders is necessary to consider their perspectives and needs [40].

Once identified, stakeholders can form collaborative associations that guarantee the viability of urban regeneration processes, communicating structures of participation, management and decision-making as practices of full citizenship [43]. For these reasons, the sociogram is postulated as a social mapping tool for the identification and analysis of stakeholder networks, with the capacity to favour the creation of partnerships that allow the collaborative governance of urban regeneration processes.

The sociogram is a sociometric technique derived from the confluence of different authors and approaches, from social psychology to field theory, in conjunction with graph analysis. It is aimed to analyse networks as theoretical and methodological perspectives for social studies [44–46]. The sociogram is, therefore, descriptive in terms of relationships or links between pairs that establish a social network [44,47,48]. They are a graphic representation of groups and individuals and their relationships, projecting the complex social structure of a community while tracing and analysing their interconnections [13]. Participatory and dialogic analysis [49] is used to study these relationships between social actors in a given context [13] incorporating the discursive interpretations of the participants, projecting them to the functional needs of the context.

The sociogram has proved to be a very useful tool for establishing participatory diagnoses, proposing actions and identifying related groups with whom to expand networks by accessing other less related groups [50]. Likewise, the sociogram identifies sets of actions around specific issues or themes that connect them [36]. However, the literature on the ability of the sociogram to mobilize collaborative governance processes at the neighbourhood level is scarce. This study aims to fill this gap by demonstrating how sociograms can be used effectively to enhance democratic participation and collaborative governance, providing useful insights for the development of more inclusive and equitable urban policies. The need for this research is underpinned by the limited amount of comparative studies in this specific area, making the findings particularly valuable for both academics and practitioners. This work deepens the potential of the sociogram for community collective action through the identification of stakeholders in three disadvantaged neighbourhoods with incipient urban regeneration processes.

This article is structured as follows. The presentation of the theoretical framework is followed by a description of the methods, the materials and the selection of cases. Then, we present the results obtained using density, dispersion and average analysis. Finally, the results are discussed in the context of the conceptual framework to draw the appropriate conclusions.

2. Materials and Methods

The objective of this research is to examine, through a comparative case study, the understanding that residents of three socially excluded neighbourhoods have about the stakeholder networks existing in their territories [51]. This research was carried out at the beginning of a grassroots initiative urban regeneration process. The ultimate goal is to promote networks that facilitate collaborative governance processes in sustainable and endogenous urban regeneration processes.

Thus, this is an instrumental case study [52,53] of a comparative nature, in which each sociogram created by the participants in each neighbourhood is taken as the object of study. The sociogram is adequate for interpretation but has too few data for a statistical analysis [54]. Consequently, we use a comparative method and a logical and systematic procedure to identify similarities, dissimilarities, contrasts and patterns [55,56].

The selection criteria for the case studies were contextual similarities in time and space [55]. The cases are three neighbourhoods in the city of Córdoba (Spain) with similar indicators of poverty and social exclusion. They also have similar historical trajectories as they are immersed in the same community intervention process for urban regeneration.

To gather the data, we rely on participant observation and document analysis [57]. Participant observation was documented in a diary, registering all 19 specific sessions with both reflexive and participative components [58,59]. Document analysis examined all documents generated from the application of the sociogram in the three contexts: minutes, graphs in process and final graphs. Written data were transcribed for later content analysis, and visual data were translated to Microsoft Excel to generate graphs for visual data analysis.

The generated graphs were projected as a Cartesian axis with "X" being the degree of affinity and "Y" being the degree of influence, assigning a value to each actor. To preserve the confidentiality of both the participants and the actors they identify, we assign a letter to each neighbourhood (A, B and C) and we standardize actors to generic definitions (i.e., Civil Society Organisation (CSO), local entity, etc.). Once systematized and standardized, all information was analysed to assess the density of stakeholders by using the aforementioned three techniques: heatmaps, scatter charts and average analysis. Heatmaps express numerical data that are represented graphically in a grid by means of colour grading [60,61]. They are recognized as the main method to determine distribution patterns, identifying concentrations and determining areas of action [62]. For heatmaps, we categorize actors in power groups, formal groups and informal groups. Scatter plots allow the measurement of trends between two correlated variables in datasets. The sociogram matrix behaves like a scatter graph, allowing comparisons of trends [63]. Finally, for the study of averages, balloon graphs are used, making it possible to correlate influence with affinity in relation to density. For average analysis, we sub-categorize the former groups: power groups include public administration, public service units; formal groups include private service providers, religious entities, local associations and platforms; and informal groups include informal leaders, self-organized citizenry and unorganized citizenry [35,36,64].

However, we recognize certain methodological limitations associated with our comparative case study approach. The unique nature of the contexts examined calls for caution in generalizing findings to other urban realities. In addition, while data collection was comprehensive, the qualitative and subjective nature of the participants may mediate the interpretation of the data. These limitations should be considered when applying our findings to other settings.

2.1. *Presentation of the Cases*

2.1.1. Córdoba, a Heritage City of Delight and Exclusion

Córdoba, located in Andalusia in the south of the Iberian Peninsula, is a city with a long historical and patrimonial heritage. It holds four UNESCO World Heritage designations due to a thousand-year-old cultural confluence, and, like other European cities, urban development has occurred in each era. However, this development has maintained deep socio-spatial segregation in such a way that four of the fifteen poorest neighbourhoods in Spain are in Córdoba according to the National Institute of Statistics (INE) [65]. They all share similar indicators of urban vulnerability, and the three neighbourhoods selected for the study are among the four poorest neighbourhoods in the city. Their socioeconomic characteristics (Table 1) show high social vulnerability and a poor urban configuration, which qualifies them as disadvantaged urban areas [2,66].

Table 1. Socioeconomic characteristics of each neighbourhood.

	Neighbourhood A	Neighbourhood B	Neighbourhood C
Population	10,457	5689	2195
Housing stock	2930	2215	719
Per capita income by year (in Euro)	€6920	€6682	€4565
Unemployment rate	36.8%	45.2%	55.5%
Illiterate population rate	13.5%	17.9%	29.9%

Source: own elaboration based on [67,68].

All these neighbourhoods also share their history. They were built between the 1960s and the 1980s, a time characterized by the suburban growth of the city under urban renewal and urban redevelopment policies. They are therefore peripheral neighbourhoods, with working class characteristics, absorbed by the metropolis and affected by the growing gentrification and touristification of the historic centre [69,70]. They have been the object, since the 1990s and to the present, of various urban regeneration programmes [21], which have consistently proven ineffective.

In this context, the research team, together with citizen representatives, developed a shared methodology for a grassroot urban regeneration initiative in each neighbourhood. The aforementioned socioeconomic conditions explain the existence of a broad and heterogeneous group of local partnerships already existent, who led the process.

To constitute this neighbourhood-based group, a series of steps was designed that entailed the application of the sociogram as a participatory technique for the analysis of networks [35,36]. This allowed mapping the local reality and analysing the complex network of stakeholders with neighbours in order to identify possible transformative partnerships.

2.1.2. Implementation of the Sociogram

The implementation of the sociogram conveys a group of participants that, through discussion, collectively agree on a visual representation of the actors in the territory and their mutual relations. It is graphically represented as a Cartesian axis that visually depicts two indicators or factors that are deemed relevant for participants. In our case, the X axis represent the different degrees of influence each actor has in the territory. It is how they can exert power or influence others to exert power. The Y axis was used to represent the degree of affinity with the principles of the collective action proposed to transform the territory: horizontality, the neighbour's protagonism, etc. Once all the relevant actors of the territory have been identified and consequently categorized as power, formal and informal actors, they are placed across the Cartesian axis by consensus. Finally, relevant actors are connected according to their relations, distinguishing if relevant whether these relations are strong or conflictive [36] (Figure 1).

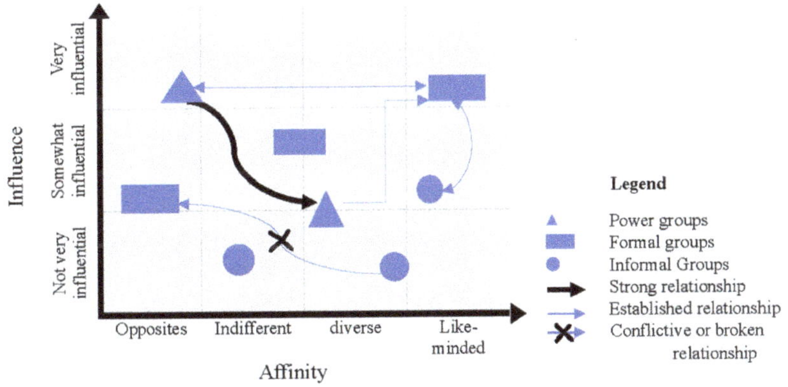

Figure 1. Areas and axes of the sociogram.

The three cases studied followed the expected sequence: identification of stakeholders; organization in categories; location in the matrix; relationship tracing; group analysis of the diagram; and planning for action [36]. Subsequently, each sociogram was analysed using different criteria: density, proximity, isolation, verticality and conflict [35].

The sociogram has a projective dimension that allows reflection on appropriate actions so that social networks are rearticulated into action sets, enabling conflict resolution and the creation of new networks [71–73]. Hence, a participatory approach was chosen in which the community is involved as a subject of knowledge, action and transformation because as the community reveals the network, it rebuilds it [9,71].

The technique was implemented in each neighbourhood between October 2020 and January 2022. Given the dates, the implementation was disrupted by the COVID-19 pandemic, which limited the techniques and instruments used but did not invalidate the results obtained. The study complied with ethical standards and was approved by the ethics committee of Universidad Loyola Andalucía on 28 April 2021, ensuring the confidentiality and anonymity of the data.

In neighbourhood A, face-to-face and virtual sessions were held. For this, a free video-conferencing application was used, and the *GenoPro* 2007 (version 2.0.1.6) program, software for generating genograms, was adapted as a participatory tool. In a total of 5 sessions, 44 people participated and 48 social actors were identified. In neighbourhood B, face-to-face and online sessions were also implemented using the same computer programs. Six sessions were held with a fluctuating participation of 61 people, and 50 social actors were identified. Neighbourhood C already had an approved community development plan. The sociogram was applied in the process of the reactivation of the plan. Eight face-to-face sessions were held in which the network map was configured, the strategic analysis was carried out and the partnerships were projected. There were a total of 37 people participating, and 42 social actors were identified. All this information is summarized in Table 2.

Table 2. Characteristics of the implementation of the sociogram in each neighbourhood.

	A	B	C
No. of Sessions	5	6	8
Duration:	3 months	3 months	5 months
Participants	44	61	37
Modality	Face to face/Online	Face to face/Online	Face to face
No. of Agents	48	50	42

Source: own elaboration.

3. Results

3.1. Stakeholder Concentration Analysis

The analysis of the concentration of the different types of actors is especially relevant when identifying stakeholders. It provides a first look at the positions of each group without actually entering into relationships. The comparative analysis of these positions allows the identification of patterns, which in turn will guide decision-making on issues such as the prospect of success or the exploration of structural conflicts [62]. Although the analysis of network density can also be measured by the proportion of the links established [47,69,74], in this case, the network was not measured, but rather the semantic location of each stakeholder was considered from a structural perspective [36] to determine the correlation between influence and affinity. Heatmaps were used for this analysis.

Figure 2a presents the heatmap of the sociogram of neighbourhood A, observing an apparent balance between 52% of the opposites and indifferent balance and the 48% of the different and like-minded. However, the graph's tendency towards extremes outlines a sociogram where a situation of opposing forces prevails. In the absence of an analysis of the relationships, these results suggest a confrontational scenario that offers limited prospects for successful outcomes.

Figure 2b highlights the imbalance in the heatmap for neighbourhood B. On the left side (opposites and indifferent) is 65%. The graph is also unbalanced horizontally, with substantial weight in the lower zone of less influence (83%). The participants consider that legitimacy in the neighbourhood is in the possession of very few. This, together with the excessive weight of the lower left quadrant, which conveys little affinity with the process, points to a vision of inaction in the neighbourhood.

Figure 2c shows a substantially different situation in neighbourhood C. Although the densest column is that of indifferent (38%), the weight of the graph is biased towards the right side (the diverse and like-minded represent 58%). The column for opposites appears depopulated (5%) and with little influence, and 45% of the groups identified are in the upper right quadrant (diverse/like-minded and somewhat or even very influential), presenting a homogeneous distribution. The density and homogeneity of this area suggest a model with viable prospects.

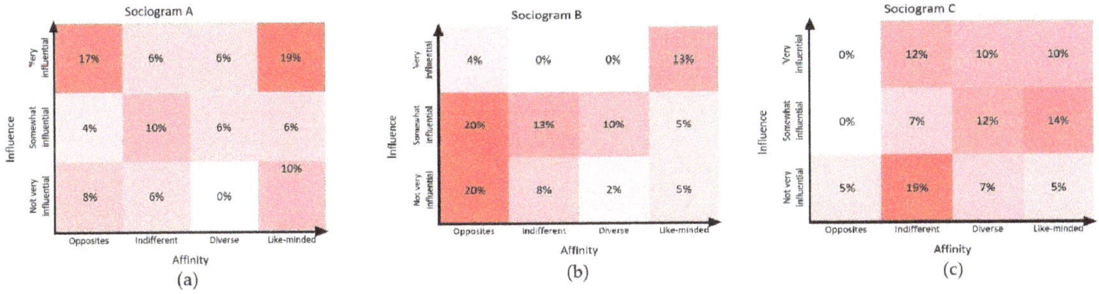

Figure 2. Heatmap of (**a**) neighbourhood A; (**b**) neighbourhood B; (**c**) neighbourhood C. The degree of colour intensity corresponds to the concentration of stakeholders, darker meaning a higher concentration.

3.2. *Analysis of the Concentration of Stakeholders Based on Organizational Structure*

After individual analysis, the stakeholders were organized based on their initial organizational structure into three categories: (1) power groups, which are organizational structures that have the competence and capacity to make direct decisions about public policy or the transformation of a neighbourhood; (2) formal groups, which are formally constituted organizations, whether public or private; and (3) informal groups, which are groups of people concerned about everyday issues, as well as neighbourhood leaders. Each group was assigned a graphic representation: triangles (power groups), squares (formal groups) and circles (informal groups) [36].

The density and location of stakeholders based on their organizational structure allow analysing the perceptions that participants have about them [36] and expand the search for patterns that indicate the organizational structure of neighbourhoods in social exclusion [47,74].

For the analysis, the neighbourhoods were grouped by each type of stakeholder to obtain a graph for each type of actor.

3.2.1. Power Groups

Figure 3 represents the positioning of the power groups, where 42% are located in the column for opposites and mostly in the area of little influence (30.43%). It is here that neighbourhoods A and B have the bulk of stakeholders who hold formal power in the neighbourhood. The image shows a diagonal trend, with neighbourhoods A and B in the lower left and neighbourhood C in the upper right (although one group is separated from it, representing provincial administration with little power in the neighbourhood).

The power groups identified are mainly institutions and public services, that is, organizations with decision-making capacity and power over the neighbourhood, because they legislate, standardize or apply laws and regulations. Importantly, despite this role, they are perceived by the participants from neighbourhoods A and B as having little influence, displacing effective leadership in the neighbourhood towards other stakeholder profiles. They are perceived as groups opposed to the community process, projecting an image of confrontation. In contrast, neighbourhood C places most of these groups in the very influential row, with a balanced weight in each grid. For this neighbourhood, 75% of these groups are located between the diverse and like-minded columns, indicating a certain proximity of these institutions to the neighbourhood.

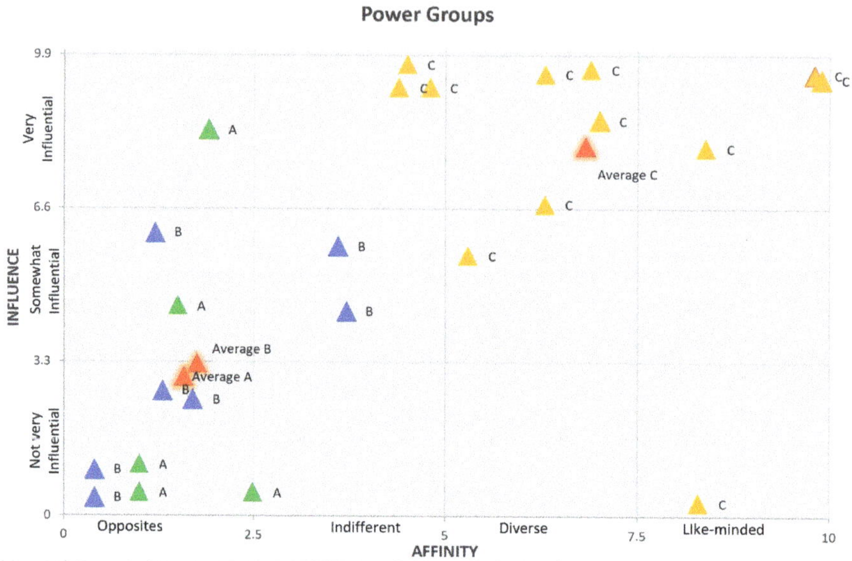

Figure 3. Positions of the power groups. In green the stakeholders in neighbourhood A, in blue those in neighbourhood B, in yellow those in neighbourhood C. Red represents the average.

3.2.2. Formal Groups

As Figure 4 shows, the formal groups present greater dispersion. The function of formal groups in the neighbourhood is broad, from constructing the social and associative fabric to conveying the public administration.

Figure 4. Positions of the formal groups. In green the stakeholders in neighbourhood A, in blue those in neighbourhood B, in yellow those in neighbourhood C. Red represents the average.

In neighbourhoods A and B, the groups are distributed throughout the map. However, certain patterns emerge. In neighbourhood A, formal groups are given high weight, with 19 organizations being identified as very influential. Of these, a considerable number are identified as either related or contrary, suggesting a slight polarization. In neighbourhood B, on the other hand, formal groups are given less weight (not very/somewhat influential), with a certain tendency to the left side (opposites and indifferent). In the extreme columns (affinity and opposites), the same high weight indicates an opposing struggle, polarizing the positions towards the extremes. This reinforces the blocking and opposing positions already detected in the general analysis. Finally, in neighbourhood C, a certain concentration is observed. In fact, in this neighbourhood, the formal groups present a high average (6.8, 4.5). Of these, up to seven groups are identified in the affinity column, and only one is found in the area of high influence. This illustrates, on the one hand, the perception that one organization drives the process and the perception of a certain lack of power and influence of the social fabric.

3.2.3. Informal Groups

Informal groups are the most diverse, ranging from individuals to semiorganized groups or groups without any organization. Despite their relative disarticulation, these groups are subject to very specific times, themes and places that weave a web across daily life in the neighbourhoods [27]. In this regard, only four groups appear in the very influential row, all from neighbourhood A (see Figure 5).

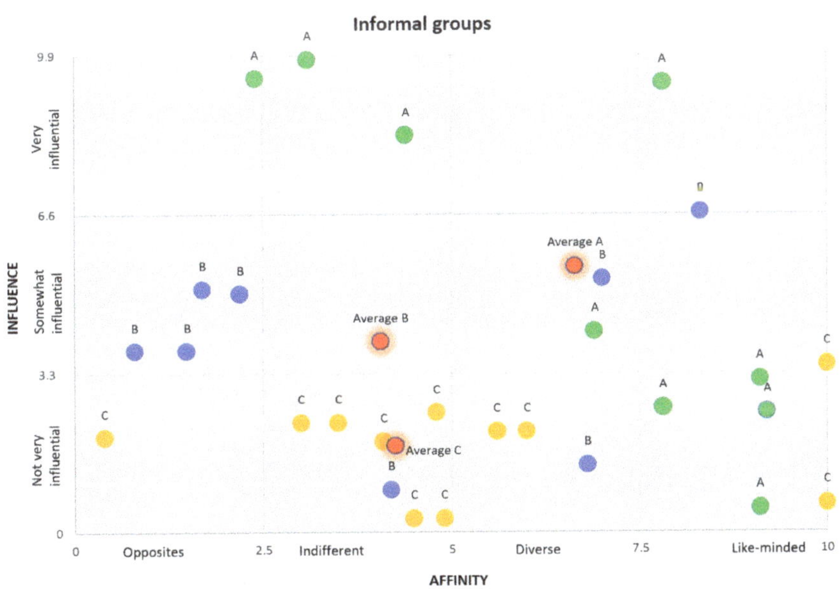

Figure 5. Positions of the informal groups. In green the stakeholders in neighbourhood A, in blue those in neighbourhood B, in yellow those in neighbourhood C. Red represents the average.

By neighbourhood, the positions of the groups in neighbourhood A indicate confrontation, but with relevant peculiarities: the groups located in the like-minded column have little influence, and the others with greater influence remain outside the process. This inverse correlation between influence and affinity indicates difficulties in achieving collective action in the neighbourhood. Contrast this with neighbourhood C, where numerous informal groups are identified, but all of them lack influence. This view of informal movements as having little influence and relatively little affinity in neighbourhood C is noteworthy despite the affinity shown by powerful and formal groups with the endogenous development process launched in the neighbourhood.

3.3. Analysis of the Concentration of Stakeholders Based on the Nature of Their Activity

For a better characterization of the stakeholders and given the different roles they play in the neighbourhoods depending on their nature, referencing Alberich [71], each typology has been subcategorized (Table 3).

Table 3. Description of the stakeholders in the territory according to their nature.

Groups	Type of Organization	Description
Power	Public administration	Public institutions at different levels: city councils, district councils, etc.
	Public service units	Public personnel and organizations that offer educational, health or other services to citizens and the community to meet the rights of citizens managed by the public administration
Formal	Private service providers	Private personnel and organizations that offer services to citizens and the community in a delegated manner to meet the rights of citizens managed by the public administration
	Religious entities	Groups with a formal link to a religious institution
	Local associations	Neighbourhood associations to meet particular or community objectives (educational, cultural, sports, etc.).
	Platforms	Conglomerate of entities that associate with each other to have a greater impact on the achievement of common objectives
Informal	Informal leaders	Groups formed by a formal link to a religious institution
	Self-organized citizenry	Non-formalized groups of citizens who organize themselves to meet some common and temporary goal
	Unorganized citizenry	Groups of citizens who are grouped together from time to time with a common, particular or community objective

Source: own elaboration.

Once categorized, Table 4 shows the prevalence of each group of actors by territory.

Table 4. Prevalence of stakeholders in neighbourhoods based on the nature of stakeholder activity.

	Organization Type	Neighbourhood A	Neighbourhood B	Neighbourhood C	Total
Power	Public administration	4.17%	10%	9.5%	7.69%
	Public service units	25.00%	20%	23.8%	23.08%
Formal	Private service providers	29.17%	20%	16.7%	22.31%
	Religious entities	6.25%	5%	4.8%	5.38%
	Local associations	12.50%	20%	19.0%	16.92%
	Platforms	4.17%	5%	4.8%	4.62%
Informal	Informal leaders	14.58%	17.5%	14.3%	15.38%
	Self-organized citizenry	4.17%	0%	7.1%	3.85%
	Unorganized citizenry	0%	2.5%	0%	0.77%

Source: own elaboration. The highest values in each column are marked in ochre and the lowest values in each column are marked in blue.

In general, the group most present in the neighbourhoods is public service units, although in neighbourhood B the perceived presence of public service units is equal to that of local associations and service provider entities. In neighbourhood A, these entities, which normally provide public services in a delegated manner, are the second largest group. Neighbourhood B is the only neighbourhood that identified a group of unorganized citizens, and neighbourhoods A and C are the only neighbourhoods that identified self-organized citizenry groups; both categories, which play roles in the daily life of the neighbourhoods, are significantly the least represented.

To provide information on the capacity to promote collaborative governance processes, the average position of each subcategory and its density are calculated for comparative purposes. They are graphically depicted on three maps (Figure 6) representing the situation in each neighbourhood.

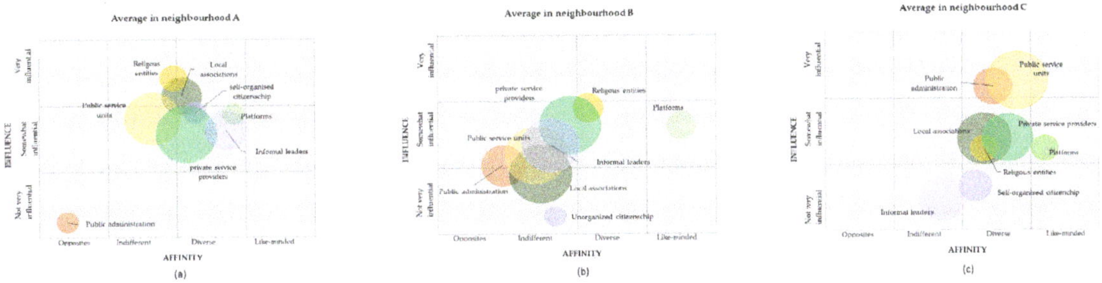

Figure 6. (**a**) Average in neighbourhood A; (**b**) average in neighbourhood B; (**c**) average in neighbourhood C.

The graphs show the average trend of each cluster in each neighbourhood. This calculation of averages implies that the more dispersed the sample, the more centrality the cluster will have. This allows identifying certain tendencies when a conglomerate moves away from centrality.

Although visually similar, Figure 6a–c present key differences in the disposition and relevance of stakeholders in each neighbourhood. In Figure 6a, most of the groups are in the central areas, where service-providing entities and distanced public administration dominate. Figure 6b, corresponding to neighbourhood B, shows a more dispersed distribution, with a tendency towards less affinity and influence, distributing the weight between public services, associations and local service provider entities. Finally, Figure 6c shows that neighbourhood C extends the influence from top to bottom, highlighting the administration and public services in the areas of greatest influence and affinity.

This comparative analysis reveals that, in general, (1) the actors in neighbourhood A are concentrated in the central and superior areas, with the notable exception of public administration; (2) the actors in neighbourhood B are concentrated in areas of little influence and opposed to or indifferent to the process, with the notable exception of neighbourhood platforms; and (3) the actors in neighbourhood C are dispersed in a diagonal that clearly tends towards the areas of greater influence and affinity, stratifying into informal, formal and power groups.

The analysis by type of actor yields very interesting results regarding the positioning of actors in disadvantaged neighbourhoods. Groups related to public administration are generally seen as opposing and with little influence (neighbourhoods A and B), although they are perceived as having much power and greater affinity in neighbourhood C. In neighbourhood A, it is noteworthy that their position is at the lower extreme (opposite/not very influential). Public service units, which are local tools of the public administrations to territorialize activity, are normally associated with public administration, except in neighbourhood A, where criticism of the administration is notable.

Service providers remain mostly indifferent to the process, with positions somewhat closer to public institutions but with a medium capacity for influence. Comparatively, neighbourhood-based local associations are perceived significantly differently in each neighbourhood with regard to the axis of influence, showing little influence in neighbourhood B but substantial influence in neighbourhood A.

A couple of entities require a detailed analysis because of their specificity in the Spanish reality but whose analysis is extendable to other contexts. Platforms, which are groups of civic organizations with a common interest, usually arise in neighbourhoods where collective action processes are initiated, thus explaining their related positions even with medium or low influence in the graphs. Religious entities, whose presence in these neighbourhoods has historical weight in movements and their welfare, maintain substantial influence, although they are located in the indifferent column because of the ageing of their members.

Finally, the social bases, represented by groups of informal leaders and self-organized and unorganized citizenry, move from indifferent positions in neighbourhoods C and B to relatively opposite positions in neighbourhood A. All show relatively little influence and are not present in some neighbourhoods.

4. Discussion

As a dialogic methodology, the sociogram has a descriptive component of reality but also a prospective purpose, allowing the incorporation of a strategic function [36]. In this line, the analysis provides a visualization of the starting positions of groups of stakeholders in each neighbourhood and indicates the feasibility and sustainability of collective action.

Starting with the heatmaps, a common inter-neighbourhood pattern is not observed. However, the graphs could indicate emerging problems in programme implementation in each neighbourhood. The starting positions indicate a playing field that, depending on the territory, points to confrontation (neighbourhood A), opposition and blockade (neighbourhood B) or collaboration (neighbourhood C). These starting positions coincide, in prospective terms, with what Villasante proposes [36,75] in his chart of positions and with Gardner and his colleagues [76]. In the most extreme cases, a prospective interpretation of the sociogram allows predicting the feasibility of development processes, as in neighbourhood B, in which the urban regeneration process that has actually started has not been implemented continuously.

Comparing the organizational structure of the stakeholders, regarding the power groups, the analysis shows that neighbourhoods A and B perceive them with suspicion and that neighbourhood C perceives them with affinity. These results demonstrate how the perceptions of the participants towards the decision-makers transition from discouragement and confrontation to collaboration and advocacy. As Rawlins poses [77] through the Grunig and Hunt bonding model [78], power groups are stakeholders associated with enabling and normative links. These types of links allow an organization to have the resources and autonomy to act. Therefore, initial positions opposed to the process could limit the possibilities of endogenous development, while projecting an aligned stance would convey a more feasible visualization of the process.

Regarding the power groups, even though there is no causal evidence, the comparative analysis points to a possible correlation between the perception in the neighbourhoods that the groups are related and an urban regeneration plan with collaborative governance exists. This is the case for neighbourhood C, a finding that supports results previously reported in the literature: for example, an increase in the importance of the interaction between public actors and nonstate actors [79] and the increased benefits when institutional context (power) encourages collaboration [42]. This result also supports the idea that when public policies are participatory, a greater proximity to public institutions is established, ameliorating general disaffection, especially in these neighbourhoods [80].

Regarding perceptions of formal groups, they integrate into the associative network of neighbourhoods and serve as functional links between public administration and the neighbourhood [77]. As seen in the graphs, the positions of these groups vary based on the situation in each neighbourhood, not finding generalized patterns that allow establishing links between positions and territorial exclusion. However, following Ran and Qi's approach [42] regarding the relationship between sharing power and the effectiveness of collaborative processes, a certain connection is observed between the expectations of success and the location of these formal groups. The farther away the groups are from the process, the more resistance to implementation there will be, perhaps affecting motivation for action. Thus, the distribution of the formal groups suggests possible positions in neighbourhood C but confrontation and blocking in neighbourhoods A and B.

The importance of informal groups lies in the fact that they weave into the network of relationships in daily life, in many cases activating endogenous resources in dialectical processes of resistance and adaptation [81]. In these spaces of connection with informal groups, neighbourhood power is activated and on appropriate occasions can give rise to the "unprecedented viable" [49], that is, processes that, being innovative, facilitate transformative action by overcoming extreme situations [82]. However, in the cases analysed, in general, informal groups are observed with caution, as they are perceived with little influence and relatively low affinity. Despite the use of participatory methodologies by the most established groups in the neighbourhoods, the existing gap with the social bases of the disadvantaged neighbourhoods is evident.

Finally, according to the nature of the social actors, the comparative analysis yields significant information about their initial positioning in the neighbourhoods. The first important element is the predominance of public service units (a power group), service providers and local associations (both formal groups) in the neighbourhoods. The reticular structure of the neighbourhoods shows a network of connections between power groups and highly organized and structured civic organizations. These connections are dominated by functional and normative links [77], projecting a managerial or even clientelist image [6,36] of the dynamics of the neighbourhoods.

Additionally, there is low perception of informal groups, i.e., those that make up the social bases. The complete absence of so-called urban social movements stands out [83,84]. The low presence of these groups is relevant because, on the one hand, they constitute diffuse links in daily life [77], and, on the other hand, they have a strong political capacity and thus play key roles in forging alliances with an impact mediated by the conflict and structural characteristics of the context [85].

Furthermore, as the administration is perceived with less influence and affinity, religious centres are perceived with greater affinity and influence. This is associated with the origin of the peripheral neighbourhoods, where the absence of the administration allowed the development of ecclesiastical services whose influence is still present.

Consistent with the principles set out by [86] in his critical analysis of case studies, our work in the three neighbourhoods shows how specific methodological approaches can reveal generalizable patterns. Although each neighbourhood has a unique stakeholder positioning scheme, the application of the sociogram has discovered distinctive models that are applicable in other urban contexts. For example, in neighbourhood A, administrative disconnection and struggles between formal and informal groups hinder progress. In neighbourhood B, the prevalence of actors with opposing positions and little influence leads to inactivity and blockade of the regeneration process. Meanwhile, in neighbourhood C, an effective collaboration between formal and power actors contrasts with the distancing of the citizen bases, highlighting the need to reconnect with the community. These findings underline the usefulness of the sociogram for identifying intervention strategies in similar settings, aligning with Flyvbjerg's perspective on the valuable contribution of case studies in qualitative research. These characterizations suggest a trend towards different models of governance in the territory, a finding that is consistent with proposals by authors such as Villasante [36] and Bradley [6]. The verification of similar models in other territories can, on the one hand, predict the drift of the governance of urban regeneration processes and, on the other hand, help establish strategies and decision-making that consider all stakeholders, creating sustainable governance coalitions [24].

5. Conclusions

The comparative study of the sociogram for three disadvantaged neighbourhoods has allowed us to draw some general conclusions about the configuration of collaborative governance networks in sustainable and endogenous urban regeneration processes, as well as the capacity of the sociogram not only to promote such processes but also to predict their viability. The sociogram is a useful tool to identify stakeholders and their starting positions and to provide inputs for strategic planning.

In general, the results are consistent with the literature; i.e., there is a relationship between the effectiveness of processes and an express will to share power. The assessment of actors' positions has allowed the classification of their roles as enabling, confrontational or blocking. This suggests that some settings, such as neighbourhood B, face significant challenges to collaborative governance, while others, such as neighbourhood C, present greater opportunities for the success of these processes

Another general conclusion regarding the stakeholders in disadvantaged neighbourhoods pertains to the nature of their positions, even in the absence of establishing relationships between them. Thus, the shared difficulty in identifying informal groups, which constitute the social bases and weave into the daily fabric of the neighbourhood, has been verified. Notably absent are the groups closest to urban social movements. This absence of community fabric demonstrates how managerial and even clientelist actors prevail in neighbourhoods, especially when organizations linked to public administration predominate.

In parallel to the conclusions, it is necessary to highlight some limitations. This study was carried out during a specific phase of process implementation; therefore, the location of the stakeholders in the sociogram matrix could have been affected. The analysis of the networks established between the actors could shed light on the interconnection processes between actants, complicating the understanding of the social processes in the territory. Furthermore, as Gutiérrez states [35,87], a sociogram is static and descriptive. Knowing that networks are by definition contingent [26], stakeholder positions and projections may vary over time, which is important to consider.

Despite these limitations, the similarities and contrasts found in the different neighbourhoods support the importance of identifying the profiles and positions of the stakeholders so as to develop sustainable and endogenous processes of urban regeneration. Therefore, addressing these limitations in future lines of research can reinforce the potential of using the sociogram as a tool to legitimize urban regeneration processes through the participation and commitment of all stakeholders, their alliances and coalitions. The analysis of neighbourhoods with successful urban regeneration processes could offer valuable reference points. Future research could explore the application of the sociogram in various urban contexts, including comparative studies between disadvantaged and non-disadvantaged neighbourhoods, or between different regions or cities. Furthermore, exploring the effect of the combined use of quantitative and qualitative data could enrich our understanding and provide practical guidelines. Given the dynamic nature of the sociogram, it seems relevant to implement longitudinal studies that monitor changes over time.

Author Contributions: All authors contributed equally. All authors have read and agreed to the published version of the manuscript.

Funding: This research received no external funding.

Data Availability Statement: Data are contained within the article.

Conflicts of Interest: The authors declare no conflicts of interest.

References

1. Ansell, C. Collaborative Governance. In *The Oxford Handbook of Governance*; Oxford University Press: Oxford, UK, 2012. [CrossRef]
2. Sianes, A.; Vela-Jiménez, R. Can Differing Opinions Hinder Partnerships for the Localization of the Sustainable Development Goals? Evidence from Marginalized Urban Areas in Andalusia. *Sustainability* **2020**, *12*, 5797. [CrossRef]
3. Oslender, U. Decolonizing Cartography and Ontological Conflict: Counter-Mapping in Colombia and "Cartographies Otherwise". *Polit. Geogr.* **2021**, *89*, 102444. [CrossRef]
4. Olarte-Riaño, J.R.; Mundo-Hernández, J.J.; Santiago Azpiazu, G.C. Diseño Metodológico Para El Ordenamiento Territorial Participativo: Una Propuesta Para Valle Del Paraíso En Puebla, México. *Cult. Científica* **2020**, *18*, 74–94. [CrossRef]
5. Habegger Lardoeyt, S. *La Cartografía del Territorio como Práctica Participativa de Resistencia. Procesos en Metodologías Implicativas, Dispositivos Visuales y Mediación Pedagógica para la Transformación Social*; Universidad de Málaga: Málaga, Spain, 2008.
6. Bradley, S.; Mahmoud, I.H.; Arlati, A. Integrated Collaborative Governance Approaches towards Urban Transformation: Experiences from the CLEVER Cities Project. *Sustainability* **2022**, *14*, 15566. [CrossRef]
7. Suša, R.; de Andreotti, V.O. Social Cartography in Educational Research. In *Oxford Research Encyclopedia of Education*; Oxford University Press: Oxford, UK, 2019. [CrossRef]
8. Caballero Ferrándiz, J.; Martín Gutiérrez, P.; Villasante, T.R. Debatiendo Las Metodologías Participativas: Un Proceso En Ocho Saltos. *Empiria Rev. Metodol. Cienc. Soc.* **2019**, *44*, 21–45. [CrossRef]

9. Martí, J. Diagnósticos Comunitarios y Participación Local: El Diagnóstico Comunitario de La Zona Ponent de Tarragona. Observatorio Internacional de Ciudadanía y Medio Ambiente Sostenible (CIMAS). 2017. Available online: http://www.redcimas.org/wordpress/wp-content/uploads/2012/08/m_JMarti_DIAGNOSTICOS.pdf (accessed on 30 April 2024).
10. Wihofszky, P. Die Praxis der Gesundheitsförderung Zwischen Top-Down und Bottom-Up: Ansätze Partizipativen Planens und Forschens in der Konzeptentwicklung. *Pravent. Gesundheitsford.* **2013**, *8*, 181–190. [CrossRef]
11. Gareis, M.C.; Miguel, R.E. Contribución del Sociograma al Enfoque Nexo Agua, Energía y Alimentación. Avances en el Estudio del Valle Antinaco—Los Colorados (La Rioja, Argentina). *Rev. Latinoam. Metodol. Cienc. Soc. Relmecs* **2022**, *12*, 4, ISSN 1853-7863. [CrossRef]
12. Vidalou, J.-B. *Ser Bosques*, 1st ed.; Errata Naturae Editores: Madrid, Spain, 2020; ISBN 978-84-17800-48-2.
13. Carlos, A.F.A. El Concepto de "Producción Del Espacio" y La Dinámica Urbana Contemporánea Bajo El Dominio Del Capital Financiero. *Rev. Geogr. Norte Gd.* **2022**, *2022*, 89–107, ISSN 0718-3402. [CrossRef]
14. Lefebvre, H. *El Derecho a la Ciudad/Henri Lefebvre*; Prólogo de Mario Gaviria; González-Pueyo, J., Translator; Península: Barcelona, Spain, 1969.
15. Harvey, D. *Urbanismo y Desigualdad Social*, 6th ed.; Siglo XXI: Madrid, Spain, 2007; ISBN 9788432302527.
16. Castells, M. *La Cuestión Urbana; Arquitectura y Urbanismo*, 2nd ed.; Siglo Veintiuno de España Editores: Madrid, Spain, 1976; ISBN 84-323-0139-6.
17. Alsayel, A.; de Jong, M.; Fransen, J. Can Creative Cities Be Inclusive Too? How Do Dubai, Amsterdam and Toronto Navigate the Tensions between Creativity and Inclusiveness in Their Adoption of City Brands and Policy Initiatives? *Cities* **2022**, *128*, 103786. [CrossRef]
18. Tellería Herrera, I. Neoliberalización y Gobernanza En Ciudades Del País Vasco. *Inguruak Soziologia Eta Zient. Polit. Euskal Aldizkaria Rev. Vasca Sociol. Cienc. Política* **2019**, *66*, 1–19, ISSN 0214-7912.
19. Vela-Jiménez, R.; Sianes, A.; López-Montero, R.; Delgado-Baena, A. The Incorporation of the 2030 Agenda in the Design of Local Policies for Social Transformation in Disadvantaged Urban Areas. *Land* **2022**, *11*, 197. [CrossRef]
20. Moore, S.; Trudeau, D. New Urbanism: From Exception to Norm—The Evolution of a Global Movement. *Urban Plan* **2020**, *5*, 384–387. [CrossRef]
21. Fernández García, M. Regeneración Urbana En Perspectiva Andaluza. Los Programas ARB (1996–2012). *Anduli* **2018**, *17*, 67–88. [CrossRef]
22. De Gregorio Hurtado, S. Políticas Urbanas de La Unión Europea Desde La Perspectiva de La Planificación Colaborativa. *Cuad. Investig. Urbanística* **2015**, *98*, 1–74. [CrossRef]
23. Healey, P.; Hillier, J. Reflecting on the BOM: The Transformative Potential of Neighbourhood Initiatives. In *Social Innovation as Political Transformation: Thoughts for a Better World*; Edward Elgar Publishing Ltd.: Cheltenham, UK, 2019; pp. 62–67. ISBN 9781788974288.
24. Navarro Yáñez, C.J.; Echaves García, A.; Guerrero Panal, G.; Mateos Mora, C.; Rodríguez García, M.J.; Moya Alfonso, R.; Zapata Moya, Á.R. *Mejorar La Ciudad Transformando Sus Barrios. Regeneración Urbana En Andalucía (1990–2015)*; Navarro, C.J., Ed.; Centro de Sociología y Políticas Locales—Universidad Pablo de Olavide: Sevilla, Spain, 2016; ISBN 978-84-608-8550-4.
25. Portela, M.; Errandonea, L.P. The Role of Participatory Social Mapping in the Struggle of the Territory and the Right to the City: A Case Study in Buenos Aires. In Proceedings of the 8th International Conference on Communities and Technologies, Troyes, France, 26–30 June 2017; ACM International Conference Proceeding Series. ACM Press: New York, NY, USA, 2017; pp. 100–104, Part F128532. [CrossRef]
26. Latour, B. Network Theory | Networks, Societies, Spheres: Reflections of an Actor-Network Theorist. *Int. J. Commun.* **2011**, *5*, 1–15.
27. Encina, J.; Ezeiza, A. *De Los Modelos Participativos a La Construcción Colectiva: Culturas Populares, Ilusionismo Social y Desempoderamiento*; Ikaskideta Sarea-Diputación foral de Gipuzkoa: Donostia, Spain, 2015; Volume 8, ISBN 978-84-15602-11-8.
28. Latour, B. On Actor-Network Theory. A Few Clarifications, Plus More Than a Few Complications. *Logos* **2017**, *27*, 173–200. [CrossRef]
29. Ortiz Ocaña, A. Fritjof Capra y La Teoría Social. *Rev. Logos Cienc. Tecnol.* **2017**, *9*, 35–42. [CrossRef]
30. Capra, F. Complexity and Life. *Syst. Res. Behav. Sci.* **2007**, *24*, 475–479. [CrossRef]
31. Ledesma González, O.; Dates, F. Redes de Gobernanza y Complejidad de La Planificación Territorial. *Cuad. Geográficos* **2021**, *60*, 91–104. [CrossRef]
32. Bone, J. The Social Map: Cohesion, Conflict and National Identity. *Natl. Ethn. Politics* **2007**, *12*, 347–372. [CrossRef]
33. Harley, J.B. Cartography, Ethics and Social Theory. *Cartogr. Intemational J. Geogr. Inf. Geovisualization* **2006**, *27*, 1–23. [CrossRef]
34. Ujaque, D.G.; Jeyasingham, D. El Contra-Mapeo En El Grado de Educación Social: Una Práctica Crítica y Atmosférica Para Repensar El Espacio Urbano. *Scr. Nova Rev. Electrónica Geogr. Cienc. Soc.* **2021**, *25*, 97–122. [CrossRef]
35. Gutiérrez, P.M. El Sociograma Como Instrumento Que Desvela La Complejidad. *Empiria Rev. Metodol. Cienc. Soc.* **1999**, *2*, 129–152. [CrossRef]
36. Villasante, T.; Gutiérrez, P.M. Redes y Conjuntos de Acción: Para Aplicaciones Estratégicas En Los Tiempos de La Complejidad Social. *Redes Rev. Hisp. Para Análisis Redes Soc.* **2006**, *11*, 1–22. [CrossRef]
37. Rhodes, R.A.W. Understanding Governance: Ten Years On. *Eur. Group Organ. Stud.* **2007**, *28*, 1243–1264. [CrossRef]

38. Barandiaran, X. Collaborative Governance for Public Social Innovation: The Case of Gipuzkoa Basque Country. *Eur. Public Soc. Innov. Rev.* **2021**, *6*, 78–96. [CrossRef]
39. Villasante, T.R. Para Avanzar Con Las Metodologías Participativas, Usando La Socio-Praxis. In *De sur a Norte: Metodologías Participativas Desde la Sociopraxis*; Editora Faith: São Borja, Brazil, 2015.
40. Healey, P. *Collaborative Planning: Shaping Places in Fragmented Societies*, 2nd ed.; Palgrave Macmillan: London, UK, 2006; ISBN 9781137086006.
41. Rodríguez, R.M. Desarrollo Local y Análisis de Redes Sociales: El Valor de Las Relaciones Como Factor Del Desarrollo Socioeconómico. *Redes Rev. Hisp. Para Análisis Redes Soc.* **2010**, *18*, 278–304.
42. Ran, B.; Qi, H. Contingencies of Power Sharing in Collaborative Governance. *Am. Rev. Public Adm.* **2018**, *48*, 836–851. [CrossRef]
43. Lima, V. The Limits of Participatory Democracy and the Inclusion of Social Movements in Local Government. *Soc. Mov. Stud.* **2019**, *18*, 667–681. [CrossRef]
44. Requena Santos, F. Análisis de Redes Sociales: Orígenes, Teorías y Aplicaciones. *Monografías* **2003**, *198*, 497.
45. Tubaro, P.; Ryan, L.; D'angelo, A. The Visual Sociogram in Qualitative and Mixed-Methods Research. *Sociol. Res. Online* **2016**, *21*, 180–197. [CrossRef]
46. Drahota, A.; Dewey, A. The Sociogram: A Useful Tool in the Analysis of Focus Groups. *Nurs. Res.* **2008**, *57*, 293–297. [CrossRef] [PubMed]
47. Requena Santos, F. El Concepto de Red Social. *Rev. Esp. Investig. Sociol.* **1989**, *48*, 137–152. [CrossRef]
48. Rapoport, A.; Horvath, W.J. A Study of a Large Sociogram. *Behav. Sci.* **1961**, *6*, 279–291. [CrossRef] [PubMed]
49. Freire, P. *Pedagogía del Oprimido*, 18th ed.; Siglo XXI: Montevideo, Uruguay, 1977.
50. Alberich Nistal, T. Iap, Redes y Mapas Sociales: Desde La Investigación a La Intervención Social. *Portularia Rev. Trab. Soc.* **2008**, *8*, 131–151, ISSN 1578-0236.
51. Harrison, K.L.; Taylor, H.A. Healthcare Resource Allocation Decisions Affecting Uninsured Services. *J. Health Organ. Manag.* **2016**, *30*, 1162–1182. [CrossRef] [PubMed]
52. Creswell, J.W. *Qualitative Inquiry and Research Design: Choosing Among Five Approaches*, 3rd ed.; Knight, V., Ed.; Sage: Los Angeles, CA, USA, 2013; ISBN 9781412995306.
53. Stake, R. *The Art of Case Study*; Sage Publications: Los Angeles, CA, USA, 1995; ISBN 9780803957671.
54. Perez-Linan, A. El Método Comparativo: Fundamentos y Desarrollos Recientes. In *Política Comparada Sobre América Latina: Teorías, Métodos y Tópicos*; Castiglioni, R., Fuentes, C., Eds.; Universidad Diego Portales: Santiago, Chile, 2007; pp. 47–78.
55. Tonon de Toscano, G. La Utilización Del Método Comparativo En Estudios Cualitativos En Ciencia Política y Ciencias Sociales: Diseño y Desarrollo de Una Tesis Doctoral. *Kairos Rev. Temas Soc.* **2011**, *7*, 1–12, ISSN 1514-9331.
56. Mills, A.J.; Durepos, G.; Wiebe, E. *Encyclopedia of Case Study Research*; SAGE Publications: Los Angeles, CA, USA, 2010; ISBN 978-1-4129-5670-3.
57. Yin, R.K. *Qualitative Research from Start to Finish*; Guilford Press: New York, NY, USA, 2011; ISBN 9781606239773.
58. Bolger, N.; Davis, A.; Rafaeli, E. Diary Methods: Capturing Life as It Is Lived. *Annu. Rev. Psychol.* **2003**, *54*, 579–616. [CrossRef] [PubMed]
59. Marincowitz, G.J.O. How to Use Participatory Action Research in Primary Care. *Fam. Pract.* **2003**, *20*, 595–600. [CrossRef] [PubMed]
60. Babicki, S.; Arndt, D.; Marcu, A.; Liang, Y.; Grant, J.R.; Maciejewski, A.; Wishart, D.S. Heatmapper: Web-Enabled Heat Mapping for All. *Nucleic Acids Res.* **2016**, *44*, W147–W153. [CrossRef] [PubMed]
61. Bertens, L.C.M.; Mohabier, K.S.C.; van der Hulst, M.; Broekharst, D.S.E.; Ismaili M'hamdi, H.; Burdorf, A.; Kok, R.; de Graaf, J.P.; Steegers, E.A.P. Complexity and Interplay of Faced Adversities and Perceived Health and Well-Being in Highly Vulnerable Pregnant Women—The Mothers of Rotterdam Program. *BMC Public Health* **2023**, *23*, 43. [CrossRef] [PubMed]
62. Gomez, N.J. Mapa de Calor Del COVID-19 En Santa Fe: Análisis Provincial y Metropolitano. *Rev. Del Inst. De Investig. Geográficas* **2021**, *5*, 1–15.
63. Raidou, R.G.; Eduard Groller, M.; Eisemann, M. Relaxing Dense Scatter Plots with Pixel-Based Mappings. *IEEE Trans. Vis. Comput. Graph.* **2019**, *25*, 2205–2216. [CrossRef] [PubMed]
64. Alberich, T. *Desigualdad, Clases Sociales y Sociedad de Los Tres Tercios (2ª Parte de Tres)*; Pensamientos Talberich: Madrid, Spain, 2017.
65. INE Instituto Nacional de Estadística. *Indicadores Urbanos Edición 2022*; INE Instituto Nacional de Estadística: Madrid, Spain, 2022.
66. García-Araque, J. Inclusive Capacity of Public Space in Disadvantaged Urban Areas: Case in the City of Valladolid. *OBETS* **2022**, *17*, 101–120. [CrossRef]
67. Ministerio de Vivienda y Agenda Urbana. Catálogo de Barrios Vulnerables de España 2011. 2021. Available online: https://portalweb.mitma.es/aplicaciones/portalweb/BarriosVulnerables (accessed on 15 November 2021).
68. Ayuntamiento de Córdoba—Estadísticas 2021. Available online: https://www.cordoba.es/la-ciudad/cifras-estadisticas/estadisticas-de-poblacion/estadisticas-2021 (accessed on 10 April 2023).
69. Serrano, L.; Sianes, A.; Ariza-Montes, A. Using Bibliometric Methods to Shed Light on the Concept of Sustainable Tourism. *Sustainability* **2019**, *11*, 6964. [CrossRef]

70. Egea Jiménez, C.; Antonio, J.; Calmaestra, N.; Domínguez Clemente, J.; Rego, R.A.G.; Egea, C.; Nieto, A.; Domínguez, J.; González, R.A. Viejas y Nuevas Realidades Urbanas. Identificación de Zonas de Habitabilidad Desfavorecida En La Ciudad de Granada. *Cuad. Geográficos* **2009**, *45*, 83–105.
71. Alberich, T.; Belmonte, R.; Espinar, C. *Metodologias Participativas: Sociopraxis para la Creatividad Social*; Dextra Editorial: Madrid, Spain, 2015; ISBN 9788416277506.
72. Gutiérrez, P.M. Balance de Diferentes Técnicas Participativas Para Integrar Metodologías Creativas. In *Prácticas Locales de Creatividad Social. Construyendo Ciudadanía 2*; Villasante, T.R., Montañés, M., Martíne, P., Eds.; El Viejo Topo: Madrid, Spain, 2001; pp. 41–66. ISBN 84-95224-18-6.
73. Serrano, M.M.; Prieto, T.R.-V.; Gutiérrez, P.M. Prácticas Locales de Creatividad Social. In *Prácticas Locales de Creatividad Social. Construyendo Ciudadanía 2*; El viejo Topo: Madrid, Spain, 2001; pp. 123–136.
74. García Hernández, A. Las Redes de Colaboración Científica y Su Efecto En La Productividad. Un Análisis Bibliométrico. *Investig. Bibl. Arch. Bibl. Inf.* **2013**, *27*, 159–175. [CrossRef]
75. Villasante, T.R.; Garrido, F.J. *Metodologías y Presupuestos Participativos*; Garrido, M., Ed.; IEPALA: Madrid, Spain, 2002; ISBN 84-89743-24-x.
76. Gardner, J.R.; Rachlin, R.; Sweeny, H.W.A.; Richards, A. Handbook of Strategic Planning. *RD Manag.* **1989**, *19*, 201. [CrossRef]
77. Rawlins, B. Prioritizing Stakeholders for Public Relations. Institute for Public Relations. Published March 2006. Available online: https://www.instituteforpr.org/wp-content/uploads/2006_Stakeholders_1.pdf (accessed on 5 January 2022).
78. Grunig, J.E.; Hunt, T. *Managing Public Relations*; Holt, Rinehart and Winston: Austin, TX, USA, 1984; ISBN 978-0-03-058337-7.
79. Ansell, C.; Gash, A. Collaborative Governance in Theory and Practice. *J. Public Adm. Res. Theory* **2008**, *18*, 543–571. [CrossRef]
80. Villasante, T.R. Las Democracias Como Los Bosques. *Liminar Estud. Soc. Humanísticos* **2005**, *3*, 26–38, ISSN 1665-8027/2007-8900.
81. Encina, J.; Rosa, M. El Uso de Espacios y Tiempos Cotidianos Como Constructor de Democracias Ambientales en el Sur. In *Democracia Ecológica. Formas y Experiencias de Participación en la Crisis Ambiental*; Atrapasueños: Madrid, Spain, 2006; pp. 1–13. ISBN 9788461115709.
82. Puiggrós, A. Lecturas de Freire. *Perfiles Educ.* **2021**, *43*, 11–21. [CrossRef]
83. López, M.M. Los Movimientos Sociales Urbanos. Un Análisis de La Obra de Manuel Castells. *Rev. Int. Sociol.* **2003**, *61*, 81–106. [CrossRef]
84. Castells, M. Movimientos Sociales Urbanos. In *Arquitectura y Urbanismo*, 3rd ed.; Siglo Veintiuno de España: Madrid, Spain, 1977; ISBN 84-323-0153-1.
85. Martínez, M.A.; Wissink, B. Urban Movements and Municipalist Governments in Spain: Alliances, Tensions, and Achievements. *Soc. Mov. Stud.* **2021**, *21*, 659–676. [CrossRef]
86. Flyvbjerg, B. Five Misunderstandings about Case-Study Research. *Qual. Inq.* **2006**, *12*, 219–245. [CrossRef]
87. Gutiérrez, P.M. Mapas Sociales: Método y Ejemplos Prácticos'. *Diálogos Educ. Form. Pers. Adultas* **2001**, *25*, 26–36, ISSN 1134-7880.

land

MDPI

Article

Mining versus Indigenous Protected and Conserved Areas: Traditional Land Uses of the Anisininew in the Red Sucker Lake First Nation, Manitoba, Canada

Chima Onyeneke [1], Bruce Harper [2] and Shirley Thompson [1,*]

[1] Natural Resources Institute, University of Manitoba, Winnipeg, MB R3T 2M6, Canada; onyenekn@myumanitoba.ca

[2] Red Sucker Lake First Nation and Natural Resource Institute, University of Manitoba, Winnipeg, MB R3T 2M6, Canada; harperb7@myumanitoba.ca

* Correspondence: s.thompson@umanitoba.ca; Tel.: +1-204-291-8413

Abstract: Indigenous traditional land uses, including hunting, fishing, sacred activities, and land-based education at the Red Sucker Lake First Nation (RSLFN) in Manitoba, Canada, are impacted by mining. The Red Sucker Lake First Nation (RSLFN) people want their territories' land and water to be protected for traditional uses, culture, and ecological integrity. Towards this goal, their Island Lake Tribal Council sought support for an Indigenous-protected and conserved area (IPCA) in their territory, outside of existing mining claims, but without success. The two-eyed seeing approach was adopted in this study. Traditional land use mapping and interviews were undertaken with 21 Indigenous people from the RSLFN, showing that many traditional land uses are concentrated on greenstone belts. The interviews revealed that mining exploration has resulted in large petroleum spills, noise distress, private property destruction, wildlife die-offs, and animal population declines. These issues negatively impact RSLFN's traditional land use practices, ecosystem integrity, and community health. Governments need to partner with Indigenous communities to reach their biodiversity targets, particularly considering northern Canada's peatlands, including those in the RSLFN territory, surpassing Amazon forests for carbon storage. The role of critical minerals in renewable energy and geopolitics has colonial governments undermining Indigenous rights, climate stabilization, and biodiversity to prioritize extractivism. Mining at the RSLFN has environmental impacts from exploration to decommissioning and after, as well as the massive infrastructure required that includes roads, hydro, and massive energy supplies, with a proposed multimedia national Northern Corridor to export RSLFN's resources and other resources to six ports.

Keywords: mining impacts; indigenous knowledge; traditional land use; land conservation; land-back; indigenous rights

1. Introduction

Should gold mining be favored over Indigenous protected and conserved areas (IPCA) when land use planning? Indigenous people have protected four-fifths of global biodiversity and a third of old-growth forests, despite making up only one-twentieth (6%) of the world's population [1–3]. This biodiversity concentrated on Indigenous-protected land demonstrates the effectiveness of Indigenous peoples as environmental guardians [2,4]. By observing Indigenous protocols and natural laws, Indigenous peoples' ecosystems have operated within ecological limits for millennia [5,6]. This paper documents the Red Sucker Lake First Nation's (RSLFN) traditional land uses and their efforts towards biodiversity conservation in their Indigenous territory in the Anisininew's Island Lake area of Manitoba, Canada, considering the impact of the competing land use of mining.

Indigenous peoples' knowledge is deemed critical in adapting and mitigating the biodiversity and climate change crises [3,7] for human survival on Earth. Permanently passing the 'survival target' of 1.5 °C is close at hand with record heat waves, forest fires, droughts, storms, and water scarcity in 2023. That year, temperatures averaged 1.48 °C and 1.5 °C for twelve consecutive months until February 2024 [8]. If global warming surpasses 1.5 °C, concurrent heatwaves with droughts, compound flooding, and/or fire weather in many regions are predicted [8]. Climate change and land use changes are the largest extinction threats [9]. Biodiversity loss threatens one million species with extinction, many at risk over the next few decades [3,7]. Nature is on the verge of collapse with 50% biomass reduction and 40% of plants endangered [4,7]. The United Nations' global biodiversity target has each country striving for the ambitious conservation target of 30% land and water conservation by 2030, related especially to many sustainable development goals (SDGs), particularly climate action (goal 13), life on water (goal 14), life on land (goal 15), and partnerships (goal 17) [7,9]. This conservation goal is considered to be impossible in Canada without Indigenous people actively protecting Native land through Indigenous conserved areas [10,11].

Mitigating the root causes of climate change, biodiversity, and ecological collapse requires addressing their colonial roots, according to Indigenous people [12]. Colonial power dynamics have shaped climate change and biodiversity loss so that a shift from carbon to renewables will not solve the crisis [12]. Despite an increased renewable electricity share, greenhouse gas emissions reached a new record high in 2023 [12]. Colonialism is when a foreign power controls economic, political, social, and cultural aspects of people's lives in a colonized nation [13]. Settler-colonial states largely ended during the national liberation movement era post-World War II, but not for Indigenous people in Canada and in several other countries. Indigenous people continue to struggle for self-determination and Native land protection against colonial intrusion, including critical mineral development in RSLFN's ancestral territory [14,15].

This paper discusses the role of self-determination and traditional land uses [16] for biodiversity and conservation in the RSLFN territory, through a two-eyed seeing approach [17–21]. The literature on self-determination, two-eyed seeing, resource curse, mining, Indigenous protected and conserved areas (IPCAs), and traditional land uses are explored in this paper. We examine if IPCAs can protect biodiversity and traditional land uses [22]. The impacts of mining and other extractive industries [23], including the resource curse brought to First Nation people [23], who are the Indigenous people in Canada governed under the Indian Act, are reviewed. RSLFN's remote location, population size, Anisininew culture, and economic hardships are profiled [24]. In the method, the two-eyed seeing process is adopted in collecting, mapping, and analyzing stories and traditional land use. Interviews and maps reveal the importance of the land and the impacts of mining in the findings. Finally, we analyze the intersection of mining and traditional land uses using maps, and what that means for self-determination and biodiversity. Further, the discussion considers the importance of land rights to the RSLFN people and the role of an IPCA in biodiversity conservation and climate goals, compared to mining's impacts. The limitations of this study and future research areas are identified in the discussion section before concluding.

Indigenous Self-determination

The International Covenants on both Civil and Political Rights and Economic, Social and Cultural Rights, Article 1 states: "All people have the right to self-determination" [25]. Self-determination is defined as: "the right to live a particular way of life, to practice a specific culture or religion, to use own languages, and the ability to determine the future course of economic development" [25]. Indigenous peoples possess the right to define and govern their knowledge, social, economic, cultural systems, and ecosystems [5]. Self-determination is a way of life determined by the creator, not one dictated by colonial government [6]. The RSLFN people's Anisininew word for a good life, as defined by the creator, is Mino Bimaadiziwin [6].

The United Nations Declaration on the Rights of Indigenous Peoples (UNDRIP) affirms Indigenous peoples' rights to self-determination of their lands, territories, resources, and cultural identity [26]. In 2007, UNDRIP was signed by 144 countries [27]. Three years later, in 2010, UNDRIP was signed by the settler-colonial states of Canada, the United States, Australia, and New Zealand [27]. Specifically, UNDRIP's Article 3 states: "Indigenous peoples have the right to self-determination. By virtue of that right, they freely determine their political status and freely pursue their economic, social, and cultural development" [26]. Rather than the United Nations conferring the implementation of UNDRIP to Indigenous people, problematically, UNDRIP's devolution of power is to nation-states. In Canada, the colonial state is tasked with UNDRIP's progress on Indigenous self-determination, which conflicts in many ways with national interests [28].

Free, Prior, and Informed Consent (FPIC) of Indigenous people is affirmed in UNDRIP, which reads: "States shall consult and cooperate in good faith with the Indigenous people concerned through their representative institutions to obtain their free, prior and informed consent before adopting and implementing legislative or administrative measures that may affect them [26]". Consultation with Indigenous people is to occur before development or using resources on Indigenous territory [29,30]. Articles 26-1,-2 of UNDRIP declare that Indigenous people have a right to control their traditional lands, territories, and resources: "Indigenous peoples have the right to the lands, territories and resources, which they have traditionally owned, occupied or otherwise used or acquired. . .[and] the right to own, use, develop and control the lands, territories and resources [26]".

Self-determination principles apply to Indigenous research. Research partnerships with Indigenous people and nations must apply self-determination principles in an ethical process. Self-determination requires the ownership, control, access, and possession (OCAP) of data and research by the First Nations engaged in the research [31]. Self-determination for Indigenous people, and coexistence, necessitates researching how to shift colonial policies, which benefits from a two-eyed seeing approach [32].

Two-eyed seeing

Two-eyed seeing harnesses the insights of Indigenous and Western ways of knowing in a complementary fashion [16–18]. Two L'nu Elders, Drs. Albert and Murdena Marshall, conceived of the two-eyed seeing approach in research to decolonize research and advance the self-determination of Indigenous communities [16]. Two-eyed seeing shifts the narrative from subjugating and colonizing Indigenous knowledge to the co-existence and integration of different knowledge systems [16–18]. Indigenous and Western systems have distinct values, purposes, protocols, methods, data collection, and outcomes that offer different insights [16–18].

Western knowledge has long been considered universal knowledge, obliterating other ways of seeing the world [16–18]. Western knowledge has Euro-centric values and worldviews [33,34]. Western knowledge makes philosophical assumptions guided by theories, power structures, and hierarchies. Western science tends towards a narrow view by isolating factors and disciplines, and this view is often over a short timeframe [33]. Hill [5] calls Western knowledge "industrial knowledge", with its focus on serving capital accumulation and modernization.

Indigenous knowledge operates within the Indigenous cosmovision that the world is alive and sacred [5]. Indigenous knowledge is learned from stories, language, culture, and the land [34,35]. Place-based, experiential knowledge, and spirituality are valued and considered a core aspect of Indigenous identity and health [34]. Ecosystem integrity and human well-being are not isolated within different disciplines, but are considered inseparable and a part of natural law [36]. Through ancient knowledge of their ecosystem, Indigenous land protocols and traditional land uses were aligned to follow natural laws [5]. This ecological knowledge values a stewardship relationship with land and wildlife, rather than an extractivist view [3,5].

Traditional Land Uses

Through traditional land uses, Indigenous people continue to derive many of their basic needs from the land, including food, water, and medicines [2,5,6]. Traditional land uses are undertaken in a sustainable, regenerative way, based on a strong reciprocal relationship with the land and water, and a culture of stewardship [5,6,32]. Traditional land uses include hunting, gathering, farming, fishing, ceremony, spirituality, education, and land stewardship [14,15].

Traditional land uses, such as wildlife harvesting and ceremonies, represent Indigenous spiritual and physical connection with the land. These practices heal trauma and protect biodiversity [6,36]. However, Indigenous land stewardship systems and traditional land uses have been under attack for centuries by colonial policies negatively impacting culture, ecosystem integrity, land access, and human health [5]. Elders blame the skyrocketing rates of addictions and suicides on spiritual and physical disconnection from the land [36].

Land stewardship and other traditional land uses represent Indigenous peoples' cultural identity. From the land, sustenance requires knowledge of traditional foods [6]. Teaching youth about traditional land uses, including food harvesting, storage, and processing, is typically done within the family [6]. However, this intergenerational knowledge transfer of traditional land use and food was disrupted by the Canadian government taking children away from their families, culture, and communities to residential schools [36–38]. At residential schools, Indigenous children were indoctrinated into the settler society and often abused [36–41]. Households with adults affected by residential schools have a significantly higher rate of severe food insecurity for the residential school attendees, their children, and their grandchildren, with elevated rates of 10%, 9%, and 5%, respectively [39,40]. Residential schools are partly blamed for the high prevalence of food insecurity and diminished traditional land use among First Nation people, which is creating a population-level health crisis and cultural genocide [39–41].

The drop in traditional foods eaten stems from a lack of access to land. Colonial land uses for settlement and extractive industries conflict with traditional land uses, reducing wildlife [41,42]. Their decline in traditional food use results in First Nation peoples having higher rates of heart disease, obesity, diabetes, cancer, osteoporosis, infections, and tooth decay [41,42]. These negative outcomes of the extractive industry are part of the resource curse that befalls Indigenous communities.

Resource Curse

A resource curse is defined as the largely negative impacts felt by a community proximate to abundant natural resources, when resources are extracted by outsiders [23,42]. Also known as the "paradox of plenty", a resource curse occurs when mainly negative impacts are felt by local people from resource extraction [23]. The resource curse most often befalls isolated or remote marginalized communities [42,43]. Isolated and marginalized describes many First Nation communities in Canada, including the RSLFN.

The Indian Act makes First Nation communities highly vulnerable to the resource curse [23]. The Indian Act (1876) made First Nations' people wards of the state, thereby placing Native lands and resources under the Crown's trust laws [28,44]. In this way, the Crown trustee gained legal authority over land, resources, and people [28,44]. The Crown permits industrial extraction and settler development on Native land: "Provincial and federal authorization for extraction and development on Indigenous territories take place without Indigenous consent [44] (p. 44)".

Resource wealth from Indigenous territories does not flow to First Nations people [23,45,46]. Since Canada's confederation, the Crown usurped billions in profits from Native land, timber, energy, gold, and other resources [23,45]. For example, the Crown collected $50 million in energy royalties from oil patch activity as the trustee of the Bearspaw First Nation's territory in Alberta [45]. Also, this racialized law results in inequitable human rights, services, and infrastructure. For example, most First Nations lack hospitals, with only one hospital in 63 First Nation communities in Manitoba. Further, 122 First Nation communities in Canada are without all-season access roads, including large communities of four thousand [47,48].

Canadian courts limit the power of First Nations to protect or benefit from Native land: "First Nations are radically constrained in negotiations for their rights and by the oppressive socio-economic structures of settler society, where industry interests often drive politics [44] (p. 13)". This explains why the fight of First Nation people for the environmental protection of their territory has been a losing battle. First Nations people's injunctions against the Canadian government and corporations were mostly denied, with only 18% granted, while 76% of corporation's injunctions were successful [44]. For example, in 2019, Coastal GasLink Ltd. was granted an injunction against Wet'suwet'en members protesting pipeline construction on their Native land [23]. A 2013 example occurred in the RSLFN territory when the RSLFN chief delivered an eviction notice to the mining company at Monument Bay. In response, the mine was granted a court injunction to evict RSLFN. Despite Monument Bay being part of RSLFN's territory, the Manitoba Courts authorized the mining company to arrest anyone obstructing, trespassing, or creating a nuisance or "engaging in any act which interferes with the operations of the Monument Bay project [49]".

Mining

Mining extracts non-renewable geological resources for industrial purposes. Minerals, particularly critical minerals, are considered essential for modern industrial society for the green and digital economy, which creates strong government support for their development [50–52]. Mining is connected with nation-building and wealth generation [50]. With renewable energies dependent on critical minerals for generation and storage, critical minerals are replacing oil to define geopolitics [53]. Critical minerals have become the new energy resource, with the shift to renewable energy [53].

A rapid global energy system transition to renewable energy from fossil fuels is deemed necessary to prevent catastrophic climate change [52–56]. That demands massive amounts of critical minerals. Critical minerals include tungsten, cobalt, copper, lithium, nickel, and rare earth elements. These critical minerals are required for photovoltaic cells, electric vehicles, batteries, wind turbines, and electrical grid connectivity [57–59].

Mineral demand has skyrocketed with the shift to renewable technologies from fossil fuels. Since 2010, minerals required per new unit of power generation capacity increased on average by 50% as renewable shares grow [54]. Wind farms, electric vehicles (EVs), and solar photovoltaics (PVs) take more minerals to build than their fossil fuel-based counterparts. An EV takes six times more minerals than a conventional energy vehicle and nine times for onshore wind over a gas-fired plant [57,58].

Mining requires massive development to create the necessary infrastructure to operationalize the mine, providing a burst of short-term employment [55,60,61]. Industrial mining projects generally require utility corridors, access roads, transfer stations, site preparation (e.g., draining of lakes), flying skilled workers in, and tailing ponds [62]. For example, to facilitate mining extraction and export in Canada, a northern multi-modal corridor is proposed to transport resources to six ports [63–65]. Industry and the Canadian Senate Committee support the Northern Corridor idea. The Northern Corridor is being devised to access mineral and oil deposits in Island Lake, the Ring of Fire, Saskatchewan's potash, and tar sands in northern locations to transport to six ports. Greenstone belts are a geologic term for an ore deposit associated with high concentrations of precious and critical metals. This proposed northern corridor nominal route crosses many greenstone belts and many areas proposed as Indigenous Protected Conservation Areas (IPCA), including in the Island Lake region, near the RSLFN. An IPCA was proposed by the Island Lake Tribal Council (ILTC) to protect lands.

Indigenous Protected and Conserved Area (IPCA)

An IPCA designation is described as "lands and waters where Indigenous governments have the primary role in protecting and conserving ecosystems through Indigenous laws, governance and knowledge systems" [66] (p. 4). The IPCAs differ from typical state-run parks and conservation initiatives. State-run parks have historically denied Indigenous people a role in land management decisions, resulting in Indigenous dispossession and exclusion from their territory [10,11,22].

Canada has advanced a colonial narrative about land management and conservation. This narrative disregarded Indigenous knowledge, purporting that traditional land uses harm natural habitats [11]. Oppositely, IPCAs recognize the reciprocal relationships that Indigenous people have with their lands and water [10]. The IPCAs recognize, under an Indigenous cosmovision, that traditional land uses of harvesting, hunting, ceremony, education, living, and sustainable industrial activity are beneficial [10,22]. Thus, IPCAs fulfill Indigenous people's cultural, educational, and spiritual purposes, as well as Canada's conservation goals. Youth mentored by traditional knowledge keepers are employed as Indigenous Land Guardians to monitor and manage environmental programs, thus providing jobs, biodiversity protection, and knowledge transfer [66,67].

The Intergovernmental Panel on Climate Change (IPCC) found that land rights for Indigenous peoples are vital for biodiversity, land protection, and climate mitigation [68]. The IPCC recognized that securing Indigenous tenure is highly cost-effective in reducing deforestation and improving land management [68]. Recognizing the tenure of Indigenous communities is important, "particularly those that authorize and respect Indigenous and communal tenure", [68] (p. 6), which improves the management of carbon-dense forests.

In Canada, Indigenous-led conservation initiatives have made the biggest advances in protecting land and water [68]. The Łutsël K'e Dene First Nation protected 26,376 km^2 in the Northwest Territories called Thaidene Nëné ("land of the ancestors") under Dene laws [69]. Thaidene Nëné is one of the largest protected areas in North America. This IPCA managed by the Dehcho First Nations is a partnership with Canada [69].

Canada will fall short of this global UN biodiversity goal without Indigenous leadership in biodiversity conservation. From 2010 to 2020, Canada achieved 12.2% of land and inland water in protected areas, falling short of its 17% goal [68]. Reaching 25% of protected areas by 2025 requires speeding up the process to more than double existing protected areas by 2025 [69]. Canada has recognized and turned to IPCAs to fill gaps. In June 2023, Environment and Climate Change Canada funded studies on 59 Indigenous-led conservation area proposals, but only recognized three across Canada [70]. Many more proposals than the 59 were not funded, including Island Lake Tribal Council, which includes the Indigenous territory of the RSLFN, despite having many at-risk species in an intact ecosystem with rich peatlands. Indigenous people want to commit to IPCAs and conservation, but Canada is reticent. Despite global recognition of the positive role that IPCAs play in biodiversity conservation and climate change, Canada remains slow to commit [3,4,7].

The three established IPCAs are in the boreal forest and have rich peatlands. Peatlands are critical wetlands for mitigating climate change and preserving biodiversity, among other ecological roles [70]. Peatlands constitute the largest natural terrestrial carbon store, holding more carbon than all other vegetation types in the world combined and representing up to 44% of all soil carbon [70]. Canada has 25% of global peatlands, which store at least five times more carbon than tropical forests. These peatlands store carbon for 50 to 100 times longer, at 10,000 years for peat, compared to 100 to 500 years for tropical forests. Canada is the largest peat reservoir of carbon in the world, with the richest peat stored in the northern boreal forests of Ontario and Manitoba. Canada's peatlands are a priority for protection, as damaged peatlands are a major greenhouse gas emitter responsible for almost 5% of global anthropogenic CO_2 emissions [70].

Red Sucker Lake First Nation Community Profile

The Red Sucker Lake First Nation (RSLFN) is one of four Anisininew Island Lake communities in northeast Manitoba. The 953 people living in RSLFN, a population density of 15.4/km^2, are a young, fast-growing population, with one-third of its population below the age of 15 years [24,71]. The RSLFN community is in the pristine Hayes River Watershed, which is one of two watersheds in Manitoba that flows naturally without water control structures or dams. The RSLFN territory is covered in boreal forests and peatlands [70]. The Canadian Shield is the oldest volcanic mountain range in the world, worn down by time and rich in minerals, with many greenstone belts identified in the Island Lake region. Figure 1 shows that the RSLFN is located near the Manitoba-Ontario border with many mining claims, including the very large claim at Monument Bay in Manitoba.

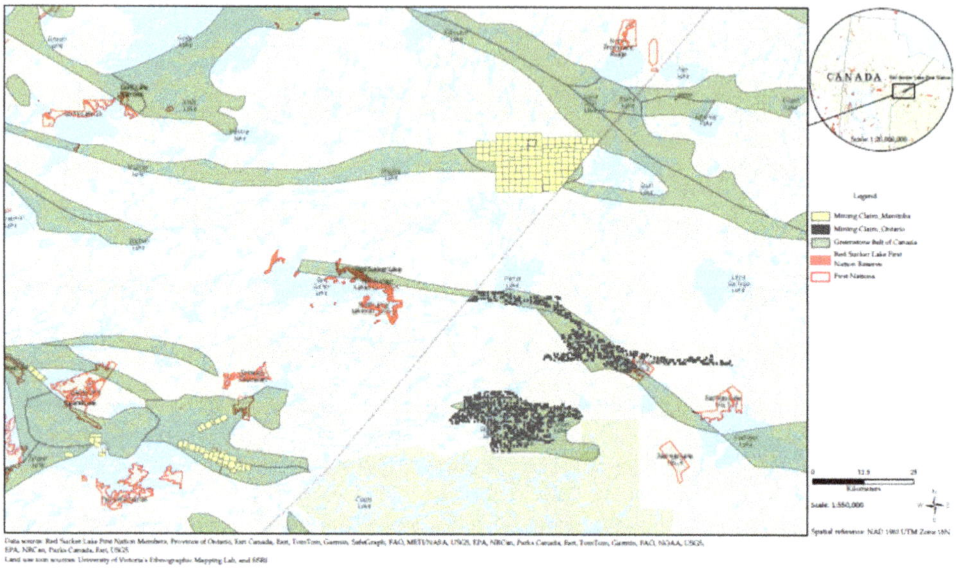

Figure 1. Map showing mining claims within the Red Sucker Lake First Nation traditional territory.

The Red Sucker Lake First Nation is isolated, without any access road to service centers. The community is located 350 air kilometers (km) northeast of Winnipeg and 285 air km southeast of Thompson. The community lacks services, without any hospitals, banks, colleges, or retail options.

The limited infrastructure and services create regional unemployment and a lack of opportunities. Table 1 shows many depressing indicators that RSLFN has, compared to Manitoba and other First Nations, despite having a billion-dollar gold mine on its territory. Table 1 shows the high rates of overcrowded housing, with the average household size in the RSLFN of 4.4 people compared to 2.5 for the total Manitoba population [71]. Houses are often overcrowded due to the colonial government's underfunding of First Nation housing. In the RSLFN, the education attainment is very low, with 5.3% of RSLFN members over 15 years of age having a post-secondary degree and 20% graduating from high school [71]. At seven times lower rates than for Manitoba, 43.5% of RSLFN people over 15 have no certificate, diploma, or degree [71]. This is blamed on the lack of funding available for post-secondary education and access barriers. The unemployment rate in the RSLFN stands at 23%, which is five times the unemployment rate in Manitoba of 4.6% [71]. Employment rates at 36% are roughly half that of Manitoba at 63% [71]. The RSLFN employment, education, and poverty rates are much worse than for the average First Nation in Canada as well.

Table 1. Comparison of economic and educational indicators for the Red Sucker Lake First Nation compared to Manitoba and all First Nations in Canada.

	Manitoba General Population	Red Sucker Lake FN	First Nations In Canada
Average household size (number of people)	2.5	4.4	3.7
Population without formal education (%)	6.8%	43.5%	28.9%
Employment rate (%)	63.1%	35.6%	46.8%
Unemployment rate (%)	4.6%	22.5%	18.0%

Data Source: [24].

2. Materials and Methods

A two-eyed seeing approach applied Indigenous and Western knowledge to the RSLFN case study of traditional land use. In Figure 2, a two-eyed seeing process displays a process to bring Indigenous and Western knowledge together, despite their differences. In this research, Anisininew knowledge keepers, primarily the late Elder Norman Wood and Bruce Harper, guided this research. They defined how research would be conducted, following an approved ethical protocol based on the Indigenous community's consent. Bruce Harper served as the community coordinator, translator, researcher, and protocol expert, participating in guiding the interviews.

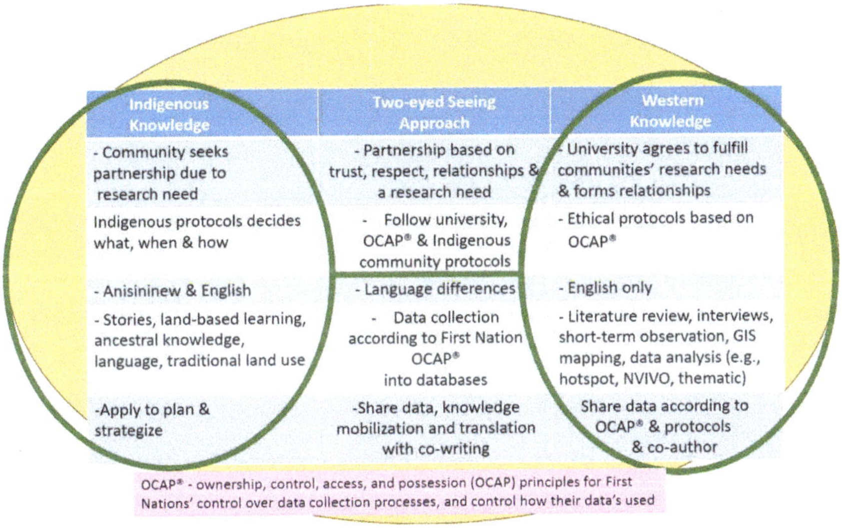

Figure 2. Two-eyed seeing process diagram to incorporate Indigenous and Western views.

The two-eyed seeing journey began with a request from Norman Wood to Dr. Shirley Thompson to help with traditional land use mapping. Dr. Thompson had worked with Island Lake communities, including RSLFN, on other projects and was able to obtain the matching funding required by the funder, Yamana Gold Inc., through MITACs. Two-eyed seeing demanded the academic researchers, who were newcomers to Canada, to undertake a large learning curve, having limited experiential knowledge of RSLFN land, culture, and language.

Ownership, control, access, and possession (OCAP) principles were applied. Control over data collection processes and controlling how their data was applied was provided to the RSLFN. Two First Nation representatives approved Onyeneke's master thesis content [72] after presentations and receiving the content, which this paper presents. The database with all the information, Excel data, and maps are on a protected, shared Teams site, accessible by community members and their GIS experts at Maawandoon Inc., located in Fort William First Nation, Canada.

Indigenous knowledge research, specifically traditional ecological knowledge (TEK) interviews and traditional land use mapping, was undertaken with the RSLFN from 22–24 September 2023. Thompson, Onyeneke, Harper, and Thapa documented the traditional land use interviews with 21 Indigenous people. Each RSLFN participant's shared stories, mapping, and photos provided Indigenous knowledge of the land. These data were recorded on maps and videos or audio tapes, based on the participant's preferences [72–74]. Territorial maps of Manitoba and Ontario at the 1:50,000 scale were used.

Interviewees were asked questions about mapping, hunting, fishing, berry picking, medicinal plant gathering, timber harvesting, community/recreational areas, cultural/sacred sites, and youth training areas. An interview guide and protocol with 64 questions were adopted, as had been approved previously by the RSLFN and the Island Lake Tribal Council (ILTC) [15,72]. The interviews were recorded and transcribed. The periods for interviews ranged from 1 to 3 h. The interviewees consisted of 16 males and five females. All participants were adults over 18 years of age, and three were Elders 65 years of age or older. Interviews were coded to provide confidentiality to participants [72].

The Canada Impact Assessment Registry [62] was reviewed on Feb 10th, 2024 to determine potential mining impacts. The new IAA, 2019 is a pertinent upgrade to the defunct EAA, 2012 of Canada, particularly in its attempt to incorporate Indigenous peoples' ideas and knowledge in the assessment process. The provinces, territories, and Indigenous jurisdictions, in collaboration with the Impact Assessment Agency of Canada, support a single impact assessment, and this reduces duplication of efforts (one project, one assessment) and promotes efficiency, rigor, certainty, and a robust registry [62]. Assessed sites similar to the RSLFN were chosen to review the impacts of mining. The sites considered were those with gold mining in the boreal forest of Northern Canada with nearby Indigenous people.

Western Science

Western science approaches were used to analyze the data. Onyeneke transcribed the audio recordings of the participants' interviews using artificial intelligence (AI) by Otter.ai to generate transcripts, which were corrected by relistening to the audio [72]. NVIVO's description-focused coding method was used with the final transcripts. During the coding process, information related to the research objectives was identified and placed into nodes/containers in NVIVO to categorize these codes into themes and sub-themes [72,75].

Interviewees' land use data were digitized on ArcGIS Pro, applying the North American Datum 1983 (NAD 83) Universal Transverse Mercator (UTM) Zone 15 projection system. We applied the ESRI geographical information system (GIS). A map biography for each interviewee was prepared using shapefiles from the digitized maps [72–74]. Data from all 21 interviews were mapped into thematic maps, hotspot maps, density maps, and summary maps [15,72]. The spatial analysis tool calculated the aerial distance for different land use locations [72,76].

We overlapped GIS layers for traditional land uses with mining greenstone belts and claims layers [72,77,78]. Further, the kernel density formula [72,79] and hotspot function [72,79,80] were applied to land use data and mapped with greenstone belts. The kernel density mapping approach was applied to analyze the density of traditional land uses. The kernel function created regular density areas from all the traditional land use point data in raster form. Mapping [72,81,82] applied the following Kernel function formula:

$$\text{Kernel function } f(x,y) = \frac{1}{nh^2}\sum_{i=1}^{n} K\left(\frac{di}{h}\right)$$

where h is the bandwidth, d_i is the distance of the variable from the center in the bandwidth, k is the kernel density function, and n is the number of observations [72,79,80].

Optimized hot spot mapping was applied, using the hotspot function to identify whether the distribution of land use spots is random or statistically significant at the 95% level. The null $\{H_0\}$ and alternative $\{H_1\}$ hypotheses were:

H_0. *The distribution of land use features of the 21 RSLFN members was random.*

H_1. *The distribution of land use features was statistically significant.*

The optimized hot spot analysis used land use spots to create a map of statistically significant high TLU and medium TLU spots using Getis-Ord Gi statistics [72,76,79,80].

3. Results

3.1. *Land Use of Red Sucker Lake First Nation Community Members*

Many traditional land use activities occur throughout the RSLFN territory. Land use activities of 21 RSLFN members include bird and egg harvesting, cultural sites, fishing, hunting, overnight stay, plants/wood/earth materials harvesting, sacred sites, and trapping. In Figure 3, the summary map biography, all traditional land uses from the 21 interviewed community members' map biographies are compiled. The many land use locations are only a tiny fraction of the entire community's land use sites, being only 21 out of the 953 community members [72,83,84]. If more people were interviewed, the result would be the identification of more traditional land use sites (TLUs), covering a wider region, as indicated by the 2018 RSLFN study [81]. The 2018 study in RSLFN involved 14 different community members and showed many TLUs on different lakes to this study. Traditional land use areas peculiar to the 2018 study include Namapanis Lake, Moose Lake, Mistune Lake, Sakwasi Lake, Robson Lake, Erin Lake, York River, Mukataysip Lake, and Jeffers Lake. Both studies recorded different TLUs around these same lakes, namely: Kistigan Lake, Pierce Lake, Seeber Lake, Rorke Lake, Sharp Lake, Richardson Lake, Stull Lake, and Lenover Lake [81].

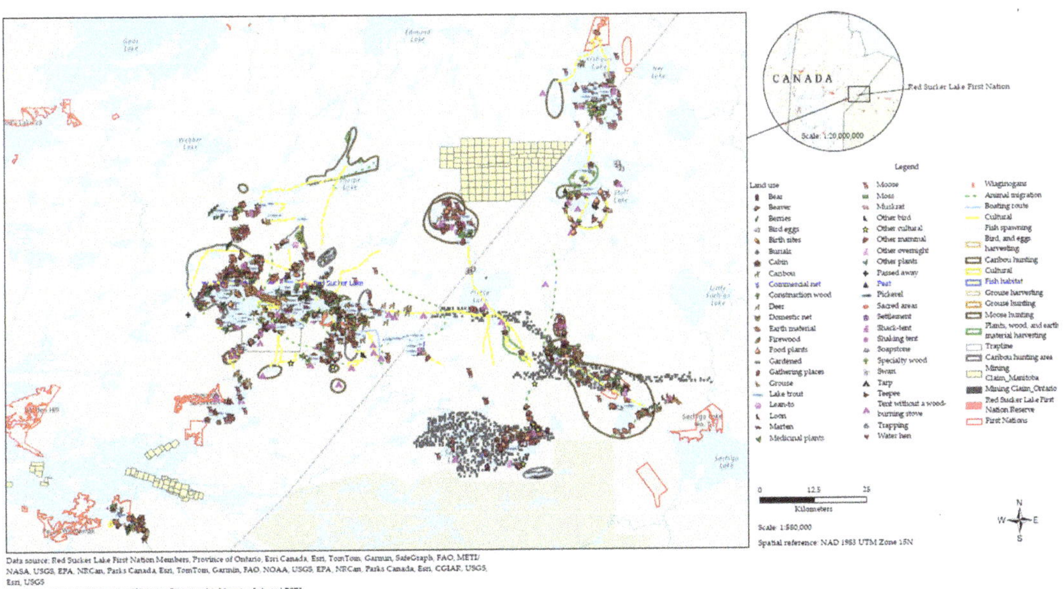

Figure 3. Summary of land use activities of 21 Red Sucker Lake First Nation members in their traditional territory and mining claims.

Hunting, fishing, and gathering is a way of life for the Anisininew people. The land was described as a source of irreplaceable medicines, food, teachings, and healing by an Elder:

> "Aside from traditional food, berries, and medicine, being out on the land has healing from the problems that we face because of Western or European influence. Being out in the wild brings healing. Healing of the mind, tranquility, if we're getting problems and you recoup, you can build yourself up. Having peaceful scenery."

Indigenous people learn from the land, which requires Indigenous-led land conservation to glean the messages from the creator to sustain their culture and well-being. The land provides both sustenance and traditional teachings that bring wisdom, according to another Elder:

> "It (The land) provides sustenance. The traditional teachings...work smart, not hard. My grandfather and uncles used to teach us how to do things, how to set snares, how to trap, and how to hunt moose. I mean you don't just go to the bush and make some noise to scare everything away. The teachings are in the land."

The land is important to the 21 RSLFN interviewees in diverse ways related to their levels of connection to the land and sustenance needs, including hunting, gathering, and other cultural activities. Many expressed having a spiritual connection to the land. The land, like a baby, was considered priceless and alive to be cared for and loved. The land, like a baby, was not to be sold or harmed. An RSLFN member talked about how the creator had given them this land to protect:

> "Because it was given to me...it was given to my family. Man did not give it to us- God gave it to us. That's why it's important to me. And it's a gift, we can't put a price on it...we can't put value on it in terms of money thinking. It's the same thing as you getting a gift. If you get a new baby from your wife, can you sell him- your baby? Can you imagine making a baby so that I can sell it? It's the same thing- it's given to us, not to sell. We have to take care of it. And all the animals. The trees are not given to us but we are entrusted as caretakers. No man gave it to me, God gave to me- my land, my language, my heritage."

Every interviewee emphasized the importance of the land for food and other sustenance. One Elder explained that living off the land was the healthiest way to live:

> "It's a way of providing sustenance, food. . .There are no organic materials that are better than the animals that are here."

Another Elder explained the community's dependence on traditional food, how he got all his meat from the land, not the food store:

> "I don't really buy any meat from the Northern store because I mostly use wild food. That's the number one important thing."

The TLUs of the 21 RSLFN members encircle lakes and rivers. Figure 3 signals the importance of pristine water for trapping beaver, fishing, water birds, and moose for the sustenance of the RSLFN people. Cabins on the trapline are always located adjacent to water bodies to haul water for drinking, cooking, and cleaning. Also, lakes and rivers are their primary travel routes, using canoes, motorboats, and float planes. People travel great distances to reach harvesting and cultural sites. RSLFN members travel aerial distances of 90 km within their traditional territory for traditional land uses. Due to the many bends in the river and portages, the distance travelled in canoes or motorboats is much longer.

Without access to roads, RSLFN people typically travel to their traditional home by canoe when the rivers and lakes are free from ice. One of the interviewees narrated how canoeing took several days from the RSLFN, through Pierce Lake to Ponask Lake (both in Ontario) and back home, with their canoe heavy-laden with harvests:

> "I remember when we took the boat to Rorke Lake. Oh, from Red Sucker Lake to Pierce Lake to Richardson, Twin Lakes, then to Stall Lake. . .there's Kistigan River, then to Rorke Lake. That portage is about 6 miles. We took a boat, gasoline, food, guns, and our clothing. It took three days to get to our destination."

The ability of community members to traverse the land in their territory without access to roads indicates their deep knowledge of the land. Their long trips to harvest food and visit often required overnight stays. On the land, they would stay in cabins, tents, other camping structures, or under the stars.

Traditional land uses of RSLFN people cross the Ontario border. The spatial distribution of land use sites extends beyond RSLFN's reserve areas, traplines, and provincial boundaries to the North Prominent Ridge. The territory goes into Ontario beyond Monument Bay and Stull Lake in the north-east direction, and beyond Ponask Lake near Sachigo Lake area in the south-east direction. Banksian River runs towards Island Lake in the south-west direction. Figure 3 shows the spatial extent of traditional land uses, going 84 km aerially to Sachigo Lake in Ontario. These measurements show the large traditional land use area, with RSLFN harvesting fish "all over" their traditional territory.

3.1.1. Traditional Land Uses Heavily Impacted by Exploration and Mining Activities

Mapping the geological layer of greenstone belts with traditional land uses shows many overlapping areas. Figure 3 shows many traditional land uses nearby mining claims. This proximity or overlap shows a conflict of land uses between mining and traditional land uses of Indigenous people. This conflict is evident in RSLFN participants' descriptions of wildlife and themselves being negatively affected. Figure 4 shows the proximity of distances to the current major mining or exploration areas in operation.

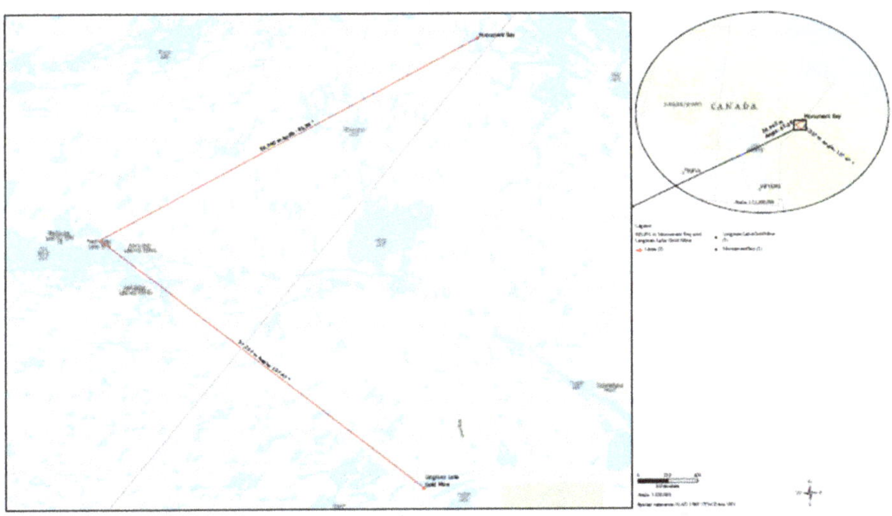

Figure 4. The Red Sucker Lake First Nation community's proximity to the gold exploration site at Monument Bay (60 km away) and the gold mine at Lingman Lake, owned by Signature Resources Ltd. (57 km away).

The RSLFN people reported that mining reduces wildlife populations, undermining wild food procurement. The frequent flights of helicopters and float planes for mining purposes at the RSLFN airport and mining explosives disrupt wildlife behavior and traditional land uses. People complained about moose and other wildlife being driven away by the constant noise of the helicopters flying back and forth shipping goods, fuel, and people.

This affects the availability of traditional food, according to an RSLFN Elder:

> "So much disturbing the land by this mining. Yamana (now Agnico Eagle), they had choppers going from here to Lingman and Twin Lake. I remember that winter all the moose were coming from the north side, heading south. That trail to Pierce Lake, there was a track of moose... They were heading south...away from the sound. I remember last year when they were flying from here to Lingman Lake I don't think they killed anything there when they went moose hunting. Then at Pierce Lake, Irene's camp, there's been a lot of disturbance from the choppers hauling their equipment for the mining."

Explosives from mining exploration and operations are also disrupting hunting. A community member complained about the large noise and disruption made by mining explosives, which waged war with animals:

> "There was a lot of prospecting, and they would fly [explosive] materials there. There is a lot of it I've seen, even in the deep-water areas, and the problem is when they're heated. And then explode, they make a loud noise. And that would scare the animals away. And it's just all over."

A male Elder, when asked about the impacts of mining in his community, described the great loss of wildlife due to mining impacts:

> "So, all that scares away the animals- moose. So, there's not much. There used to be a lot of moose there before. ...The birds too- ducks, geese, and beavers die there on the water. We pull them [the dead carcuses] out of the water because they'll damage the river."

Another respondent opined that mining would result in the land being destroyed and lost:

> "Well, like if this mine starts up. . .I know for a fact that we are going to lose the whole area. People are going to come in and destroy...So, it'll be flights in and out. It will be oil and gas. It'll be maybe hydro development."

Talking about the environmental pollution impact of mining on the community, another respondent talked about a spill of oil he discovered, and that oil spills were common:

> "Yes, and you don't know where they [mining explorers] left maybe gas or other materials and it's leaking out into the land...Animals take that up. . .. I remember we were hunting...and we checked and there were [gas] barrels there. Rechecked next day, and there's a spill there, needing clean up."

Another RSLFN member shared his experience of how diesel from mining contaminates whole areas:

> "I used to work for a mine. Twin Lake. I do not want any kind of mining or development in this area, because I've seen how they do things. They bring in these big bladders for diesel. They fly in these big bladders and sometimes those bladders are on the ice...these little barriers....it wouldn't contain the spill. It will just contaminate the whole area. One of those bladders ruptured. No, I don't want anything like that around here."

Mining in the RSLFN territory is changing intergenerational use of land and wildlife abundance. The people of RSLFN are impacted by Monument Bay, which is 60 km northeast and Lingman Lake, which is 57 km southeast (Figure 4). Moose used to be abundant but are no longer. An Elder described killing his first moose at age 17, compared to his son at age 35. The Elder attributed this difference in the 'first moose kill age' to a decrease in wildlife abundance due to industrial development. As killing a moose in Anisininew culture is a sign of manhood and maturity, these rights of passage are slowing or dying. The interviewees' perspectives concerning land protection priorities were unanimous. All 21 community members wished that their entire traditional lands be protected from all forms of industrial development, which is clearly stated by one RSFLN member saying:

> "In future, where my grandchildren.., I would like to see the protection of all this- all around Red Sucker. All this territory."

Another person said he wanted 'everywhere' in the RSL territory to be protected. RSLFN people's perspectives towards mining varied from person to person, especially between age groups. However, nobody wanted mining companies or exploration on their RSLFN traditional territory. This is summarized by one interviewee saying:

> "I don't think anybody wants their traditional lands to be disturbed, you know, to be destroyed, or altered in any way."

The Elders unanimously disapproved of mining, concerned about the negative impacts of mining. Youths had mixed views of mining, without awareness of their impact on the land [85]. The respondents' opinions on the impacts of mining reveal different levels of land knowledge.

3.1.2. Overlap of Traditional Land Use Hotspots and Greenstone Belts

Figure 5 shows the highest density regions for TLUs as yellow to lime green, with moderate density shown as purple. This map shows that most high-density TLU spots are located on greenstone belts. All but one of these medium to high-density TLU spots are on greenstone belts. The high-density TLU spots are located around Red Sucker Lake, Rorke Lake, Lenover Lake, and Ponask Lake. Moderate land use density of TLUs exist around Richardson Lake, Stull Lake, Sachigo Lake, Banksian River, Angling Lake, Seeber River, Pullan Lake, and Durell Lake.

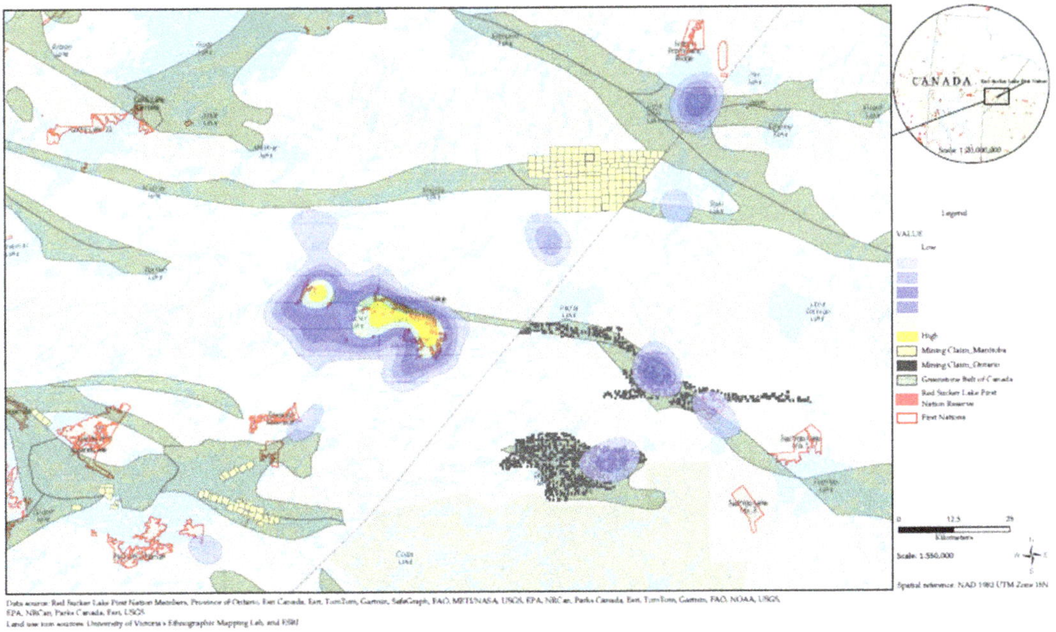

Figure 5. Summary heat map of the land uses of 21 Red Sucker Lake First Nation members.

The optimized hot spot map in Figure 6 identified statistically significant land use locations on/within greenstone belts. Significance levels with 90%, 95%, and 99% confidence, whether high or medium-density TLU spots, imply non-random land use. Statistically significant clusters of high and low incident counts of land uses are identified by lime green output features, while fuchsia pink output features represent medium TLU spots.

Figure 6 shows that statistically significant TLUs were all located on lakes in greenstone belts. These lakes with statistically significant TLUs were Red Sucker Lake (high), Pierce Lake (medium), Stull Lake (medium), Seeber Lake (medium), and Angling Lake (medium). High-density TLU spots were statistically significant for bird/egg harvesting, fishing, hunting, plants/wood/earth materials harvesting, and trapping.

All land and traditional land use areas are considered to be culturally and environmentally significant. 'Not statistically significant' does not mean 'not traditionally important or significant'. The summary land use map (Figure 3) shows that many culturally and traditionally important land uses occur in areas not considered to be significant hotspots. The 'not significant' spots refer to incident counts (of land use activities other than travel routes) that are not statistically significant based on False Discovery Rate (FDR) correction for multiple testing and spatial dependence.

The small sample size of this research means the traditional land uses do not represent the entire community and miss many hotspots. Also, we cannot be sure that the research captured all of the traditional land uses for the 21 interviewees, as community members may withhold information on certain landscapes/land use [83,84]. Documenting the land uses of more community members would have produced higher land use densities and much more statistically significant output clusters, as evidenced by the 2018 TLU study, which involved 14 community members different to those selected for this research [81]. While both studies recorded some TLU areas in common, each study also recorded distinct TLU areas, as community members' traplines differ.

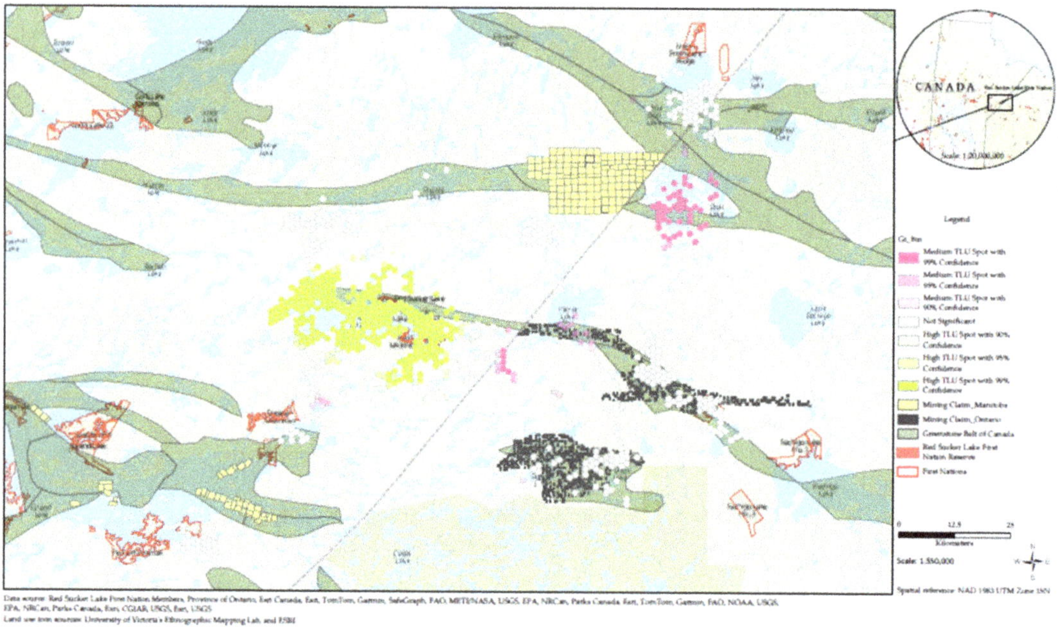

Figure 6. Traditional land use hotspots of 21 Red Sucker Lake First Nation members on greenstone belts.

3.2. *Mining versus Protection*

The RSLFN territory and land uses include the many mining claims in Monument Bay. Monument Bay in the Island Lake region has high-grade gold-tungsten, with tungsten being a critical mineral [59,60]. High-grade gold-tungsten ore at Monument Bay is worth an estimated one billion dollars [59]. The mining claims at Monument Bay are owned by Agnico Eagle Mines Limited (previously Yamana Gold Inc. until 2023), a Canadian-based multinational company.

Island Lake contains many greenstone belts. Canadian mineral tenure law and Manitoba's Planning Regulations dictate that mining is the only land use designation possible for greenstone belts [82] and neighboring land. Even with no mining claims, greenstone belt areas and their surrounding lands are deemed ineligible for IPCA funding or protection by Canada's colonial governments.

Referring to the territory of RSLFN's traditional land that he wishes to protect, one of the interviewees simply said- 'everywhere'. Community members also revealed that the mining companies, contractors, explorers, and colonial governments did not respect the duty to consult. According to the UNDRIP [27,28], RSLFN community members have the right to free, prior, and informed consent (FPIC) regarding developmental activities in their traditional territory, but that right is not being respected, creating conflict and discord [82].

An Elder also commented on prospectors' camps in the RSLFN territory as having many negative impacts. The respondent called prospectors 'invasive', as prospectors invade their territory:

> "Invasive... that's prospecting. We don't like it. It's not appreciated. It [prospecting] is like walking into somebody's house and sitting down...turning on the TV, without permission. This trapline...nobody should be there at all when we are not there unless they ask. They need to get permission first."

When asked if the community was consulted before such prospecting activities, he answered "no":

> "No, they don't [ask or consult] but we see those camps. There is supposed to be a consultation... But they don't do that."

Mining claims and greenstone belts overlap TLU hotspots at Monument Bay and Lingman Lake. As a result, mining is in direct conflict with RSLFN's traditional activities and the livelihoods of the RSLFN people and already created exclusion zones. At Monument Bay, RSLFN Indigenous people have a court injunction keeping them from using this territory. As RSLFN's territory reaches over two provinces with different regulators, the policy implications extend to both the provincial and federal levels of the Canadian government. Mining puts RSLFN's traditional land uses, culture, and ecosystem integrity at risk.

The RSLFN people want their land unspoiled by industrial developments, including hydro development and mining. The view is that the land is perfect the way the creator made it. The RSLFN people intend to fulfill their sacred role as guardians of the pristine nature of their forests and lakes. Their wish is to create land-based education to teach youth to live on the land, while monitoring wildlife abundance, the quality of environmental media, and natural cycles.

A review of the Impact Assessment Agency of Canada [62] database listed mining impacts for gold mines near First Nation communities. The many negative environmental impacts of mining provide the potential risks for mining development at the RSLFN [62]. Mining impacts watersheds during site preparation and operation with heavy water usage, causing groundwater drawdown and impacting surrounding wetlands. Mining worsens water quality through sediment loading, erosion of suspended solids, acid mine drainage, and metal leaching, including leaching of radioactive metals, such as uranium. Mining operations use massive equipment that contributes to noise and toxic pollution, which lowers air quality. The reduced abundance of wildlife in the area, especially moose, is an example of traditional food availability affected by mining operations. Potential spills from mining operations pose safety and health challenges to humans and wildlife, altering the availability and acceptability of traditional food [62].

Mining destroys natural habitats [62]. Mining causes environmental pollution, biodiversity loss, and the decimation of wildlife, including species at risk of extinction [62]. Mining also reduces the abundance of valuable species to Indigenous people by destroying habitat, important breeding sites, migration paths, and poisoning wildlife [62]. Changes in terrestrial and wetland environments alter landscapes, upset ecosystem dynamics, reduce biodiversity, and diminish the abundance of species at all levels of the food web, including species at risk [62].

Mining development and operations negatively impact human health and Indigenous rights in many ways [62]. Mining encroaches on Indigenous rights by curtailing or limiting fishing, hunting, trapping, medicinal plant harvesting, cultural, educational, and economic aspects with ceremonies and local language aspects [63]. Mining eliminates the availability of nearby freshwater bodies and land for traditional food production and harvesting [62]. Opportunities for Indigenous knowledge sharing and community interactions diminish due to habitat destruction and land use changes from mining-related activities [62]. Mines also result in an increasing loss of traditional food choices [63]. An inability to maintain food sustenance undermines Indigenous food sovereignty and leads to a continual decrease in food security [62]. Mining also creates inequity between genders by employing mainly settler males, which has resulted in gender violence toward Indigenous women [63]. The benefits of mining projects for Indigenous communities are often conditional, including potential employment, training, and some funding to the First Nation, if a negotiated benefit agreement occurs [62].

Mining requires massive supportive infrastructure, including rail, roads, and reliable energy access, with a big ecological footprint. A proposed multi-modal right-of-way, encompassing road, rail, pipelines, and transmission lines [63], is proposed that is connected with mining and trade [63–65]. The Canadian Senate Committee on Banking, Trade and Commerce endorses, based on the enhancement of commerce and trade, this Northern Corridor as indicated by their title 'National Corridor Enhancing and Facilitating Commerce and Internal Trade'. The nominal route shown in Figure 7 links greenstone belts with ports to facilitate their export after extraction. The route is not designed to provide access roads to those 122 First Nation communities lacking access to roads, being nearby to only seven First Nations [63]. This route is 25 km from three Island Lake First Nations (a little farther to RSLFN), North Spirit Lake, and Cat Lake First Nations, proximate to the greenstone belts and many mining claims.

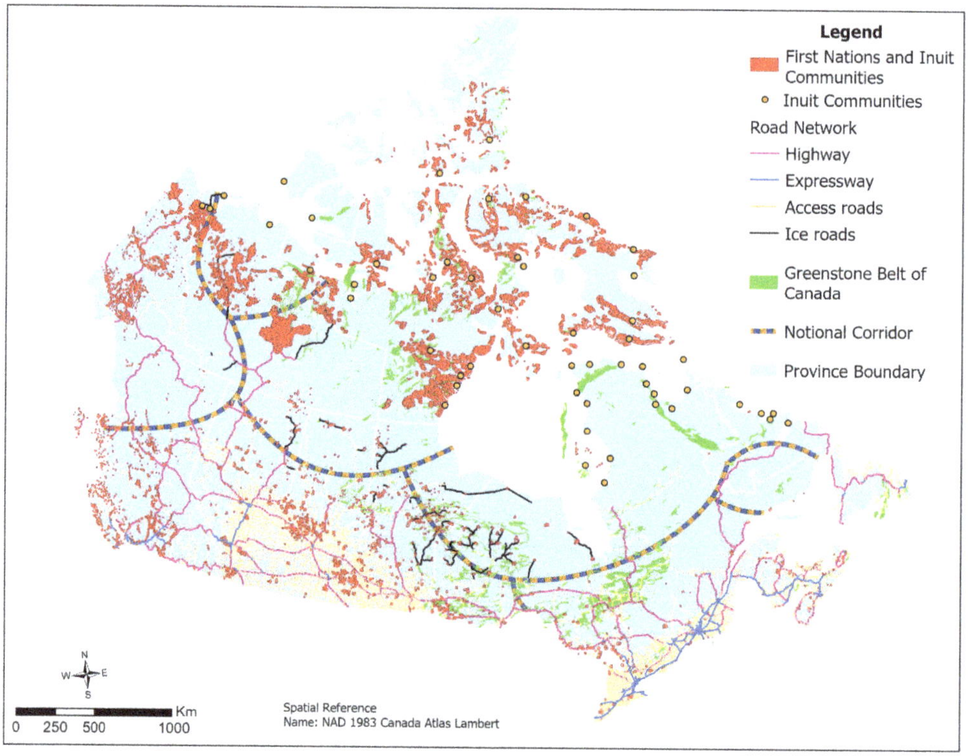

Figure 7. National northern corridor notional route linking greenstone deposits to ports through the Island Lake region. Data Source: [23].

The Island Lake Tribal Council for the RSLFN and the other three First Nations in Island Lake applied for funding towards making 500,000 hectares of their territory an IPCA. This area proposed for an IPCA is shown in Figure 8 marked in green. This area is all within the Anisininew traditional territory marked in blue. Their IPCA proposal purposely avoided mining claims to abide by the only criteria listed, which is that the land must be free of claims. This IPCA funding application [83] reads:

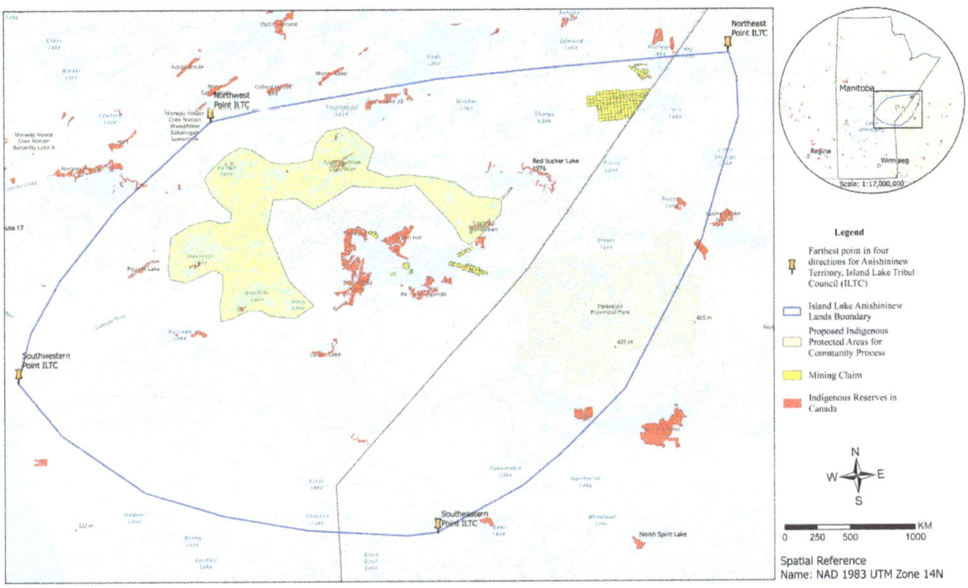

Figure 8. Island Lake Tribal Council's proposed 500,000 hectares of Indigenous protected conservation area in their Anisininew Territory. Data Source: [83].

> "The Island Lake Anisininew leaders and communities are dedicated to keeping their aki (land) sacred as the Creator made it. We want to protect the Hayes Watershed in the Island Lake region (3 million hectares) but focus this proposal on ecosystem conservation of 500,000 hectares for preserving our aki, culture, biodiversity, sustainable livelihoods, and threatened species. The overall aim is to protect aki according to Anisininew ways, with Indigenous-led land-based conservation education and protocols. Through this project, the four Island Lake First Nations will educate to protect their traditional territory to sustain the benefits of conservation and traditional land use for future generations."

The IPCA aimed to foster land stewardship by Elders teaching youth the ways of the land, according to their IPCA application, which states:

> "Elders and elders-in-training will teach the practice and theory of land guardianship, Anishininew culture, language, respect for aki, traditional protocols, and traditional land uses. The community youth hired will be called land guardians and taught to survive on the land, monitor ecosystems, feed the community, and build permaculture camps to monitor and educate others.... [The teachings] will provide Anishininew worldview programming that considers holistic traditional territory protection to achieve mino bimaadiziwin (the good life) for the present and future generations of Island Lake First Nations."

4. Discussion

Indigenous people in the RSLFN and Island Lake want their traditional land uses and land protected. To this end, the Island Lake Tribal Council for the RSLFN and the other Island Lake communities carefully devised an IPCA area without including existing mining claims or other incumbrances. Given the pressing issues of climate change and biodiversity loss, the proposed IPCA offers global benefits. Supporting IPCAs for Island Lake and other IPCAs designated by Indigenous governments would advance Canada's target for biodiversity, preserve peatlands, ensure traditional land uses, protect endangered species, and restore ecosystem balance [7,86]. More research is required but this Island Lake IPCA appears to meet the requirements for a protected area. However, mining interests are interfering with this IPCA. Although Manitoba and Canada proclaim wanting to conserve biodiversity and peatlands, the Island Lake IPCA was not funded to conduct IPCA research due to mining interests [53].

Critical minerals are considered of national importance. Mining, as a result, typically trumps other land uses. For example, in Manitoba, regulations dictate that greenstone belts can only be used for mining. Mining is the polar opposite of an IPCA. Mining creates environmental destruction and obliterates traditional land uses, being an incompatible land use. Most high- and medium-density spots were found to be on greenstone belts, with these land uses in conflict.

Many negative impacts occur from mining, including pollution and habitat loss. Large-scale development of roads and power lines is required as part of a large mining operation in a remote area [3,62,87]. These impacts are felt by the local community, even at the exploration phase. Many RSLFN people complained about mining destroying their livelihoods, wild food supply, traditional land uses, and displacement [3,62,87], without economic benefits. Large spills and noise impacting wildlife in the exploratory phase foreshadow larger impacts when the mine starts.

Maps and statistics clearly show the overlap of TLUs with greenstone belts in RSLFN's territory. RSLFN members complain about mining's many negative impacts on their TLUs. Mining supplies continuously fly in and out of the airport adjacent to the RSLFN in helicopters and planes. The constant noise is disruptive to wildlife and requires some no-fly zones or times to protect wildlife. Amidst these adverse impacts of mining activities on the community, RSLFN members are denied meaningful consultation and reasonable economic benefits during the exploration phase for Monument Bay.

Colonial governments support mining, undermining Indigenous people's self-determination. These colonial governments evicted the RSLFN chief and members from their Native land at Monument Bay and sacked the RSLFN-supported IPCA application, due to mining interests. The Manitoba provincial laws restrict greenstone belts for mining land uses only [82]. Manitoba's Land Use Planning Act Regulation 81/2011 dictates that mining comes before all other interests, stating: "the best and only use of greenstone belts is mining and greenstone belts . . . must be identified and protected from conflicting surface land uses that could interfere with access to the resources" [82] (pp. 39–40). The rich deposits in the RSLFN, indicated by greenstone belts, and the Indian Act resulted in a natural resource curse on RSLFN. This resource curse sank their IPCA proposal, which was to safeguard their land for traditional land uses and give their youth jobs as land guardians.

Canada has laws to reduce the impacts of mining and other developments. Canada's Impact Assessment Act provides a forum to discuss and mitigate mining impacts, but seldom prevents unwanted projects. Further, the Canadian Environmental Protection Act (1999) has a legislative basis for a range of federal environmental and health protection programs that require managing industrial risks.

Mining impacts are expected to ramp up after the exploration stage in the RSLFN, causing more extensive ecosystem damage and impacts to traditional land uses. An onslaught of exploration and mining is expected globally, with an estimated six-fold increase in mineral resources (e.g., lithium, graphite, cobalt, etc.) from 2020 levels to 'net zero' by 2050 [52].

This study expands on Onyeneke's master's thesis with a limited sample size (n = 21). Due to the limitations of this study, further research to determine the coexistence possibilities of industrial mining and IPCAs is needed.

5. Conclusions

Two-eyed seeing provides a way for Indigenous knowledge to guide land use planning to consider IPCAs, biodiversity, and climate mitigation. A two-eyed seeing approach to research prioritizes Indigenous self-determination, with RSLFN wanting to protect its territory in the pristine Hayes watershed. The Island Lake Tribal Council (ILTC) with the RSLFN proposed an IPCA to protect their sacred relationship with the land. The 500,000 hectares of land for protection were carefully chosen around existing mining claims to balance the protection of land in Island Lake with other interests. This IPCA is a way for Canada to meet biodiversity and climate change commitments, through Indigenous traditional ways of environmental stewardship [7,67,88,89] in an area rich in peat.

Climate stabilization and biodiversity require careful planning to balance the economic engine of renewables, which require critical minerals, with land protection for biodiversity, climate stabilization, and Indigenous peoples' livelihoods. Renewables are part of the solution to ecological collapse, if done sustainably, but so are IPCAs. The IPCAs provide a pathway for self-determination, traditional land uses, climate change stabilization, and biodiversity, considering that mining is already heavily impacting their way of life [13].

Mining is an unsustainable activity. Yet, Manitoba's laws state that mining is the best and only 'sustainable' use of greenstone belts. This blanket approach to planning development proximate to greenstone belts undermines the self-determination and IPCA proposal of the RSLFN and Island Lake. This unsustainable broad approach of prioritizing mining counteracts Canada's stance as a global leader in the green economy [90]. Land use planning on greenstone belts need not obliterate the sacred, cultural, and livelihood sites in the RSLFN homeland. More careful planning laws need to consider TLUs, IPCAs, and Indigenous self-determination, not only the mining of precious and critical minerals.

Author Contributions: Conceptualization, C.O. and S.T.; methodology, C.O., B.H. and S.T.; validation, B.H.; formal analysis, C.O. and S.T.; investigation, C.O., B.H. and S.T.; resources, C.O., B.H. and S.T.; data curation, C.O. and S.T.; writing—original draft preparation, C.O.; writing—review & editing, S.T. and C.O.; visualization, C.O. and S.T.; supervision, S.T.; project administration, S.T.; funding acquisition, S.T. All authors have read and agreed to the published version of the manuscript.

Funding: This research was funded by MITACS Accelerate grant number IT24601 and the SSHRC partnership grant number 895-2017-1014 REF 47534—the Mino Bimaadiziwin partnership.

Data Availability Statement: The original contributions presented in the study are included in the thesis, further inquiries can be directed to the corresponding author/s.

Conflicts of Interest: The authors declare no conflicts of interest. The funder, MITACS, in conjunction with Yamana Gold, had no role in the design of the study; in the collection, analyses, or interpretation of data; in the writing of the manuscript; or in the decision to publish the results.

References

1. United Nations Environment Programme. The Convention on Biological Diversity. Available online: https://www.cbd.int/convention/text (accessed on 14 May 2024).
2. Raygorodetsky, G. Of Swiddens, snakes, and honey. *Biodiversity* **2018**, *19*, 244–247. [CrossRef]
3. Vogel, B.; Yumagulova, L.; McBean, G.; Charles Norris, K.A. Indigenous-Led Nature-Based Solutions for the Climate Crisis: Insights from Canada. *Sustainability* **2022**, *14*, 6725. [CrossRef]
4. Dawson, N.; Bhardwaj, A.; Coolsaet, B.; Scherl, L.M.; Massarella, K.; Ndoinyo, Y.; Oliva, M.; Suich, H.; Worsdell, T. *Journeys to More Equitable and Effective Conservation: The Central Role of Indigenous Peoples and Local Communities*, 23rd ed.; IUCN (Policy Matters): Gland, Switzerland, 2023. Available online: https://portals.iucn.org/library/sites/library/files/documents/Policy-Matters-Issue-23_0.pdf (accessed on 14 May 2024).
5. Hill, S.L. The Autoethnography of an Ininiw from God's Lake, Manitoba, Canada: First Nation Water Governance Flows from Sacred Indigenous Relationships, Responsibilities and Rights to Aski. Ph.D. Thesis, University of Manitoba, Winnipeg, MB, Canada, 2020.
6. Thompson, S.; Harper, V.; Whiteway, N. *Let's Keep Our Land Sacred as the Creator Taught Us: Wasagamack First Nation Ancestral Land Use*; Manitoba First Nations Education Resource Centre: Winnipeg, MB, Canada, 2020; pp. 1–63. Available online: http://ecohealthcircle.com/book-lets-keep-our-land-sacred-as-the-creator-taught-us/ (accessed on 13 July 2022).

7. Webb, J. Indigenous-Led Conservation in the Amazon: A Win-Win-Win Solution. Available online: https://www.amazonfrontlines.org/chronicles/indigenous-conservation-amazon/ (accessed on 22 September 2021).
8. Malcolm, M. A Deeply Troubling Discovery: Earth May Have Already Passed the Crucial 1.5 °C Warming Limit. Available online: https://www.theconversation.com/a-deeply-troubling-discovery-earth-may-have-already-passed-the-crucial-1-5-c-warmking-limit-222601 (accessed on 23 March 2024).
9. Whyte, K. Too late for indigenous climate justice: Ecological and relational tipping points. *Wiley Interdiscip. Rev. Clim. Change* **2020**, *11*, e603. [CrossRef]
10. Mansuy, N.; Staley, D.; Alook, S.; Parlee, B.; Thomson, A.; Littlechild, D.B.; Munson, M.; Didzena, F. Indigenous protected and conserved areas (IPCAs): Canada's new path forward for biological and cultural conservation and Indigenous well-being. *FACETS* **2023**, *8*, 1–16. [CrossRef]
11. Nitah, S. Indigenous peoples proven to sustain biodiversity and address climate change: Now it's time to recognize and support this leadership. *One Earth* **2021**, *4*, 907–909. [CrossRef]
12. Huseman, J.; Short, D. A slow industrial genocide: Tar sands and the indigenous peoples of northern Alberta. *Int. J. Hum. Rights* **2012**, *16*, 216–237. [CrossRef]
13. Kohn, M.; Reddy, K. Colonialism. In *The Stanford Encyclopedia of Philosophy*, 8th ed.; Edward, N.Z., Uri, N., Eds.; Metaphysics Research Lab, Stanford University: Stanford, CA, USA, 2023. Available online: https://plato.stanford.edu/archives/spr2023/entries/colonialism/ (accessed on 20 March 2023).
14. Thapa, K.; Thompson, S. Applying Density and Hotspot Analysis for Indigenous Traditional Land Use: Counter-Mapping with Wasagamack First Nation, Manitoba, Canada. *J. Geosci. Environ. Prot.* **2020**, *8*, 285–313. [CrossRef]
15. Thompson, S.; Thapa, K.; Whiteway, N. Sacred Harvest, Sacred Place: Mapping harvesting sites in Wasagamack First Nation. *J. Agric. Food Syst. Community Dev.* **2019**, *9*, 251–279. [CrossRef]
16. Iwama, M.; Marshall, M.; Marshall, A.; Bartlett, C. Two-eyed seeing and the language of healing in community-based research. *Can. J. Nativ. Educ.* **2009**, *32*, 1–6.
17. Martin, D.E.; Thompson, S.; Ballard, M.; Linton, J. Two-eyed seeing in research and its absence in policy: Little Saskatchewan First Nation Elders' experiences of the 2011 flood and forced displacement. *Int. Indig. Policy J.* **2017**, *8*, 6. [CrossRef]
18. Peltier, C. An Application of Two-Eyed Seeing: Indigenous Research Methods with Participatory Action Research. *Int. J. Qual. Methods* **2018**, *17*, 1609406918812346. [CrossRef]
19. Kimmer, R.W. *Braiding Sweetgrass: Indigenous Wisdom, Scientific Knowledge, and the Teachings of Plants*; Milkweed Editions: Minneapolis, MN, USA, 2013.
20. Kovach, M. *Indigenous Methodologies: Characteristics, Conversations, and Contexts*; University of Toronto Press: Toronto, ON, Canada, 2009; pp. 2–16.
21. Wilson, S.; Hughes, M. Why Research is Reconciliation. In *Research and Reconciliation: Unsettling Ways of Knowing through Indigenous Relationships*; Wilson, S., Breen, A.V., Dupre, L., Eds.; Canadian Scholars: Toronto, ON, Canada, 2019; pp. 5–19.
22. Indigenous Circle of Experts' Report and Recommendation. We Rise Together: Achieving Pathway to Canada Target 1 through the Creation of Indigenous Protected and Conserved Areas in the Spirit and Practice of Reconciliation. Available online: https://publications.gc.ca/collections/collection_2018/pc/R62-548-2018-eng.pdf (accessed on 22 November 2023).
23. Thompson, S.; Hill, S.; Salles, A.; Ahmed, T.; Adegun, A.; Nwankwo, U. The Northern Corridor, Food Insecurity and the Resource Curse for Indigenous Communities in Canada. *Sch. Public Policy Publ.* **2023**, *15*, 1–44.
24. Statistics Canada. Census Profile. 2021 Census of Population. Statistics Canada Catalogue no. 98-316-X2021001. Ottawa. Census of Population. Released 27 April 2022. Available online: https://www12.statcan.gc.ca/census (accessed on 11 September 2022).
25. International Covenant on Civil and Political Rights. Archived 3 March 2012. Available online: https://www.web.archive.org/web/201203030011412/https:/www2.ohchr.org/english/law/ccpr.htm (accessed on 9 April 2024).
26. Eide, A. Economic, social and cultural rights as human rights. In *Economic, Social and Cultural Rights*, 2nd ed.; Brill Nijhoff: Leiden, The Netherlands, 2001; pp. 9–28. [CrossRef]
27. United Nations. United Nations Declaration on the Rights of Indigenous Peoples (UNDRIP). 2007. Available online: https://www.un.org/development/desa/indigenouspeoples/wp-content/uploads/sites/19/2018/11/UNDRIP_E_web.pdf (accessed on 23 March 2024).
28. Indigenous Foundations. UN Declaration on the Rights of Indigenous Peoples. Available online: https://indigenousfoundations.arts.ubc.ca/un_declaration_on_the_rights_of_indigenous_peoples/ (accessed on 10 April 2024).
29. Blacksmith, C.; Thompson, S.; Hill, S.; Thapa, K.; Stormhunter, T. The Indian Act Virus Worsens COVID-19 Outcomes for Canada's Native People. In *Canadian Yearbook on Human Rights' Special Issue on COVID-19*; University of Ottawa Centre for Human Rights Press: Ottawa, ON, Canada, 2021.
30. Fontana, L.B.; Grugel, J. The politics of Indigenous participation through "free prior informed consent": Reflections from the Bolivian case. *World Dev.* **2016**, *77*, 249–261. [CrossRef]
31. United Nations for Indigenous People. Free, Prior, and Informed Consent: An Indigenous Peoples' Right and Good Practice for Local Communities. Available online: https://www.un.org/development/desa/indigenouspeoples/publications/2016/10/free-prior-and-informed-consent-an-indigenous-peoples-right-and-a-good-practice-for-local-communities-fao/ (accessed on 14 October 2023).

32. Schnarch, B. Ownership, control, access, and possession (OCAP) or self-determination applied to research: A critical analysis of contemporary First Nations research and some options for First Nations communities. *Int. J. Indig. Health* **2004**, *1*, 80–95.
33. Porter, L.; Barry, J. *Planning for Coexistence? Recognizing Indigenous Rights through Land-Use Planning in Canada and Australia*; Routledge: New York, NY, USA, 2016.
34. Mazzocchi, F. Western science and traditional knowledge. *EMBO Rep.* **2006**, *7*, 463–466. [CrossRef] [PubMed]
35. Aikenhead, G.S.; Ogawa, M. Indigenous knowledge and science revisited. *Cult. Stud. Sci. Educ.* **2007**, *2*, 539–620. [CrossRef]
36. McGregor, D. Coming full circle: Indigenous knowledge, environment, and our future. *Am. Indian Q.* **2004**, *28*, 385–410. [CrossRef]
37. Lines, L.A.; Jardine, C.G. Connection to the land as a youth-identified social determinant of Indigenous Peoples' health. *BMC Public Health* **2019**, *19*, 176. [CrossRef] [PubMed]
38. Marquina-Márquez, A.; Virchez, J.; Ruiz-Callado, R. Postcolonial Healing Landscapes and Mental Health in a Remote Indigenous Community in Subarctic Ontario, Canada. *Polar Geogr.* **2016**, *39*, 20–39. [CrossRef]
39. Kuhnlein, H.V.; Receveur, O. Dietary Change and Traditional Food Systems of Indigenous Peoples. *Annu. Rev. Nutr.* **1996**, *16*, 417–442. [CrossRef]
40. National Report of the First Nations Regional Health Survey. First Nations Information Governance Centre 3(1) (March). Available online: https://www.fnigc.ca/wp-content/uploads/2020/09/53b9881f96fc02e9352f7cc8b0914d7a_FNIGC_RHS-Phase-3-Volume-Two_EN_FINAL_Screen.pdf (accessed on 9 April 2024).
41. Laurie, C.; Batal, M.; Sadik, T.; Tikhonov, C.; Schwartz, H.; Fediuk, K.; Ing, A.; Marushka, L.; Lindhorst, K.; Barwin, L.; et al. FNFNES Final Report for Eight Assembly of First Nations Regions: Draft Comprehensive Technical Report. Assembly of First Nations, University of Ottawa, Université de Montréal. 2019. Available online: https://www.fnfnes.ca/docs/FNFNES_draft_technical_report_Nov_2__2019.pdf (accessed on 26 March 2024).
42. Chan, L.; Malek, B.; Tonio, S.; Tikhonov, C.; Schwartz, H.; Fediuk, K.; Ing, A.; Marushka, L.; Lindhorst, K.; Barwin, L.; et al. Corrigendum to the FNFNES Final Report for Eight Assembly of First Nations Regions: Draft Comprehensive Technical Report. November 2021. FNFNES. Available online: https://www.fnfnes.ca/researcher-docs/FNFNES_National_Report_Corrigenda_2021-10-27.pdf (accessed on 28 March 2024).
43. Havranek, T.; Horvath, R.; Zeynalov, A. Natural Resources and Economic Growth: A Meta-Analysis. *World Dev.* **2016**, *88*, 134–151. [CrossRef]
44. The Indian Act. *An Act to Amend and Consolidate the Laws Respecting Indians*; Government of Canada: Ottawa, ON, Canada, 1876; sec 12; Available online: https://nctr.ca/wp-content/uploads/2021/04/1876_Indian_Act_Reduced_Size.pdf (accessed on 29 March 2024).
45. King, H. *Land Back: A Yellowhead Institute Red Paper*; Yellowhead Institute: Toronto, ON, Canada, 2019; pp. 8–44. Available online: https://redpaper.yellowheadinstitute.org/wp-content/uploads/2019/10/red-paper-report-final.pdf (accessed on 28 March 2024).
46. Bakx, K. Alberta's Bearspaw First Nation Fighting Federal Government for Right to Manage Own Savings. *CBC News*. 27 July 2021. Available online: https://www.cbc.ca/news/canada/calgary/bakx-bearspaw-first-nation-government-savings-1.6117818 (accessed on 27 July 2021).
47. The National Aboriginal Economic Development Board. Recommendations on First Nations Access to Indian Moneys. 2017; p. 16. Available online: http://www.naedb-cndea.com/reports/recommendations-on-first-nations-access-to-indian-moneys.PDF (accessed on 8 April 2024).
48. Adegun, A.; Thompson, S. Higher COVID-19 rates in Manitoba's First Nations compared to non-First Nations linked to limited infrastructure on reserves. *J. Rural Community Dev.* **2021**, *16*, 4.
49. The Free Press: Judge Nixes Stop Work Order; Miner Forges on. Available online: https://www.winnipegfreepress.com/business/2013/07/05/judge-nixes-stop-work-order-miner-forges-on (accessed on 30 March 2024).
50. IPCC. *AR6 Synthesis Report: Climate Change 2023*; AR6 Synthesis Report: Climate Change 2023—IPCC; IPCC: Geneva, Switzerland, 2023. Available online: https://www.ipcc.ch/report/sixth-assessment-report-cycle/ (accessed on 3 January 2023).
51. USEPA. *Using a Total Environment Framework (Built, Natural, Social Environments) to Assess Life-Long Health Effects of Chemical Exposures Request for Applications (RFA)*; U.S. Environmental Protection Agency: Washington, DC, USA, 2017. Available online: https://www.epa.gov/research-grants/using-total-environment-framework-built-natural-social-environments-assess-life (accessed on 13 April 2023).
52. IEA. *World Energy Outlook 2022*; IEA: Paris, France, 2022. Available online: https://www.iea.org/reports/world-energy-outlook-2022 (accessed on 3 January 2023).
53. Thompson, S. Strategic Analysis of the Renewable Electricity Transition: Power to the World without Carbon Emissions? *Energies* **2023**, *16*, 6183. [CrossRef]
54. IEA. *The Role of Critical Minerals in Clean Energy Transitions*; IEA: Paris, France, 2021. Available online: https://www.iea.org/reports/the-role-of-critical-minerals-in-clean-energy-transitions (accessed on 8 April 2024).
55. Davies, M.; Swilling, M.; Wlokas, H.L. Towards New Configurations of Urban Energy Governance in South Africa's Renewable Energy Procurement Programme. *Energy Res. Soc. Sci.* **2018**, *36*, 61–69. [CrossRef]
56. Viebahn, P.; Soukup, O.; Samadi, S.; Teubler, J.; Wiesen, K.; Ritthoff, M. Assessing the Need for Critical Minerals to Shift the German Energy System Towards a High Proportion of Renewables. *Renew. Sustain. Energy Rev.* **2015**, *49*, 655–671. [CrossRef]

57. Seibert, M.K.; Rees, W.E. Through the Eye of a Needle: An Eco-Heterodox Perspective on the Renewable Energy Transition. *Energies* **2021**, *14*, 4508. [CrossRef]
58. Michaux, S.P. *The Mining of Minerals and the Limits to Growth*; Geological Survey of Finland: Espoo, Finland, 2021.
59. Manitoba's Mineral Industry: Exploration and Development Highlights 2014. Available online: https://www.gov.mb.ca/iem/industry/exp-dev/highlights2014.html (accessed on 23 March 2024).
60. Tükenmez, M.; Demireli, E. Renewable Energy Policy in Turkey with the New Legal Regulations. *Renew. Energy* **2012**, *39*, 1–9. [CrossRef]
61. Ackah, I.; Graham, E. Meeting the Targets of the Paris Agreement: An Analysis of Renewable Energy (RE) Governance Systems in West Africa (WA). *Clean Technol. Environ. Policy* **2021**, *23*, 501–507. [CrossRef]
62. Impact Assessment Agency of Canada. Government of Canada. Available online: https://www.canada.ca/en/impact-assessment-agency.html (accessed on 23 March 2024).
63. Sulzenko, A.; Fellows, G.K. Planning for Infrastructure to Realize Canada's Potential: The Corridor Concept. *Sch. Public Policy Publ.* **2016**, *9*, 1–84. [CrossRef]
64. Fellows, G.K.; Tombe, T. Gains from Trade for Canada's North: The Case of a Northern Infrastructure Corridor. *Sch. Public Policy Publ.* **2018**, *10*, 1–25. [CrossRef]
65. Tombe, T.; Munzur, A.; Fellows, G.K. Implications of an Infrastructure Corridor for Alberta's economy. *Sch. Public Policy Publ.* **2021**, *14–17*, 1–36. [CrossRef]
66. Indigenous Circle of Experts. Pathway to Canada Target 1 Indigenous Circle of Experts (ICE) Regional Gatherings Reports MARCH. Available online: https://static1.squarespace.com/static/57e007452e69cf9a7af0a033/t/5ab9504c0e2e7246a9551a5a/1522094157137/Regional+Gathering+Reports+EN.pdf (accessed on 3 April 2024).
67. O'Bryan, C.J. The importance of Indigenous peoples' lands for the conservation of terrestrial mammals. *Conserv. Biol.* **2021**, *35*, 1002–1008. [CrossRef] [PubMed]
68. IPCC. Climate Change and Land: An IPCC Special Report on Climate Change, Desertification, Land Degradation, Sustainable Land Management, Food Security, and Greenhouse Gas Fluxes in Terrestrial Ecosystems. 2019. Available online: https://www.ipcc.ch/srccl/ (accessed on 2 April 2024).
69. Ginger, G.; Ford, R.; Firelight Group. Implementing Indigenous Protected and Conserved Area Agreements in Canada: A Review of Successes, Challenges, and Realities. Available online: https://makeway.org/wp-content/uploads/2023/05/Firelight-Makeway-IPAs-April-28-2023.pdf (accessed on 2 April 2023).
70. Wildlife Conservation Society Canada. Northern Peatlands in Canada. An Enormous Carbon Storehouse. Available online: https://storymaps.arcgis.com/stories/19d24f59487b46f6a011dba140eddbe7 (accessed on 10 April 2024).
71. Indigenous and Northern Affairs Canada. Red Sucker Lake First Nation Detail. AANDC. Available online: https://fnp-ppn.aadnc-aandc.gc.ca/fnp/Main/Search/FNMain.aspx?BAND_NUMBER=300&lang=eng (accessed on 25 February 2024).
72. Onyeneke, C. Mining Impact and Indigenous Protected and Conserved Areas. Master's Thesis, University of Manitoba, Winnipeg, MB, Canada, 2023. Available online: https://mspace.lib.umanitoba.ca/server/api/core/bitstreams/a15289fe-a9b1-427c-b922-b4c974808d62/content (accessed on 2 February 2024).
73. Olson, R.; Hackett, J.; DeRoy, S. Mapping the digital terrain: Towards Indigenous geographic information and spatial data quality indicators for indigenous knowledge and traditional land-use data collection. *Cartogr. J.* **2016**, *53*, 348–355. [CrossRef]
74. Thapa, K. Indigenous Land Rights and Indigenous Land Use Planning: Exploring the Relevance and Significance to Wasagamack First Nation, Northern Manitoba, Canada. Master's Thesis, University of Manitoba, Winnipeg, MB, Canada, 2018.
75. Blair, E. A reflexive exploration of two qualitative data coding techniques. *J. Methods Meas. Soc. Sci.* **2015**, *6*, 14–29. [CrossRef]
76. Esri Canada. Optimized Hotspot Analysis. Optimized Hot Spot Analysis (Spatial Statistics)—ArcGIS Pro | Documentation. Available online: https://www.pro.arcgis.com/en/pro-app/latest/tool-reference/spatial-statistics/optimized-hot-spot-analysis.htm (accessed on 26 February 2023).
77. Ministry of Northern Development and Mines. Available online: https://www.ontario.ca/page/ministry-mines (accessed on 20 May 2023).
78. Integrated Mining and Quarrying System. Available online: https://web33.gov.mb.ca/imaqs/ (accessed on 11 October 2022).
79. Shafabakhsh, G.A.; Famili, A.; Bahadori, M.S. GIS-based spatial analysis of urban traffic accidents: Case study in Mashhad, Iran. *J. Traffic Transp. Eng. (Engl. Ed.)* **2017**, *4*, 290–299. [CrossRef]
80. Grubesic, T.H.; Murray, A.T. Constructing the divide: Spatial disparities in broadband access. *Pap. Reg. Sci.* **2002**, *81*, 197–221. [CrossRef]
81. Thompson, S.; Thapa, K. *Red Sucker Lake First Nation Ancestral Land Use and Traditional Activities Report*; Personal Communication; Natural Resources Institute, University of Manitoba: Winnipeg, MB, Canada, 2018.
82. Manitoba Government. Manitoba's Provincial Planning Act (81/2011). Available online: https://web2.gov.mb.ca/laws/regs/current/_pdf-regs.php?reg=81/2011 (accessed on 13 July 2023).
83. Wood, N.; Ahmed, T.; Thompson, S. Island Lake Tribal Council's Expression of Interest (EOI) for an Indigenous-Led Area-Based Conservation (ILABC) in the Island Lake Region of Manitoba, Canada, Environment Canada Funding Application. 2023. Available online: https://ecohealthcircle.com/important-docs-to-read/expression-of-interest-final-ilabc/ (accessed on 12 October 2022).

84. Rezende, C.L.; Scarano, F.R.; Assad, E.D.; Joly, C.A.; Metzger, J.P.; Strassburg, B.B.N.; Mittermeier, R.A. From hotspot to hopespot: An opportunity for the Brazilian Atlantic Forest. *Perspect. Ecol. Conserv.* **2018**, *16*, 208–214. [CrossRef]
85. Huff, A. Indigenous land rights and the new self-determination. *Colo. J. Int. Environ. Law Policy* **2005**, *16*, 295.
86. Catovsky, S.; Bradford, M.A.; Hector, A. Biodiversity and ecosystem productivity: Implications for carbon storage. *Oikos* **2002**, *97*, 443–448. [CrossRef]
87. Zakharchenko, N.V.; Hasanov, S.L.; Yumashev, A.V.; Admakin, O.I.; Lintser, S.A.; Antipina, M.I. Legal Rationale of Biodiversity Regulation as a Basis of Stable Ecological. *J. Environ. Manag. Tour.* **2018**, *9*, 510–523. [CrossRef]
88. Wardle, D.A.; Jonsson, M.; Bansal, S.; Bardgett, R.D.; Gundale, M.J.; Metcalfe, D.B. Linking vegetation change, carbon sequestration and biodiversity: Insights from island ecosystems in a long-term natural experiment. *J. Ecol.* **2012**, *100*, 16–30. [CrossRef]
89. MacArthur, J.L.; Hoicka, C.E.; Castleden, H.; Das, R.; Lieu, J. Canada's Green New Deal: Forging the socio-political foundations of climate resilient infrastructure. *Energy Res. Soc. Sci.* **2020**, *65*, 101442. [CrossRef]
90. Hayden, A. *When Green Growth Is Not Enough: Climate Change, Ecological Modernization, and Sufficiency*; McGill-Queen's Press-MQUP: Kingston, ON, Canada, 2014; pp. 1–5.

land

MDPI

Article

Civic Engagement in Urban Planning and Development

Liisa Horelli * and Sirkku Wallin

Department of the Built Environment, Aalto University, 02150 Espoo, Finland; sirkku.wallin@aalto.fi
* Correspondence: liisa.horelli@aalto.fi

Abstract: The scientific literature has revealed that there is often a discrepancy between urban planning initiatives and the actual development of cities. Urban development is a complex process strongly affected by the self-organisation of citizens and entrepreneurs, who produce services and events, create new public places, and shape urban regeneration through everyday practices. However, the actors of self-organised urban development are rarely met in participatory urban planning practices. The gap raises the research problem of how to recognise and facilitate civic engagement in urban development. The authors of this article have been investigating the ontology and methodology of urban planning and development over the past two decades, culminating in the framework of expanded urban planning. The article aims to present multiple approaches to civic engagement in urban planning and development, based on an updated longitudinal study conducted in a Helsinki neighbourhood. The authors argue that urban planning should be updated with expanded urban planning (EP), which extends the focus of civic engagement from public participation to self-organisation and everyday life practices.

Keywords: expanded urban planning; urban development; public participation; self-organisation; civic engagement

1. Introduction

Studies and empirical evidence demonstrate that urban planning often fails to achieve the desired outcomes of urban development. This is primarily due to the complex nature of urban development[1], which is heavily influenced by the self-organising actions of citizens and entrepreneurs. These actors are crucial in generating services, events, and public spaces, thereby shaping urban regeneration through everyday practices. Furthermore, urban development occurs continuously and independently of the formal planning system, sometimes even in opposition [1].

The definitions of urban planning tend to vary according to planning theory and the planning system of the country [2]. Definitions range from those that focus mainly on physical form to ones that include development and governance. Leonie Sandercock [3] presents a holistic view of urban planning as "An unfinished social project whose task is to manage our coexistence in the shared spaces of cities and neighbourhoods in such a way as to enrich human life and to work for social, cultural and environmental justice". However, planning often seeks to guide urban development through land-use planning and targeted initiatives. Additionally, other administrative sectors implement urban policies to foster social and economic development. However, the effective implementation of these policies requires localised actions that account for the polymorphic and contradictory nature of urban change [4].

The communicative turn in planning since the 1960s provided an alternative to the rational comprehensive planning approach by enhancing public participation [5–7]. Public participation has become an integral part of planning procedures in the EU countries, like in many other democratic nations. Participatory processes in statutory planning are expected to empower people and enhance pro-environmental attitudes. Consequently, planners receive richer planning data, which enables them to improve plans and urban space.

However, traditional participatory urban planning has produced disappointing results. Boonstra and Boelens [1] argue that the failure lies in the fact that participatory planning proposals remain controlled by public government, which is not adaptive to initiatives that emerge from the dynamics of civil society itself. Consequently, the actors of self-organising urban development are rarely met in participatory urban planning practices. The gap raises the research problem of how to recognise and facilitate self-organising urban development.

Over the years, participatory planning research has evolved from advocacy planning [8] to various supportive technological systems, for example, PPGIS [9], and to actor networks and their self-organised actions, which have been recognised as crucial in urban development. According to the post-structural approach, urban space and other socio-cultural phenomena can be produced and fostered outside formal governance and the actual planning system [1,9–12].

Our research group[2] has conducted extensive research on participatory planning and development over several decades, culminating in the conceptualisation of expanded urban planning (EP), which focuses on civic engagement in various forms. EP is a post-structural approach to urban development [1,20] that can potentially generate an architecture of opportunities (see more in the next section).

This article aims to present multiple approaches to civic engagement in urban planning and development, based on an updated longitudinal study conducted in a Helsinki neighbourhood. Although we have written several articles, including three doctoral dissertations partly based on the same longitudinal study, this article comprises a conceptual elaboration of civic engagement in planning research, which has mostly concentrated on public participation within the planning process.

We argue that urban planning should be updated with expanded urban planning, which extends the focus of civic engagement from public participation to self-organisation and everyday life practices. The article begins by outlining the evolution of EP as a theoretical framework, followed by a description of the methodology employed and the research questions addressed. The subsequent sections present and discuss the research findings, leading to conclusions and recommendations for future endeavours.

2. Evolution of Expanded Urban Planning

The origins of EP can be traced back to the student revolts of the late 1960s, which sparked a widespread adoption of participation in various fields, including planning and design. In Finland, architect Heikki Kukkonen, PhD [21], pioneered the concept of self-planning language, enabling citizens to participate in the co-creation of their living environments in Helsinki. This approach to empowerment through self-organised planning and design was also explored among different demographics, including children, youth, the elderly, and people with long-term mental health conditions, in the regions of Northern Karelia and Kainuu, Finland. While the international trend of communicative planning [5,7] influenced Finnish cases, self-organisation played a central role due to the absence of legislation until 2000.

Eve Mitleton-Kelly [22], with her coevolving, adaptive systems approach, inspired the development of EP, officially named expanded urban planning ten years later [13]. Ontologically, EP adopts pluralist realism without limitations. Drawing from complexity theories, EP is sensitive to the temporal dynamics inherent in urban development, such as the emergence of phenomena, self-organisation, and the transformative nature of urban processes. EP transcends traditional strategic and statutory planning, embracing the co-production, monitoring, and evaluation of spatial and socio-cultural appropriation (see Figure 1). As a result, participation and self-organisation can occur at all stages and adapt to contextual factors dynamically. EP has a dialectical relationship with public planning and self-organisation, incorporating rhizomatic and arborescent characteristics [17].

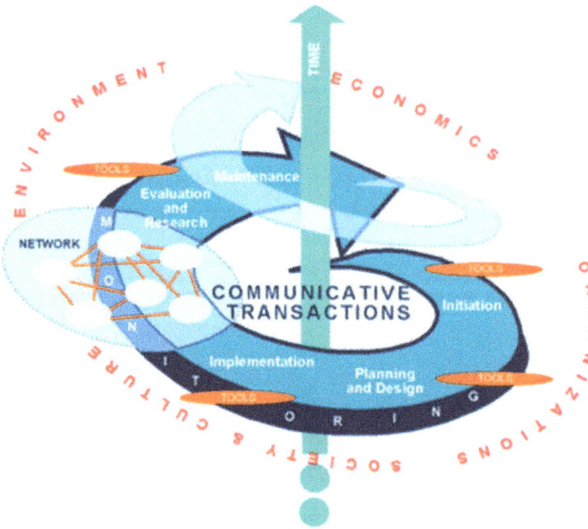

Figure 1. The learning-based network approach to urban planning and action research [23].

EP has adopted an ecosystem of digital and non-digital tools for empowerment and analysis [24] which resonate with its relational and dynamic epistemology. The methodological framework encompasses various enabling methods and future-oriented research tools such as urban informatics, city information modelling, and artificial intelligence [25,26].

This diverse methodology facilitates collaboration among local and global agents, including public, private, and community partnerships, to co-create participatory events and self-organised initiatives for adaptive neighbourhoods [27].

EP also acknowledges the dynamic role played by the spatial and material infrastructure in constructing the quality of everyday life experiences. As a result, EP integrates ecofeminist criteria that pay attention to the substance of planning [28] (see Figure 2).

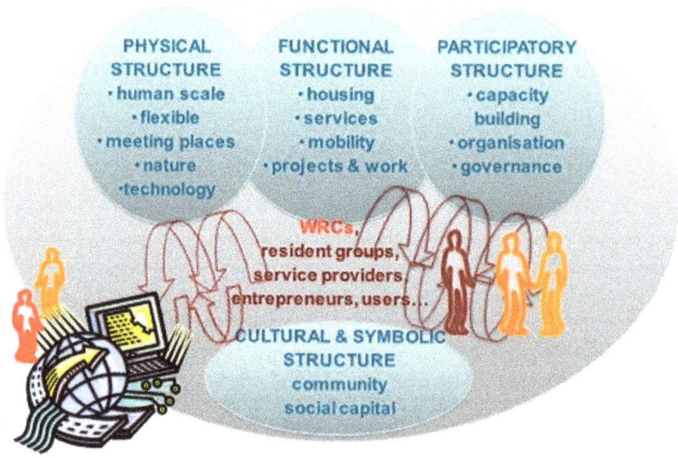

Figure 2. The supportive infrastructure of everyday life is based on the appropriation of the physical, functional, and participatory structures by a variety of actors to co-create the cultural and symbolic structure supported by digital and non-digital tools [20].

The expansion of participation to everyday practice was theorised by Sirkku Wallin [11] (see Figure 3).

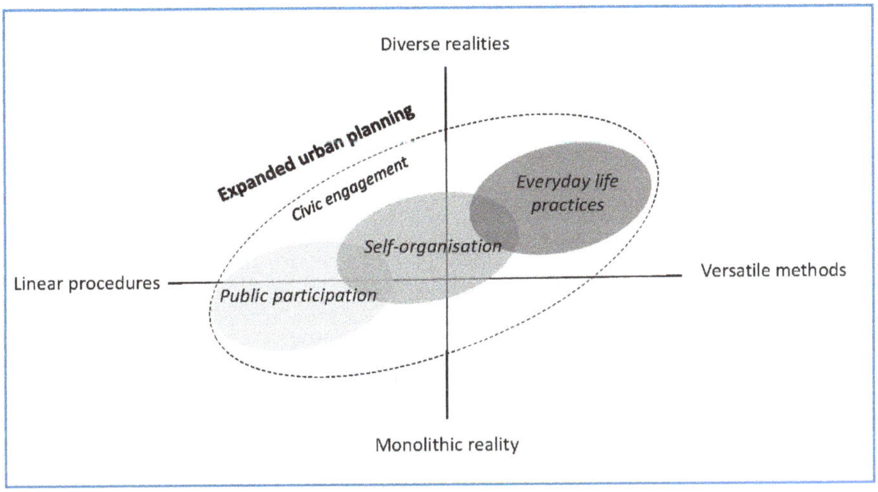

Figure 3. Expanded urban planning acknowledges the different types of civic engagement in diverse realities and applies varying methodological approaches to urban planning and development [11] (p. 30).

Thus, the forms of civic engagement comprise the following:

- Public participation [5,7] that takes place in the formal urban planning and decision-making processes;
- Self-organisation, in which people take action outside formal institutions, including non-governmental organisations NGOs [1,29,30];
- Everyday life practices which make changes in the physical and social construction of urban space [15,31].

Everyday life and its practices serve as a framework for understanding subjective and intersubjective interventions within a structural context. According to Heller [32] and Bech-Joergensen [33], the foundation of everyday practices lies in the reproductive actions that generate psychosocial forces enabling individuals to transform material and socio-cultural conditions into tangible and phenomenal experiences. The physical environment is shaped by social construction and serves as a tool for individual self-regulation [34]. While the environment may impose certain constraints on everyday activities, the way people shape their surroundings can also initiate potential changes at various levels. However, there is a need for places where intersubjective realities can be generated, and the "psychosocial powers of everyday lif". can be unleashed [33]. Therefore, the intermediary spaces of everyday life [35] become significant "free living space". that bridge the gap between individual private lives and the formal public realm, enabling the collective building of community and the neighbourhood.

EP, which incorporates extended civic engagement, builds on and enhances current planning disciplines. The traditional planning system endorses linear procedures and a monolithic reality that provides a "fixed orde". [10]. Public participation is part of the linear procedure, the main objective of which is to produce a plan or a statutory blueprint which is necessary from the administrative perspective. However, urban reality is diverse and complex, necessitating a variety of flexible approaches. EP emphasises citizen engagement that reflects and accommodates the diverse realities of urban settings in a versatile manner.

In recent years, the importance of civic engagement has been acknowledged not only in planning and development but also in co- or hybrid governance [30]. Co-governance integrates formal, semi-formal, and informal networks and public spheres to establish a deliberative system [36]. Semi-formal mixed spheres, facilitated by community informatics, act as intermediaries bridging formal and informal domains, enhancing political activities and decision-making processes. This fosters communication between grassroots-level participation and formal decision-making, facilitating the formation of new partnerships among the public, private, and people partnerships [17,37].

However, governance in its various forms is tied to the concept of public good or interest. According to the study by Puustinen, Mäntysalo, and Jarenko [38], planners tend to have a variety of approaches to the concept of public interest, depending on whether it is individually or collectively based, regulated (by authority) or non-regulated. Thus, we agree with Puustinen et al. that the interpretation of public interest and its application in specific planning situations depends on the relationship of the context to its discursive framework.

In conclusion, EP advocates for a broader understanding of civic engagement that goes beyond traditional notions of public participation, encompassing self-organisation and everyday life practices within urban planning and development.

3. Methodology and Material

The research questions are as follows:

1. What is self-organisation in urban development?
2. What is the role of civic engagement in urban development?
3. How do we enhance civic engagement in urban planning and development?

The design of the study comprised the following:

(a) A longitudinal case study in 2004–2009 in the Herttoniemi neighbourhood and its follow-up in 2023–2024; and in the spring of 2024;
(b) A conceptual analysis of the literature during the past ten years;
(c) A discussion on civic engagement based on a and b.

The article is based on a data set gathered during the decade-long case study on the Herttoniemi neighbourhood, Helsinki, by our study group [11] and its recent follow-up in 2023–2024. A case study always has limitations, but on the other hand, we believe, like Bent Flyvbjerg [39], that the strength of the case study is not in its power to comprehensively explain the world but to give an example that makes the world more comprehensible.

The first phase of the case study in 2004–2009 was based on action research that followed the principles of advocacy planning [8]. Thus, the urban planning and community development interventions were intertwined with the study of different actors and their activities in the neighbourhood. The action research followed the learning-based network approach (Figure 1), which integrated the shared practices of community development into the urban planning process by using enabling tools and various research methods [23]. The enabling tools comprised local gatherings and workshops, which were organised with community workers, local NGOs, and activists. The traditional data-gathering methods consisted of a literature review, surveys, interviews, focus groups, and observations.

The action research aimed to examine the urban development of the neighbourhood and to find out how to support people's everyday lives and opportunities to participate in urban planning. Therefore, the objectives of the data gathering were 1. to collect views of different actors concerning the ways to develop the neighbourhood and to disseminate information for deliberation and 2. to enable the monitoring of the evolving context (see Table 1).

The analysis of the action research data was based on a systems evaluation method called CATWOE [6,40]. The method provided the possibility to perform the following:

1. Conduct an analysis, which defined development patterns that took place in the neighbourhood (urban development processes and the performance of the plans);
2. Evaluate complications and progress that occurred;
3. Interpret the different modes of civic engagement versus administration [11] (p. 43).

Table 1. The data from Herttoniemi research (see also Appendices A and B).

Data Source	Phase I: Action Research 2004–2009 [11] (p. 41)	Phase II: The Follow-Up 2022–2024
Primary data Data from local gatherings, produced by the actors and stakeholders (memos, drawings, photographs)	Memos of local assemblies (23 cases), memos from the formal urban planning participation events in Herttoniemi (4 items) Memos from semi-formal planning workshops and the gatherings of local stakeholders (5 cases)	Planning documents produced by the Helsinki planning department (2018–2023) Participatory budgeting documents (2023)
Secondary data Data gathered by the researchers (surveys, interviews, and data from observation)	Survey I (2004) Survey II (2007) Group interviews with local stakeholders (3 items) Group and personal interviews with administrators (5 items) (2009–2010) Observation memos and graphics from the local events and happenings (over 200 items during 2003–2012)	Observation memos (2022–2024) Local newspaper articles (2019–2023) Articles on the Herttoniemi and Roihuvuori websites (2010–2023)
Supporting data Planning documents and formal announcements related to the participatory processes (plans, strategies, statistics, etc.)	Planning documents produced by the Helsinki planning department (2004–2017) Statistical data of Helsinki (2001–2017)	

The follow-up to the study was carried out focusing on the planning process that went on in Herttoniemi in 2023–2024. The statutory urban development initiative continued with a new planning process, which aimed to improve the metro station again, as well as to create novel housing and services. In this phase, the analysis was based on the planning documents, official data on public participation in the planning process, articles in the local newspaper, and observation in situ.

The data of the study, both the action research (Phase I) and the follow-up (Phase II), comprised plans and planning documents made by the City of Helsinki and local news describing and commenting on the Herttoniemi initiatives (Table 1).

A literature analysis of self-organising urban planning and development research was conducted to comprehend and position the case study. A Scopus search on major concepts, such as self-organisation, civil engagement, and urban development, revealed that only a few new, relevant articles were found in addition to the list of publications we were already aware of (cf. introduction and references). Our method of analysis was based on the research questions. The outcome of the case study and literature analysis are dealt with in the discussion.

4. Results

We will answer the research questions considering the case study and the follow-up in Herttoniemi.

4.1. Statutory Planning and Self-Organising Urban Development

The Herttoniemi neighbourhood has provided a rich context for studying participatory planning and urban development, reflecting the overall urban transformation in the Helsinki metropolitan area[3].

As one of the oldest suburbs of Helsinki, Herttoniemi has been subject to various internal and external factors that have influenced its development.

Since the late 1990s, significant changes have occurred in the Herttoniemi harbour site and neighbourhood. These include transformations in terms of functions, population, townscape, and its position in the metropolitan area's central hierarchy. Gentrification, like in many other neighbourhoods near city centres, has been prominent. Social housing apartments have been converted into owner-occupied dwellings, resulting in a concentration of new urbanism and green activism in what was once a working-class neighbourhood.

Over the past twenty years, the City of Helsinki has undertaken a large-scale urban development initiative aimed to rebuild the Herttoniemi metro station area with a shopping centre and housing project (see Figure 4). This endeavour has involved several planning processes, occurring from 2005 to 2007, 2009 to 2011, and again from 2018 to 2024[4].

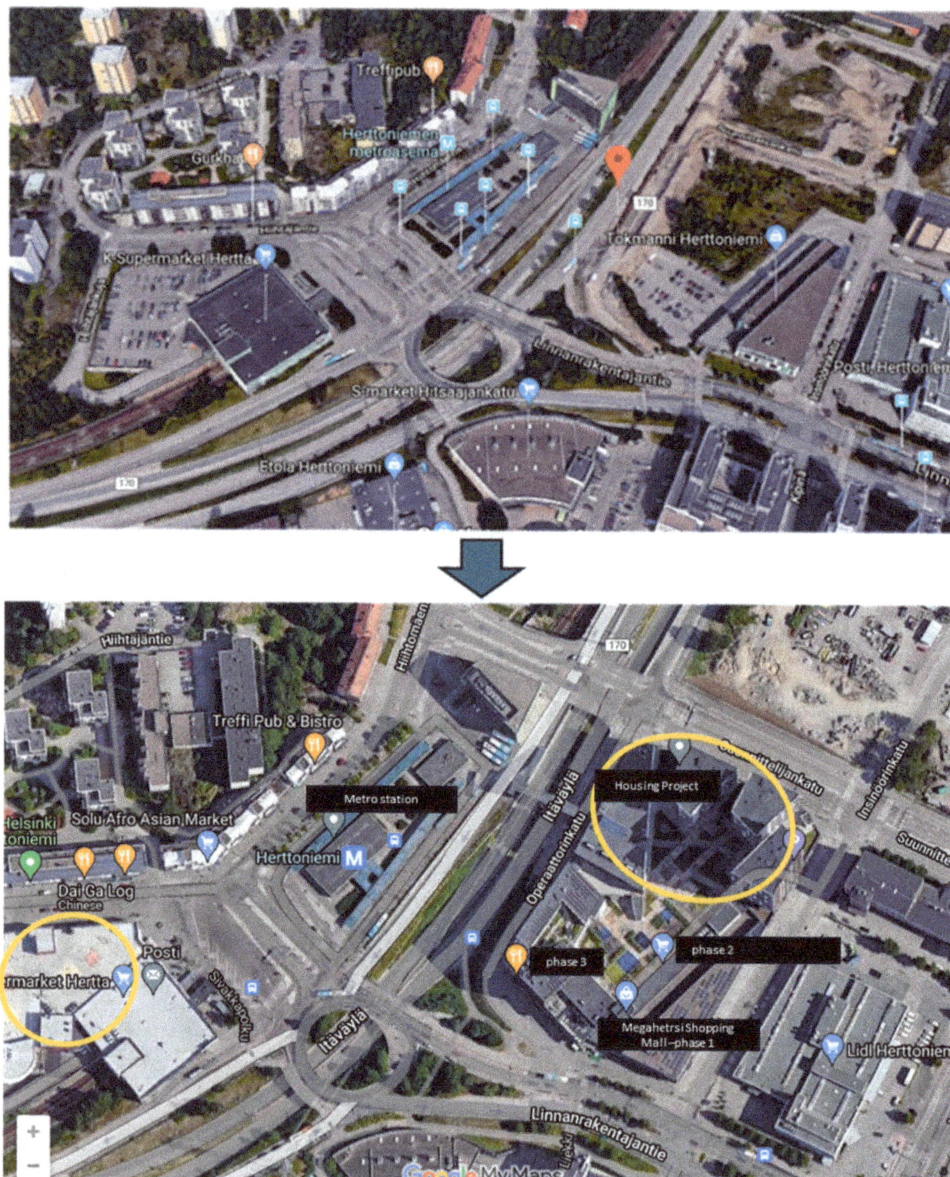

Figure 4. Urban development of Herttoniemi Centre from 2009 to 2024. Based on a comparison between Google Street Maps in June 2018 and on 4 April 2024.

During these processes, statutory planning has successfully developed the east side of Herttoniemi Centre. The shopping mall (Megahertsi) has undergone two expansions since its opening in 2003. Real estate developers have constructed three new housing blocks adjacent to the mall. On the west side of the centre, K-Market Hertta has also capitalised on the opportunity for expansion. However, the planning process for the metro station itself is still ongoing.

Figures 5 and 6 show how the city has been persistent with the idea of creating high-rise dwelling blocks and a bridge between the east and west sides of the centre in all different planning processes during the past two decades.

Figure 5. The visualisation of the local detailed plan in 2011. A plan of Herttoniemi Centre and Megahertsi, Helin & co Arkkitehdit/Arkkitehtuuritoimisto B&M Oy.

Figure 6. The visualisation of the local detailed plan in 2023. A plan of Herttoniemi Centre and Megahertsi, Arkkitehtuuritoimisto B&M Oy.

The plan for Herttoniemi Centre and the associated planning process faced criticism from residents, as well as from the local library and youth services, when it began in

2005. Participants in the statutory planning process put forward an alternative plan that focused on the renovation of the metro station. This alternative plan did not propose additional square meters beyond the metro plot, which the city required for its financial contribution to the development. The number of participants in planning meetings and local debates ranged from 30 to 60 people, indicating that most residents were unaware of the administrative intentions for the neighbourhood.

Throughout the 2010s, urban activism grew more prominent in Helsinki, with various web-based communities advocating for urban planning and development agendas. Some groups proposed urban densification, while others pushed for improved bicycle infrastructure or the preservation of green spaces and natural sites [12,16]. The City of Helsinki took note of this increased interest in urban issues and emphasised participatory processes in their statutory planning efforts. It also sought legitimacy through numbers using large-scale web-based surveys like the local place-based questionnaire "Kerro Kartall". (Tell on the map) [41]. However, this data-gathering survey, designed and delivered by the administration, primarily provided data for planners without necessarily reflecting local aspirations or self-organising initiatives.

By 2020, the planning process in Herttoniemi had not delt with the metro station plot, but otherwise, the development had followed the goals set by the administration 15 years earlier. Commercial services were concentrated near the metro station, and many public services followed suit, despite the criticism from residents in the public hearings and planning surveys.

Consequently, the locals shifted their aspirations towards other issues, particularly the maintenance of the surroundings. The safety and cleanliness of public spaces, particularly around the metro station, became focal points, especially after the completion of the metro line's west section and the renovation of several older stations. The City of Helsinki began funding for local initiatives through participatory budgeting projects allowing citizens to allocate a small portion of public funds for urban development. The implementation of these projects involved surveys in which all citizens had the right to vote. In 2023, one of the winning ideas was the renovation of the Herttoniemi metro station, aligning with the self-organising actions of previous decades.

However, the actual renovation of the metro station has not taken place, as the city is anticipating a more substantial transformation through the ongoing planning endeavour. Instead, the city provided funding for planting flowers, which, although desired by many locals, fell short of the anticipated development.

Table 2 reveals the contrasting roles and actions of investors and landowners, including the city, in comparison to self-organising local actors. While the self-organising actors aimed to enhance the quality of urban spaces and services while maintaining the existing urban scale and structure, the city introduced a comprehensive plan for housing projects and a new service cluster. However, the city did not allocate resources towards supporting small initiatives that aimed to improve urban spaces.

Table 2. Comparison of statutory urban planning and self-organising development.

	Statutory Urban Planning	Self-Organising Urban Development
Stakeholders	Real estate developers, landowners, commercial enterprises, public services, city administration	Residents, small local enterprises, hobby groups
Interests	Large-scale real estate development for services and housing	Improvement of the current metro station, housing, and service structure without extensive new building rights
Actions	Development initiatives with several planning processes and partnership contracts	Initiatives to improve local surroundings, clean up, and plant flowers. Interests in developing services and new commercial spaces

4.2. Role of Civic Engagement in Urban Development

During the ongoing planning processes in Herttoniemi, various participatory initiatives took place, which the residents frequently commented on. However, the analysis revealed that citizens had limited influence on public participation regarding urban planning, while civic engagement at large played a significant role in shaping urban development on a local level [11].

The analysis showed that participants, particularly in formal public hearing procedures, had little actual impact on the processes or outcomes of the plan. While public gatherings provided information and engagement for residents, they served more as an arena for interaction between local actors and civil servants, benefiting the self-organising urban initiatives more than the planning of Herttoniemi Centre. The ladder of participation proposed by Arnstein [41][5] remained stuck in its early stages, as the public could not effectively implement their ideas.

The city had full control over both the planning procedure and substance. If the city had desired to improve the condition of the Herttoniemi metro station, it could have made that decision instantly years ago. However, the deteriorating state of the metro station, which serves as a major transportation hub for the eastern part of the metropolitan region, has remained unchanged since its opening in 1982. The development process of the metro station is still ongoing after 20 years of planning. Public participation was not able to challenge the administration and planners.

On the other hand, the case study revealed that self-organisation differs from public participation not only in terms of occurrence, dynamics, and motives for action but also in terms of legitimacy. The quality of public participation is often measured by democratic representativeness, as it requires participants to be in favour of or against the planning proposal. Self-organisation, however, is driven by the action itself and carries a progressive ethos. Actors participate based on their agendas, which are openly presented through implicit and explicit forms of action. These agendas are also acknowledged by those involved in the action.

Therefore, the sovereign nature of self-organisation is not based on the number of actors. Unlike public participation, nobody represents anyone else but themselves, even though many people are motivated by the desire to do good or work for others. Self-organisation does not require large numbers of social actors or movements. Even a small and enthusiastic action group can be effective, as seen in several local initiatives in the Herttoniemi neighbourhood[6] [11].

However, there were also shared features with traditional participation. The same individuals who represented the traditional neighbourhood association were also active in community initiatives and other urban activism efforts. The semi-professional "super participant". [42] engaged in public participation and formed a "community of practic". [43]. For example, the activists involved in local web pages and social media tools were just a handful of people, yet they were engaged in various urban issues and local identity-making. Most methods and tools of participation (e.g., social media gatherings, event-making, application of PPGIS) were found to be valuable and successful when applied outside of traditional planning procedures and integrated into everyday life practices. The transition from participatory planning to participatory co-production accelerated and enabled satisfactory outcomes, such as the maintenance and renovation of the Roihuvuori yard [15,16].

Thus, civic engagement was inherently dispersed but connected to the local realm, serving as a foundation for local development. It sometimes served as a check on administrative decisions through NIMBYs (not in my backyard), while at other times, it accelerated developments through YIMBYs (yes in my backyard) and even contributed to the creation and maintenance of new public spaces and functions.

Nevertheless, statutory planning and self-organisation existed in separate realms, not as competitors, but spatially and temporally separated. Civic engagement expanded the temporal aspect of urban development, as self-organising groups could take immediate or short-term action, while statutory planning aimed to attract long-term investments through development plans.

When the relationship between civic engagement, formal planning, and administration was not contested, it proved to be productive. In the early stages of the action research, representatives of traditional planning did not acknowledge local activists, but the situation has slightly changed in recent years (see next section).

4.3. Potential of Civic Engagement in Urban Planning and Development

Until recently, the administration did not prioritise urban initiatives with immediate local impact (such as participatory budgeting). The statutory urban processes were initiated by urban developers with external funding, and the initiatives focused on enhancing larger metropolitan development. These large-scale changes were often not well received by residents.

Despite the 20-year planning process, the large-scale metro station project did not materialise. However, the city did not support minor-scale renovations that could have coexisted with the larger development initiatives. As a landowner, the city could have explored various developmental strategies, such as revitalising the metro station and allocating some sections for small entrepreneurs, community services, or leisure activities. Thus, the official priority of urban development was not the physical surroundings of the metro station but rather the expectation of economic value through potential building on the plot.

Simultaneously, different departments within the city were enthusiastic about developing the neighbourhood through participatory planning processes. Community and youth workers played a crucial role in enhancing local networking and participating in unofficial co-governance [14]. Unofficial co-governance included local websites maintained by residents, as well as gatherings where people planned and designed spaces for family services, shops for DIY farming products, and leisure activities. However, the outcomes of this co-governance remained unclear, as the statutory planning process continued with predetermined objectives and outcomes.

For real estate developers, self-organising groups were not particularly interesting, as they operated within economically less-valued segments and had limited capacity to purchase or lease properties. Similarly, self-organising groups were not favoured when it came to city-owned properties. As a result, the space for self-organisation was limited. However, self-organising actions that contributed to a lively everyday life culture [33], such as the Hanami event in Roihuvuori, were appreciated, as they improved the neighbourhood's image and vibe.

In summary, while the administration accepted participatory processes involving residents, these processes did not align with the substance of urban planning. Statutory urban planning was and still is primarily focused on facilitating real estate development and commercial services, with limited attention given to the creation of shared urban spaces or the improvement of public service properties. The economic interests of the city seem to take precedence, overshadowing the potential for enhancing civic engagement at large, which nevertheless thrived despite the formal processes.

5. Discussion

According to our literature review, various researchers have explored the link between urban planning and self-organisation. The Complexity and Planning group of the Association of European Schools of Planning (AESOP) has been particularly active in this area since 2005 [4,44]. Their articles often focus on enhancing self-organisation from the perspective of governance and planners. One approach they propose is rule-based planning, which involves creating a few basic rules that enable citizens and collectives to initiate their projects or prevent negative effects [40,45]. Instead of striving for predefined outcomes, for example, through blueprint planning, the framework rules seek to make space for unplanned and spontaneous urban development or hinder undesirable consequences, for example in the case of tourism, gentrification, etc. [46].

As the relationship between planning and self-organisation is co-evolutionary and occurs on different levels, planners need a mix of rules and instruments to navigate self-organisation [45]. The authors suggest moving from a normative prescriptive approach, which focuses on specific goals or designs, to a proscriptive approach. Proscriptive rules set out simple guidelines or codes that shape endeavours and prevent negative outcomes [43,47,48]. For example, Cozzolino [27] suggests, besides proscriptive rules, to also adopt a moderate scale of design involving multiple designers/architects, promoting multifunctionality, incorporating temporal factors, and diversifying ownership systems. Thus, the means of enhancement are not only spatial, and the scope is multilevel.

Nevertheless, only a few articles deal with civic engagement and urban development, according to our Scopus search. One of them [49] describes how diverse citizen engagement in spatial planning is in terms of actors (individuals, collectives, forms of self-organisation), types and scales of spaces (neighbourhoods, districts of cities, from micro to macro levels), intentions (numerous), interventions (spatial adjustments, alternative uses of space, alternative plans, agenda setting), and roles (between market and social economy; political, critical). Yet the article is written from the governance perspective, ignoring the meaning of self-organisation and everyday life.

However, the landscape of civic engagement is changing in urban development, with new forms of action emerging through social media, digital methods, and web-based initiatives. People can now raise funds, design their spaces, and create new public and semi-public areas. The role of co-governance and the co-production of space and service facilities is becoming increasingly important [12].

Influential authors, such as Beitske Boonstra and Luc Boolens [1], recommend that planners actively engage with the issues at hand by immersing themselves in the field as participants. Similar advice can be found in the literature on community development, place-making, and tactical urbanism [50–52].

For example, the self-organising actions during the Roihuvuori Hanami event not only improved the neighbourhood's image in the short term but also had long-term effects on re-creating everyday life practices. Therefore, civic engagement supports and creates the so called "third space". where both physical and conceptual spaces are transformed through shared experiences [53]. Through self-organisation, space acquires both novel meanings and resources, even new events, services, and customers, which statutory urban planning cannot foresee. Thus, by supporting local space- and place-making, statutory planning can accelerate urban development.

6. Conclusions and Recommendations

The perspective of civic engagement in its expanded form is based on critical planning research which seeks to study urban development initiatives in practice, outside the formal planning procedures and administration. Our study in Herttoniemi revealed the diverse and multifaceted nature of civic engagement, which played a role in patching up statutory planning and urban development. Civic engagement is inherently dispersed, but it also fosters connections to the local, meso, and macro levels when necessary [49,50].

Theoretical approaches such as actor-based urban planning are relevant for explaining urban development [1,11]. However, traditional urban planning often fails to acknowledge the self-organisation of local actors or the significance of short-term developments [24,54]. Therefore, we draw the following conclusions:

Conclusion 1: Urban planning should be conceptualised as actor-based planning, such as expanded urban planning, which recognises and incorporates various forms of civic engagement, from public participation to self-organising initiatives and everyday life practices.

Recommendation 1: While public participation has acquired a legitimate role in urban planning processes, it is time to enhance the other forms of civic engagement that immediately connect the development process to the local realm and might co-produce supportive urban settings.

Conclusion 2: There are multiple approaches to enhancing civic engagement in urban planning and development.

Recommendation 2: The choice of methods to support civic engagement depends on the adopted theoretical, practical, and contextual perspectives. Methods such as place-making, tactical urbanism, and fieldwork (action research) can support local identity-building and the coordination of different socio-spatial phenomena that may not otherwise be incorporated into statutory planning processes. Therefore, expanded urban planning has the potential to provide an "architecture of hope".

However, this case study does not allow us to explain the mechanism between civic engagement and urban development in detail. That will certainly be one of the foci of future research.

Author Contributions: Conceptualization, L.H. and S.W.; Methodology, S.W.; Formal analysis, S.W.; Investigation, L.H.; Writing—original draft, L.H. and S.W. All authors have read and agreed to the published version of the manuscript.

Funding: This research received no external funding.

Data Availability Statement: The original contributions presented in the study are included in the article, further inquiries can be directed to the corresponding author.

Conflicts of Interest: The authors declare no conflict of interest.

Appendix A

The research material comprises the following sets of data:
Follow-up 2018–2024

(a) News/Web sites/Other documents:

City of Helsinki (2024). Karttapalvelu (Service map included all planning processes). Retrieved 6.6.2024 at https://kartta.hel.fi/?sukkaId=2018-000604 (accessed on 6 June 2024).

City of Helsinki (2023). Hankinnan keskeyttäminen, Herttoniemen metrokortteli, metroaseman ja linja-autoterminaalin peruskorjaus, kaupunkiympäristön toimiala (2023) https://paatokset.hel.fi/fi/asia/hel-2019-008560?paatos=7604b7be-2bc8-47fa-8cca-4de1e2144bfa (accessed on 6 June 2024) → Hankinnan keskeyttäminen, Herttoniemen metrokorttelin kehittäminen

City of Helsinki (2019). Hankinta, Herttoniemen metrokorttelin kehittäminen, Liikenneliikelaitos (2019). https://paatokset.hel.fi/fi/asia/hel-2019-008561?paatos=ec791975-7d99-4a4d-b3d6-01380c5d11f7 (accessed on 6 June 2024).

Roihuvuori Neihgbourhood website (2023). Herttoniemen metroaseman asemakaava valmisteilla (6-2023). Published 26.6.2023 at https://kaupunginosat.fi/roihuvuori/ (accessed on 20 May 2024).

Helsingin Sanomat (2024). Itäinen metro-asema on odottanut remonttia pienen iäisyyden—Helsinkiläis--mieheltä suorat sanat. Published 26.2.2024 at https://www.hs.fi/kaupunki/helsinki/art-2000010242720.html (accessed on 6 June 2024).

Helsingin Sanomat (2022). Herttoniemen metroaseman yläpuolella uusia taloja: Havainnekuvat näyttävät mullistuksen. Published 25.10.2022 at https://www.hs.fi/kaupunk/art-2000009155085.html (accessed on 6 June 2024).

Helsingin Sanomat (2022). Nainen nousi metrosta ja järkyttyi kaikesta mitä näki—Silloin hän houkutteli miehensä asemalle siivoamaan".maan-päällistä helvetti".. Published 1.6.2022 at https://www.hs.fi/kaupunki/helsinki/art-2000008851744.html (accessed on 6 June 2024).

Helsingin Uutiset (2023). Herttoniemeen suunnitellaan maamerkkiä, joka mylläisi metroaseman uusiksi—kaupunki julkaisi poikkeuksellisen havainnekuvan. Published 22.5.2023 at https://www.helsinginuutiset.fi/paikalliset/5944538 (accessed on 6 June 2024).

Kauppakeskusyhdistys (2014). Finnish Shopping Centers. Retrieved 6.6.2024 at https://www.kauppakeskusyhdistys.fi/media/kauppakeskusjulkaisu/kauppakeskukset-2014.pdf (accessed on 6 June 2024).

Packalén, E. (2008). Herttoniemi. Kylä, kartano, kaupunginosa. (Herttoniemi as a village and a neihgbourhood). Helsingin kaupunginmuseon tutkimuksia ja raportteja, 2. Keuruu, Otava.

(b) Planning data:

City of Helsinki (2023a). Asemakaavaehdotuksen selostus 12.6.2023. Retrieved 4.4.2024 at https://ahjojulkaisu.hel.fi/55D50879-2084-CEBB-BA3F-8870FDB00002.pdf (accessed on 6 June 2024).

City of Helsinki (2023b). Asemakaava nähtävillä: Herttoniemi, Herttoniemen metroaseman kortteli 12.6.2023. Retrieved 4.4.2024 at https://paatokset.hel.fi/fi/kuulutukset-ja-ilmoitukset/asemakaava-nahtavilla-herttoniemi-herttoniemen-metroaseman-korttelit (accessed on 6 June 2024).

City of Helsinki (2023c). Vuorovaikutusraportti 23.5.2023. Retrieved 4.4.2024 at https://ahjojulkaisu.hel.fi/48673000-256D-C530-8BA4-87C28B900002.pdf (accessed on 6 June 2024).

City of Helsinki (2017a). *Footbridge over Eastern Motorway/Jalankulkusilta Itä- väylän yli.* Diaarinumero HEL 2017-013453 yritystontit 64/14.12.2017. Retrieved 3.4.2018 at https://dev.hel.fi/paatok-set/asia/hel-2017-013453/ (accessed on 6 June 2024).

City of Helsinki (2017b). *Resigned Rental Contract for the shopping mall plot*/Vuokrasopimuksen täydennys Lähipalvelukeskus Hertsi, Linnanrakentajan tie 2. YIT rakennus Oy:n ja Hartela Etelä-Suomi Oy:n Diaarinumero HEL 2015-009664. Retrieved 3.4.2018 at https://dev.hel.fi/paatokset/asia/hel-2015-009664/ (accessed on 6 June 2024).

City of Helsinki (2017c). *Rental contract with K-Supermarket Hertta*/Piirustusten tutkiminen ja vuokrasopimuksen muuttaminen, Kesko Oyj (Herttoniemi, tontti 43172/1) Retrieved 3.4.2018 at https://dev.hel.fi/paatokset/asia/hel-2017-002458/ (accessed on 6 June 2024).

City of Helsinki (2011). The detail plan of Herttoniemi Center/Herttoniemen keskuksen Megahertsin kortteli: Asemakaavan muutoksen selostus. Hankenumero: 1083_2, HEL 2011- 004118. Helsingin kaupunkisuunnitteluviraston asemakaavaosasto.

Appendix B

Action research 2004–2009

Primary data:

- Memos of local assemblies (23 items);
- Memos from the formal participation events in Herttoniemi (4 items);
- Memos from semi-formal planning workshops and the gatherings of local stakeholders.

Secondary data:

- Survey I (Häkkinen and Wallin 2004);
- Survey II (Jarenko 2007);
- Group interviews with local stakeholders (3 items);
- Group and personal interviews with administrators (5 items) (2009–2010);
- Observation memos and graphics from the local events and happenings. (over 200 cases during 2003–2012).

Supporting planning data:

- Planning documents produced by the Helsinki planning department at Hert-toniemi, Roihuvuori, and Kalasatama;
- Statistical data of Helsinki (2001–2012);
- Strategies and decision documents of the Departments of town planning and social affairs (2003–2009).

Notes

1 Urban development can be intentional or unintentional, just as the verb "developmen". is transitive and intransitive. Thus, urban development is larger in scope than urban planning.

2 The so called Palco-group focused from various perspectives on the characteristics of participatory local communities. It comprised Liisa Horelli, adjunct professor, PhD [13]; Karoliina Jarenko, MSci [14]; Jenni Kuoppa, PhD [15]; Joanna Saad-Sulonen, PhD [16]; and Sirkku Wallin, PhD [12,17]. Members of the group have studied expanded urban planning in Finland and in other Nordic countries [18], as well as in Italy [19].

3 According to Wallin [11], Herttoniemi has functionally been a part of Helsinki, with its old suburban structure gradually changing since 1950's. Former industrial activities in the area transformed into harbour and warehouse services serving the entire city. After World War II, Herttoniemi evolved into one of Finland's first suburbs, characterised by a classical neighbourhood unit that provided its residents with public services, commercial establishments, and workplaces. The previous harbour area was replaced by a new sub-area called Herttoniemenranta, characterised by high-density urban buildings. Commercial services were relocated to the central part of Herttoniemi, near the metro station, while the former industrial area transformed into an office and retail district. The old shopping centres in West Herttoniemi and Roihuvuori have lost most of their commercial services, except for small supermarkets. Additionally, public services such as the library and youth house have moved to a more central location at Megahertsi (see Figure 4). The changing urban landscape in Herttoniemi continues, but mostly outside of the neighbourhood. In the slightly northern region near the centre, new buildings are added to the existing point-access block houses and slab blocks in Myllypuro. Furthermore, a completely new neighbourhood called Kruunuvuorenranta is currently under construction southeast of Herttoniemi. Just a few kilometres south lies Kalasatama, an even larger agglomeration with its new metro station. While Herttoniemi has served as a transportation hub for Southeast Helsinki, its prominence in this aspect has diminished as internal transformations have taken place.

4 The research group has been monitoring the planning process of the Herttoniemi Metro Centre since 2004. The initial plan, accepted in 2007, initiated progress on the eastern side of the motorway but not on the metro station plot. In 2018, the planning and development leapt forward. This time, the plan comprised three new housing units on the east side of the centre, enlarging the quarter of the Megahertsi shopping mall to the north. Currently, the City of Helsinki is proposing a plan for the metro station with new housing quarters (19,050 km^2), a service floor (4350 km^2), and open public space on the street level with an underground metro and bus station (2350 km^2). However, the decision-making process of the plan is still ongoing (see Appendices A and B).

5 The ladder of participation by Arnstein [41] has been widely used to illustrate the distribution of power over/to/in the planning processes. The ladder comprises stages of (1) manipulation, (2) therapy, (3) informing, (4) consultation, (5) placation, (6) partnership, (7) delegated power, and (8) citizen control. The highest steps are considered to ensure citizen empowerment. However, the post-structuralists consider that even the highest stages are insufficient because the ladders are staged from the governmental perspective. The model provides participation by procedural inclusion (e.g., Boonstra and Boelens [1] (p. 107)).

6 For example, the self-organising movement that initiated and produced the neighbourhood house at Tuhkimo Kindergarten comprised a handful of people. Later, the house was widely used by locals for many occasions. Also, the Roihuvuori Yard initiative improved the local surroundings and is enjoyed daily by the residents. Relatively small numbers of residents and other stakeholders have managed to develop and maintain better public spaces in Herttoniemi.

References

1. Boonstra, B.; Boelens, L. Self-organization in urban development: Towards a new perspective on spatial planning. *Urban Res. Pract.* **2011**, *4*, 99–122. [CrossRef]
2. Reimer, M.; Getimis, P.; Blotevogel, H.H. (Eds.) *Spatial Planning Systems and Practices in Europe*; A Comparative Perspective on Continuity and Changes; Routledge: New York, NY, USA, 2014.
3. Sandercock, L. Towards a planning imagination for the 21st century. *J. Am. Plan. Assoc.* **2004**, *70*, 133–141. [CrossRef]
4. De Roo, G. Introduction to the Handbook on Planning and Complexity. In *Handbook on Planning and Complexity*; de Roo, G., Rauws, W., Zuidema, C., Eds.; Edward Elgar Publishing: Cheltenham, UK, 2020.
5. Healey, P. *Collaborative Planning: Shaping Places in Fragmented Societies*; MacMillan Press: London, UK, 1997.
6. Forester, J. *The Deliberative Practitioner- Encouraging Participatory Planning Processes*; MIT Press: Boston, MA, USA, 1999.
7. Innes, J.; Booher, E. *Planning with Complexity: Introduction to the Collaborative Rationality of Public Policy*; Routledge: New York, NY, USA, 2010.
8. Davidoff, P. Advocacy and Pluralism in Planning. *J. Am. Inst. Plan.* **1965**, *31*, 331–337. [CrossRef]
9. Jacobs, J. *Death and Life of Great American Cities*; Random House: New York, NY, USA, 1961.
10. Rantanen, A.; Rajaniemi, J. Urban planning in the post-zoning era: From hierarchy to self-organisation in the reform of the Finnish Land Use and Building Act. *Environ. Plan. B Urban Anal. City Sci.* **2020**, *47*, 321–335. [CrossRef]
11. Wallin, S. Managing Urban Complexity—Participatory Planning, Self-Organization and Coproduction of Urban Space. Ph.D. Thesis, Aalto University, Espoo, Finland, 2019. Available online: https://aaltodoc.aalto.fi/handle/123456789/40200 (accessed on 6 June 2024).

12. Mäenpää, P.; Faehnle, M. *Neljäs Sektori—Kuinka Kaupunkiaktivismi Haastaa Hallinnon, Muuttaa Markkinat ja Laajentaa Demokratiaa (How Urban Activism Challenges Governance, Transforms the Markets and Expands Democracy)*; Vastapaino: Helsinki, Finland, 2021.
13. Horelli, L. (Ed.) *New Approaches to Urban Planning*; Aalto University Publication: Otaniemi, Finland, 2013.
14. Jarenko, K. Local Co-Governance in Herttoniemi: A Deliberative System. In *New Approaches to Urban Planning*; Horelli, L., Ed.; Aalto University Publication: Otaniemi, Finland, 2013; pp. 45–63.
15. Kuoppa, J. Kävelyn Lupaukset Kaupungissa (The Potential of Walking in the City: Three Case Studies on Everyday Experiences of Walking and Urban Planning in Finnish Cities.). Tampere University. Acta Electronica Universitatis Tamperensis. Retrieved 2016. Available online: https://trepo.tuni.fi/handle/10024/98653 (accessed on 6 June 2024).
16. Saad-Sulonen, J. Combining Participation: Expanding the Locus of Participatory E-Planning by Combining Participatory Approaches in the Design of Digital Technology and Urban Planning. Ph.D. Dissertation, Media Lab, Aalto University, Otaniemi, Finland, 2013.
17. Wallin, S.; Jarenko, K.; Horelli, L. Features and Consequences of Flat Ontology in Expanded Urban Planning. *disP—Plan. Rev.* **2021**, *57*, 46–59. [CrossRef]
18. Wallin, S.; Horelli, L.; Saad-Sulonen, J. (Eds.) *Digital Tools in Participatory Planning*; Series C27; Aalto University, Centre for Urban and Regional Studies: Espoo, Finland, 2010.
19. Horelli, L. Engendering urban planning successes constraints expansion. In *Expanding Architecture from a Gender-Based Perspective*; Amoroso, S., Saree, E., Novas, M., Vilaplana, A., Eds.; MORE; Didapress: Firenze, Italy, 2020; pp. 505–521.
20. Hillier, J. Strategic navigation across multiple planes: Towards a Deleuzean-inspired methodology for strategic spatial planning. *Town Plan. Rev.* **2011**, *82*, 503–527. [CrossRef]
21. Kukkonen, H. *A Design Language for a Self-Planning System*; Helsinki University of Technology: Espoo, Finland, 1984.
22. Mitleton-Kelly, E. Complexity Research—Approaches Methods: The LSEComplexity Group Integrated Methodology. In *Organisational Complexity*; Keskinen, A., Aaltonen, M., Mitleton-Kelly, E., Eds.; Finland Futures Research Centre, Turku School of Economics and Business Administration: Turku, Finland, 2003; pp. 55–74.
23. Horelli, L. A Methodology of Participatory Planning. In *Handbook of Environmental Psychology*; Bechtel, R., Churchman, A., Eds.; John Wiley: New York, NY, USA, 2002; pp. 607–628.
24. Ingold, T. *The Perception of Environment: Essays in Livelihood, Dwelling and Skill*; Routledge: London, UK, 2000.
25. Stojanovski, T.; Partanen, J.; Samuels, I.; Sanders, P.; Peters, C. Viewpoint: City Information Modelling (CIM) and Digitizing Urban Design Practices. *Built Environ.* **2020**, *46*, 637–646. [CrossRef]
26. Horelli, L.; Wallin, S. The Renewal of the Finnish Planning Legislation as a Strategy of Urban Planning and Development. *Land* **2023**, *11*, 2085. [CrossRef]
27. Cozzolino, S. The (anti) adaptive neighbourhoods. Embracing complexity and distribution of design control in the ordinary built environment. *EPB Urban Anal. City Sci.* **2019**, *47*, 203–219. [CrossRef]
28. Horelli, L.; Damyanovic, D. Evaluation of spatial development from the gender+ perspective. In *Gendered Approaches to Spatial Development in Europe—Perspectives, Similarities, Differences*; Zibell, B., Damyanovic, D., Sturm, U., Eds.; Routledge: London, UK, 2019; pp. 157–180.
29. Boonstra, B. *Planning Strategies in an Age of Active Citizenship. A Post-structuralist Agenda for Self-organization in Spatial Planning*; Ph.D. Series; InPlanning: Groningen, The Netherlands, 2015. Available online: https://hdl.handle.net/1854/LU-7257558 (accessed on 28 December 2018).
30. Faehnle, M.; Mäenpää, P.; Blomberg, J.; Schulman, H. Civic engagement 3.0—Reconsidering the roles of citizens in city-making. *J. Finn. Urban Stud.* **2017**, *55*, 70–75.
31. de Certeau, M. *The Practice of Everyday Life*; University of California Press: Berkeley, CA, USA, 1984.
32. Heller, A. *The New Everyday Life*; Routledge & Kegan Paul: London, UK, 1984.
33. Bech-Joergensen. "Hvorfor gor de ikke noge".? In *Hverdagsliv, Kultur og Subjektivitet*; Bloch, C., Hojgaard, L., BechJoergensen, B., Nautrup, B.L., Eds.; Akademisk Forlag: Copenhagen, Denmark, 1988.
34. Eleb-Vidal, M. Se Construire et Habiter, proposition d'analyse psycho-sociale clinique. Ph.D. Thesis, Universite de Paris VII, Paris, Franch, 1980.
35. Gilroy, R.; Booth, C. Building an Infrastructure for Everyday Lives. *Eur. Plan. Stud.* **1999**, *7*, 307–324. [CrossRef]
36. Hendriks, C.M. When the Forum Meets Interest Politics: Strategic Uses of Public Deliberation. *Politics Soc.* **2006**, *34*, 571–602. [CrossRef]
37. Horelli, L.; Saad-Suloinen, J.; Wallin, S.; Botero, A. When Self-Organization intersects with Urban Planning, Two Cases from Helsinki. *Plan. Pract. Res.* **2015**, *30*, 286–302. [CrossRef]
38. Puustinen, S.; Mäntysalo, R.; Jarenko, K. The Varying Interpretations of Public Interest: Making Sense of Finnish Urban Planners' Conceptions. *Curr. Urban Stud.* **2017**, *5*, 82–96. [CrossRef]
39. Flyvbjerg, B. Five Misunderstandings about Case Study Research, Corrected. In *Qualitative Research: The Essential Guide to Theory and Practice*; Savin-Baden, M., Major, C.H., Eds.; Routledge: London, UK; New York, NY, USA, 2012; pp. 165–166.
40. de Roo, G.; Rauws, W.; Zuidema, C. (Eds.) *Handbook on Planning and Complexity*; Edward Elgar Publishing: Cheltenham, UK, 2020.
41. Nummi, P. Crowdsourcing Local Knowledge with PPGIS and Social Media for Urban Planning to Reveal Intangible Cultural Heritage. *Urban Plan.* **2018**, *3*, 100–115. [CrossRef]

42. Staffans, A. Vaikuttavat asukkaat: Vuorovaikutus ja paikallinen tieto kaupunkisuunnittelun haasteina (Residents with impact). In *A Publication Serie of the Centre for Urban and Regional Studies, 29*; Helsinki University of Technology: Espoo, Finland, 2004.
43. Rijken, B.; Buitelaar, E.; van Duinen, L. Exploring the feasibility of future housing development within existing cities: Science-for-policy in the face of complexity and politicization. *Environ. Plan. B Urban Anal. City Sci.* **2020**, *47*, 336–351. [CrossRef]
44. Portugali, J. *Self-Organization and the City*; Springer: Berlin, Germany, 2012.
45. Rauws, W.; Cozzolino, S.; Moroni, S. Framework rules for self-organizing cities: Introduction. *Environ. Plan. B Urban Anal. City Sci.* **2020**, *47*, 195–202. [CrossRef]
46. Moroni, S. Complexity and the inherent limits of explanation and prediction: Urban codes for self-organizing cities. *Plan. Theory* **2015**, *14*, 248–267. [CrossRef]
47. Partanen, J. Guiding urban self-organization: Combining rule-based and case-based planning. *Environ. Plan. B Urban Anal. City Sci.* **2020**, *47*, 304–320. [CrossRef]
48. Rosner-Manor, Y.; Borghini, S.; Boonstra, B.; Silva, P. Adaptation of the urban codes—A story of placemaking in Jerusalem. *Environ. Plan. B Urban Anal. City Sci.* **2020**, *47*, 251–267. [CrossRef]
49. Horlings, L.G.; Lamker, C.; Puerari, E.; Rauws, W. Citizen Engagement in Spatial Planning, Shaping Places together. *Sustainability* **2020**, *13*, 11006. [CrossRef]
50. Sarkissian, W.; Hofer, N.; Shore, Y.; Vajda, S.; Wilkinson, C. *Kitchen Table Sustainability: Practical Recipes for Community Engagement with Sustainability*; Earthscan: Adelaide, Australia, 2008.
51. Hamdi, N. *The Placemakers Guide to Building Community*; Earthscan Ltd.: London, UK, 2010.
52. Lydon, M.; Garcia, A. A Tactical Urbanism How-To. In *Tactical Urbanism*; Lydon, G., Ed.; Island Press/Center for Resource Economics: Washington, DC, USA, 2015; pp. 171–208.
53. Soja, E.W. *Thirdspace. Journeys to Los Angeles and Other Real-and-Imagined Places*; Blackwell: Oxford, UK, 1996.
54. Shove, E.; Pantzar, M.; Watson, M. *The Dynamics of Social Practice: Everyday Life and How It Changes*; Sage: London, UK, 2012.

land

MDPI

Article

Securing Land Tenure Through Participatory Upgrading Processes: Women's Experiences in Freedom Square, Gobabis, Namibia

Tanzila Ahmed [1,*], Astrid Ley [2], Mohamed Salheen [3] and Jennilee Kohima [4]

1 University of Stuttgart, Faculty of Architecture and Urban Planning, MSc Integrated Urbanism and Sustainable Design, 70174 Stuttgart, Germany
2 University of Stuttgart, Institute of Urban Planning and Design, 70174 Stuttgart, Germany; astrid.ley@si.uni-stuttgart.de
3 Integrated Urbanism and Sustainable Design, Ain Shams University, Cairo 11566, Egypt; mohamed_salheen@eng.asu.edu.eg
4 Department of Architecture, Planning and Construction, Namibia University of Science and Technology, Windhoek 9000, Namibia; jkohima@nust.na
* Correspondence: ahmedtanzila25@gmail.com or st184792@stud.uni-stuttgart.de; Tel.: +49-15213331859

Abstract: In Namibia, informal settlements are usually upgraded by a participatory process aimed at providing inhabitants with land tenure security through the Flexible Land Tenure Act (FLTA), followed by the provision of basic services. The pilot FLTA implementation case is the upgrading of the Freedom Square settlement in Gobabis, which was a woman-led process. This research analyzes both the experiences of the women participating during the process as well as the immediate and long-term perceived impacts on their social and economic conditions. The empirical data were collected through interviews and a focus group discussion with experts and key participants in the development process. The findings highlight socio–economic challenges faced by the participating women, including insufficient support from family and limited economic opportunities. Throughout the process, stakeholders shifted their aims. This shift influenced the involvement of women, changing their roles from active participants to service recipients. While the main goals of receiving land tenure and service provision were partially fulfilled, the long-term impact of the process on women is varied from how it was initially envisioned. This was largely due to the absence of a comprehensive plan to improve their livelihoods. The results show that for participatory land tenure processes to become inclusive, a stronger gender-mainstreaming agenda becomes necessary. It also emphasizes holistic development planning as a key factor for long-term success.

Keywords: land management; collaborative planning; participatory development; urban and land planning; informal settlement upgrading; SDG localization; gender-sensitive process; participatory land tenure

1. Introduction

1.1. Problem Statement

In Namibia, unplanned rapid urbanization has led to the growth of informal settlements [1], and marginalized women living in these informal settlements are particularly vulnerable, facing inadequate sanitation, lack of security, and insufficient basic services. In these informal settlements, the inhabitants lack tenure security for the land they occupy [2]. Tenure security is a key factor for suitable housing, enabling access to other important factors like services and habitability [3]. In Namibia's informal settlements, land tenure is secured through an upgrading process involving multiple stakeholders (e.g., local authorities, NGOs) and the community contributing with 'sweat equity' (physical labor) to reduce costs. However, finalizing the land registration and delivery is often delayed due to the inhabitants' economic constraints and the limited resources of local authorities.

Therefore, the Namibian government introduced the Flexible Land Tenure Act (FLTA) in 2012 as the foundation of the Flexible Land Tenure System (FLTS), which ensures land tenure security through several titles with reduced requirements [4]. Low-income families living in informal settlements in urban areas throughout Namibia are the main recipients of this system. At the final stage of the upgrading process, the FLTS is implemented, ensuring rapid delivery of land to the inhabitants. This act adds to Namibia's many land law reforms since 1995, recognizing women as landowners since, in Namibia, the traditional communal laws and colonial statutory land systems have historically favored men as landowners.

The town of Gobabis in eastern Namibia provides a relevant case study for this research analysis. Here, FLTS was piloted in Freedom Square, one of its largest informal settlements, as shown in Figure 1 adapted from [1]. This upgrading process was initiated and continued with its female inhabitants in the leading role, resulting in tenure security and partial access to services. However, due to the absence of post-process research, it remains unclear the extent to which this process has been gender mainstreamed.

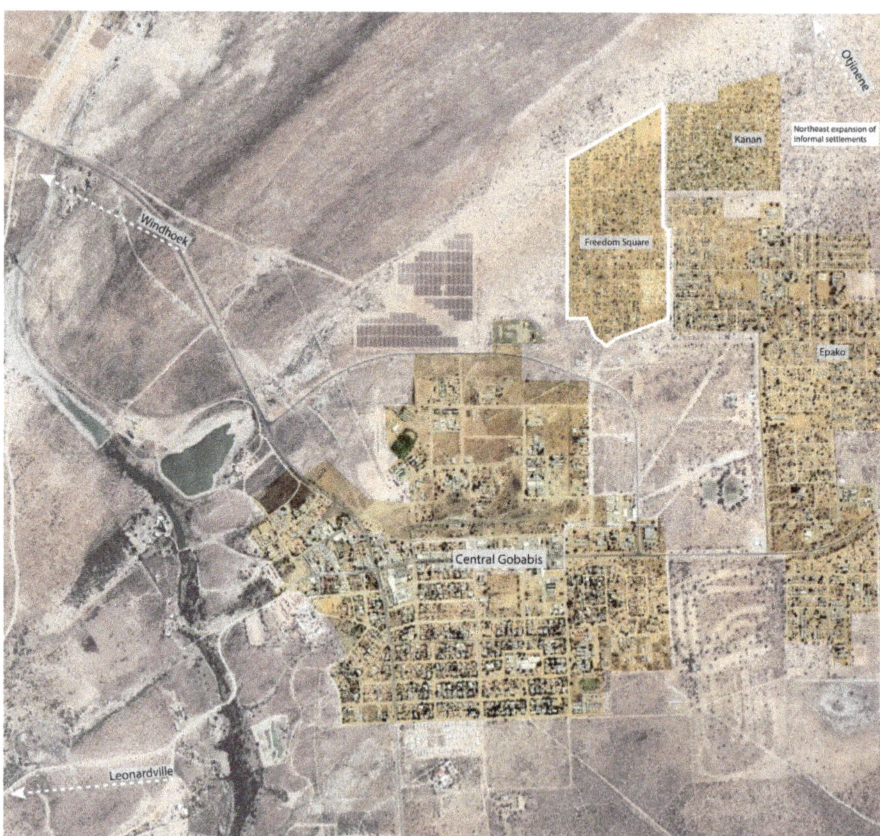

Figure 1. Freedom Square in Gobabis. Adapted from [1].

Gender mainstreaming forms an essential approach contributing to localizing Sustainable Development Goal (SDG) 11, which advocates for inclusive, safe, and sustainable urban spaces. There is limited research on the extent to which this process has been gender mainstreamed, which is crucial in informal settlements already suffering from resource imbalances. These imbalances force the local authorities to adopt innovative and unique approaches, driven by the absence of African context-specific SDG localization guidelines. Therefore, cases such as the co-production of land in Freedom Square [5] with the inhabitants and the local authority of Gobabis warrant deeper analysis. What makes this research crucial is that this process is being replicated in informal settlements across Namibia. Therefore, analyzing this case paves the way for making future approaches more gender mainstreamed. Additionally, the key findings could help monitor and improve ongoing upgrading processes across the country. This will in turn evaluate the gender-mainstreaming dimension of the FLTS and its implications in localizing SDG 11 in informal contexts. This research was conducted under the umbrella of the 'SDGs GoGlocal!' project as a trilateral partnership between the University of Stuttgart, Ain Shams University, and Namibia University of Science and Technology. The field data collection took place between April and May 2024.

1.2. Research Objectives

Given the central role of women in initiating and driving the upgrading process of Freedom Square, this research focuses on women's experiences during participation and adopts an analytical approach to evaluate them. Guided by four research objectives, this research aims to evaluate the gender-mainstreaming aspects of a locally tailored settlement's upgrading process and its contribution to localizing SDG 11.

The first objective analyzes the involvement of the stakeholders at various phases of the upgrading process, which provides insight into their influence on the participation of women. The second objective critically examines the challenges women faced during participation, the extent to which these challenges were acknowledged or mitigated, and their broader social and economic implications. The third objective evaluates the immediate outcome of women's participation in the process by assessing whether their goals were achieved. The alignment of the outcome with the initial motivations of the women results in an evaluation of the benefits of their participation in this process. The final objective is to assess the sustaining impact of the process on the participating women. Focusing on outcomes developed by actively engaged women inhabitants and the implementation of the FLTS, this research explores how land tenure has influenced these women's lives. These objectives are explored through the following research questions:

- What motivated women to participate?
- How did stakeholders influence women's participation?
- What challenges did women encounter during their participation?
- What were the immediate outputs of women's participation?
- What are the long-term impacts of women's participation?

1.3. Theoretical Background of Research

Gender is a social concept that refers to the differences in characteristics, duties, and obligations of men and women [6]. In most societies, there are differences and inequalities between women and men in responsibilities assigned, activities undertaken, access to and control over resources, and decision-making opportunities [7]. Despite Namibia's progress in closing its gender gap to 80.5% [8], gaps remain for marginalized groups living in informal settlements, constituting 25% of the population [9].

Gender mainstreaming is a central approach to closing this gender gap, while also supporting the localization of SDG 11. This approach ensures that a gender-sensitive perspective is central to all activities, such as policy development, research, advocacy, dialogue, legislation, resource allocation, planning, implementation, and the monitoring and evaluation of policies and programs [10]. Gender-sensitive approaches are required that address gender norms, roles, and gaps in resource allocation to mainstream gender [6], especially in informal settlements suffering from resource imbalances.

Ensuring tenure security is a key milestone in upgrading Namibia's informal settlements, as secure tenure rights enable impoverished people to invest in their property and livelihood without fear of eviction [11]. To ensure that the tenure securing process and its output are gender-sensitive, it must be inclusive, ensuring that resource imbalances do not continue. Ensuring tenure security for women is one primary step to discontinuing resource imbalances. As in the Global South, women are less likely to own or control land, despite performing more than half of all agricultural activities. This imbalance impacts general development activities as well. Hence, the need to close the gender gap in land ownership becomes evident [11]. The active engagement of women in the informal settlement-upgrading process and equal land rights (act and implementation) aligns with gender-mainstreaming agendas, advancing the localization of SDG 11.

Historically, in Namibia, due to a lack of recognition of rights within families and communities, African women have faced significant challenges [12]. In the pre-colonial customary or communal land tenure system, women's civil status within the community influenced their access to land, and their social roles determined which land they had access to [13]. Colonial and patriarchal systems further embedded disparities through the formalization of men's ownership of land. Since its independence, Namibia has implemented reform policies such as joint registration rights under Redistributive Reform [14] and, a more recent addition, the FLTA, which also recognizes single women-headed households as beneficiaries. Introduced under the FLTA in 2012, the FLTS is aimed towards rapid land delivery. However, the land ownership guidelines under the FLTS reveal gender ambiguities. The guide to the FLTA explains the FLTS in key points to its target users and mentions that the landowner shall be the 'Head of Household' [15]. It provides examples that seem to reinforce the idea that a woman could be the 'Head of Household' only in the absence of a man in the family. Otherwise, it suggests joint tenure for married couples. Therefore, this leaves women cohabiting with their partners without tenure security.

Furthermore, the implementation process of this system does not explicitly mention a participatory process with the community. It rather relies heavily on the existing institutions or the local authorities to conduct the process. The lack of resources and manpower of these institutions raises valid concerns regarding the implementation and outcome of FLTS even with reduced requirements. Delving deeper into cases like Freedom Square will bring out the possibilities of co-production for faster implementation of the FLTS with a sustaining impact, such as attaining SDG targets 11.1 and 11.3 and focusing on affordable housing and inclusive urbanization. Figure 2 illustrates the framework of this research.

Effectively localizing the SDGs in Africa may be complicated due to the differences between global SDG norms and Africa's diverse context, and previous studies suggest breaking down goals into local processes and acknowledging them in an integrated approach [16]. Moreover, multi-stakeholder engagement further strengthens the localization processes in these informal contexts [17]. The inclusion of marginalized groups in development, including community-based organizations, can aid in reducing challenges in achieving inclusive urbanization. For example, incorporating women's needs and experiences into tenure reforms ensures that the benefit of tenure security is equally distributed.

Thus, analyzing women's engagement in Freedom Square through these lenses is a cross-cutting approach, providing insight into the possible pathways of localizing SDG 11.

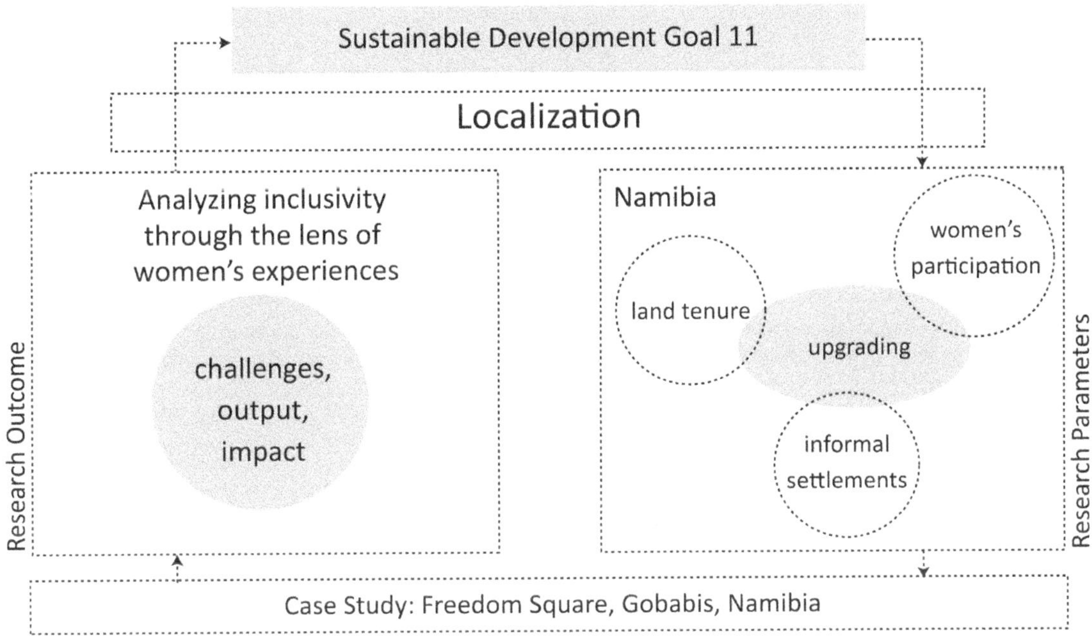

Figure 2. Framework of research.

2. Materials and Methods

2.1. Case Study Selection

In the Freedom Square settlement of Gobabis, Namibia, all inhabitants received their plots with a registration document of the 'Landhold Title', a form of land ownership under the FLTS. This process was completed in 2021, three years prior to this research, providing an opportunity to analyze the impact of land tenure on the lives of women who were actively engaged in the process. As one of the earliest informal settlement-upgrading processes, the Freedom Square process involved many international, national, academic, and government organizations throughout its nine-year lifespan, which provides a valuable opportunity to analyze the relationships between stakeholders and the output of their collaboration. Also, upon completion, this upgrading process was identified as a model to be replicated across Namibia to upgrade informal settlements. Investigating this case offers insights into the specific achievements (and pitfalls) of the process and provides learning for future gender mainstreaming in tenure upgrading. Evaluating the gender aspect throughout the phases of this process will help identify and avoid repeating mistakes. Finally, this case study is closely aligned with the broader research topic. It served as a platform to implement gender-mainstreaming initiatives, resulting in inclusive land ownership. Thus, this has been a significant step forward in localizing SDG 11, inviting further research on the process.

2.2. Data Collection

This research followed a qualitative method to answer the research questions outlined in Section 1.2. The primary and secondary literature on land tenure in Namibia, informal settlements, and upgrading processes were analyzed to provide background information and validate the data. In Gobabis and Windhoek, Namibia, semi-structured interviews were conducted and later validated through a focus group discussion (FGD). Eight experts from Namibia University of Science and Technology (NUST), Namibia Housing Action Group (NHAG), and Deutsche Gesellschaft für Internationale Zusammenarbeit (GIZ) were interviewed for their expertise and direct involvement in Freedom Square's upgrading process. Their interview questions (Appendix A) consisted of six main questions centered on women and their participation in the process.

Additionally, seventeen female inhabitants of Freedom Square were interviewed to gather first-hand experiences from them as they played a key role in leading the process. The interviewees were selected with the support of Epako municipality, who contacted the active community leaders of Freedom Square. The community leaders then further assisted in choosing interviewee participants. They were chosen based on the provided interviewee sampling criteria by the researchers: (a) female inhabitants and participants of the upgrading process and (b) female inhabitants of varied civil status. During the data collection process, it became clear that at later stages of the upgrading processes, women were less involved. As a result, further interviews were conducted with three male inhabitants to obtain an understanding of the full upgrading process. The interviews were conducted in Afrikaans, the local language, and translated by an interpreter (a Freedom Square inhabitant) during the interview. The presence of an interpreter did not affect the information, which was later ensured by translating the recorded interviews with transcribing services and the varied data received from the interviewees. Their interview questions (Appendix B) focused on their family, initiation to participation, and relationship to the stakeholders. They were also asked to discuss the challenges faced during the process and what they currently face when they have land tenure. The interviews were followed by a visit to the Freedom Square settlement to observe the upgraded formal settlement. In the second phase of data collection, the FGD was conducted at the Epako municipality. The previously interviewed inhabitants of Freedom Square were invited to participate and validate the extracted data from their interviews.

2.3. Data Analysis

The collected data were analyzed in two phases. While primary and secondary data were analyzed to obtain an overall background of the project and as supporting information to validate the analyzed data, the semi-structured interviews served as the main source of data for answering the research questions. A total of 28 interviews were recorded and transcribed. Four themes were extracted from the transcriptions to code the interviews. Thirty-three codes (Appendix C) were generated manually under the themes. Atlas.ti (Version 24.0.0) was used to code the relevant quotations of each transcription, resulting in a database of interview quotations under each theme and code. Initially, a timeline of the upgrading process was created from the database and the available secondary literature. The community later validated this timeline during the FGD mentioned in the previous section (Figure 3). The timeline was refined with additional information, resulting in the current timeline. The involved stakeholders were identified from the interview database and the secondary literature available before being validated by the workshop participants. Then, from the 17 interviewed women, a storyline of their participation was generated. This gave distinctive results for each of the participants. Their challenges were identified from there and later validated in the workshop. Women's motivations to participate were identified from the entry points and the interviews. They were validated in the same workshop. Finally, the current situation of the settlement, its built environment and services, and the socio–economic situation of the participating inhabitants were considered.

Figure 3. Data collection through FGD with female inhabitants (authors).

Due to the time limitations, the research could not consider all stakeholders' perspectives. The Freedom Square upgrading process started twelve years before this research was conducted. Therefore, the complete timeline of the process and actions were derived from participants' memories of the events, as documentation had only been performed by stakeholders during each of their involvement phases.

3. Results

3.1. The Upgrading of Freedom Square

Gobabis, the capital of the Omaheke region of Namibia, is located approximately 200 km east of Windhoek. Around 19,000 people reside in Gobabis [18], and according to the settlement enumeration called Community Land Information Program (CLIP) conducted in 2012, 47% of this population lived in its four informal settlements. The council of Gobabis planned to establish a designed formal township in the location of Freedom Square, it initially intended to relocate the inhabitants. However, due to the large size of the settlement and possibly so as not to lose political ground, the council eventually surrendered to the inhabitants' demands and agreed to sell the land to them at a subsidized price. With the support of NHAG and the Shack Dwellers Federation of Namibia (SDFN), the upgrading process started.

In Freedom Square, tenure security was identified as a means to achieve all the other services, such as water and sanitation. The upgrading started in 2012 with settlement enumeration followed by identifying and prioritizing the needs of the inhabitants. The inhabitants directly involved themselves in this process, with a few selected inhabitants traveling abroad for learning exchanges within the international network Slum Dwellers International (SDI), of which SDFN is a member. Afterwards, in 2014, they collaborated with NUST students to re-block their settlement. Figure 4 demonstrates the resulting settlement plans.

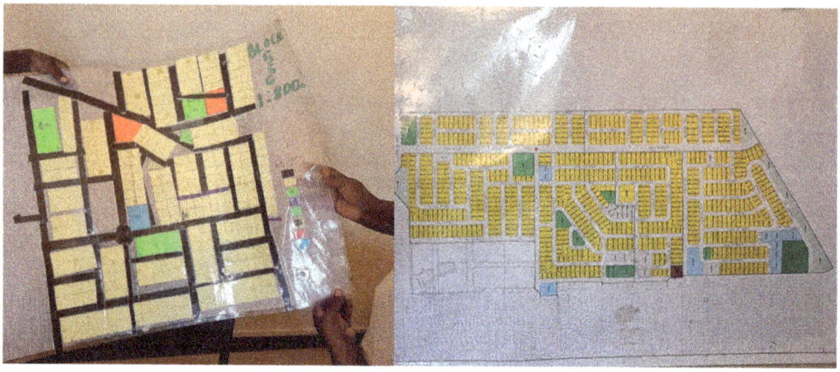

Figure 4. Re-blocked settlement of Freedom Square (authors). Legends for image on left: black: roads, yellow: residential plots, green: agricultural plots, blue: business plots and red: unusual plot sizes. Legends for image on right: white: roads, yellow: residential plots, green: agricultural plots and blue: business plots.

As things proceeded, the inhabitants moved into their re-blocked plots and provided physical labor to excavate trenches for service line installation. At this point, the Ministry of Land Reform (MLR) equipped the inhabitants with training and informed them about their tenure security options to implement the FLTA. The inhabitants were also assisted in forming management committees for each administrative block. A final survey by the government provided the updated land occupancy data, and afterwards, the inhabitants were handed over the certificates of tenure and a photocopy of the land deed in early 2021. Figure 5 presents photos of the Freedom Square settlement during the time of this research in April 2024.

Figure 5. Inside the Freedom Square settlement (authors).

3.2. Motivations to Participate

The women of Freedom Square had a range of tangible and intangible goals that motivated their participation in the process. At the inception, SDFN members and NHAG motivated women to participate. Young women were also encouraged to use their time to help the community. At the initiation of the process, they had identified that having land tenure is the means of receiving their primary needs, such as water and toilets. During an interview for this research with Tapiwa Maruza from NHAG, speaking of the involvement of single mother inhabitants, she mentioned, 'I think women, especially women-headed households, are probably in need more than the average household. So they are more vulnerable. Moreover, because these women don't have access to other opportunities that perhaps are accessible to most, they become more involved'.

Although women's basic needs overpower everything else in an informal setting, during the interviews, the participants mentioned the intangible things that motivated them to participate in this process. The joy of planning with their peers and professionals and having their opinions valued motivated them to continue being in the process. For some, the image of a livable community acted as an inspiration, while others enjoyed the group interactions and kept returning for more. Even though the process had not yielded any tangible results for the first four years, the women being motivated for these various reasons gave momentum to this process to initiate and continue. As mentioned previously, acknowledging all inhabitants' needs leads to a gender-mainstreamed process, which in this case was driven by the needs of the leading women, as men were still mostly absent from the process until the very end.

Moreover, women in Freedom Square often engage in informal work, preventing them from receiving a loan to develop their housing. They can only fulfill their housing goals by participating in this process. As Tapiwa Maruza from NHAG further mentioned during her interview for this research, 'So if you look, for example, at informal economy, it's mainly women. And so already you can see that there's a big chunk of people who are working. The biggest chunk of people who are working informally are the same people who can benefit from this process, because they don't have pay slips. They don't have all of the things that would make it possible for them to get a loan at the bank or for them to join any other housing process like NHE and all of that. So I think because women are vulnerable more than men, they participate more because they see this as a final opportunity for them to actually get housing'. This statement confirms the necessity of an upgrading process like FLTS to be implemented through community participation, as this provides a chance for the inhabitants to contribute in other ways than direct payment.

During the semi-structured interviews, many participating women from Freedom Square mentioned that they had no formal commitment to any job. Therefore, they had the flexibility to manage their schedules and be present in the community gatherings. The involved stakeholders also provided lunch to the participants. To encourage women to attend, they held the meetings within the community's reach, under a large tree, so that the women did not have to travel far. Most women responsible for looking after their children could leave them with the neighbor for the time being and attend meetings. Although not directly focused on being gender-sensitive, the process required women's active participation, and therefore, it attempted to accommodate their pressing needs to bring them into the process.

During the FGD, when presented with the participation motivations from interviews and asked to select their three most important goals, it was visible that everyone selected their needs for basic services as their highest priority. The results of the validated data are summarized in Table 1. All participating women explicitly mentioned their fear of eviction driving them. Therefore, this is highlighted as the key motivation of the participants. The rest are categorized as primary, secondary, and tertiary, corresponding to the number of women selecting these goals as their initial motivations.

Table 1. Initial participation motivations identified by women inhabitants.

Level of Goals	Motivations
Key	Fear of eviction
Primary	Owning a land Water services Sanitation services
Secondary	Single mothers without help Owning a house Owning a backyard garden Income opportunities in land
Tertiary	Leaving inheritance for children A livable community

The identification of these motivations at the beginning of any development process assists in analyzing two things. One, it assists in setting parameters to monitor the outcome of the process, and two, it sets a precedent to design a gender-mainstreamed participatory process by identifying gender-specific development goals.

3.3. Stakeholders' Influence on the Participation by Women

The dynamics of stakeholders changed according to their roles throughout the upgrading process. Initially, the key stakeholders were NHAG and SDFN, who conducted the settlement enumeration according to their contract with the Gobabis Municipality. At that time, and immediately afterwards, the community was more involved and made key decisions regarding upgrading and planning the re-blocked settlement, along with the town-planning students of NUST. The next phase proceeded as the Ministry of Urban and Rural Development (MURD) allocated funds for service installation and the inhabitants, along with NHAG and SDFN, provided physical labor to relocate themselves to their new plots and excavate trenches. Enumeration data were updated, the Social Tenure Domain Model (STDM) by the Global Land Tool Network (GLTN) was followed to register the data, and selected inhabitants were recruited to assist. The next steps included the participation of MLR and the municipality as primary stakeholders, which caused a shift in stakeholder involvement, as visualized in Figure 6. The stakeholders, such as the municipality, the MLR, and GIZ, although involved from the process initiation, increased their involvement during the implementation of the FLTS. They impacted the involvement of NHAG—and, therefore, as a cascading effect, the inhabitants—by leaving these stakeholders out of decision-making meetings [5]. The impact of this shifting relationship between stakeholders influenced the outputs of the inhabitants and especially women's participation in the process.

While achieving the earlier milestones, the women inhabitants were directly involved in every step supported by multiple actively engaged NGOs, CBOs, and international organizations; thus, the outputs were in their favor and left little room for complaints, as the direct involvement created a sense of responsibility within the inhabitants. They felt valued and liable for their own development alongside the local authorities and NGOs. During an interview for this research, Tapiwa Maruza from NHAG mentioned from her experiences in Gobabis, 'Most of the time, especially when you look at these communities who are impoverished and who do not have the resources to improve their situation. . . .a participatory process actually encourages them and that it empowers them to say, okay, I actually can do something and make decisions and be an active stakeholder in my own development'.

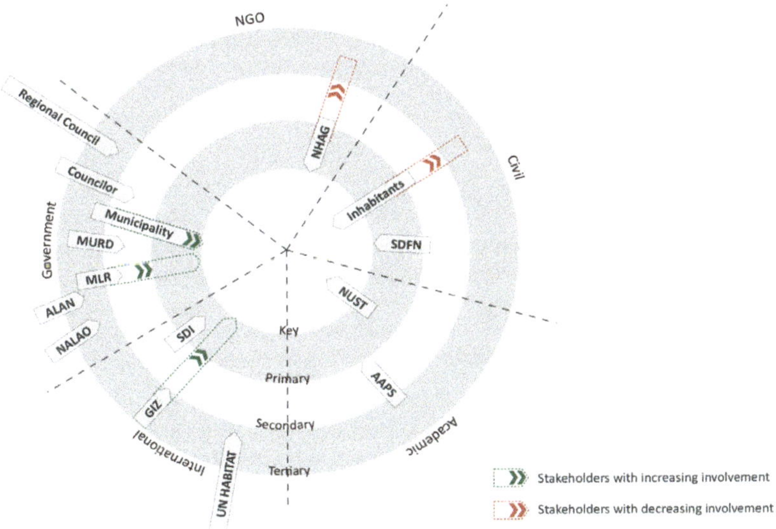

Figure 6. Stakeholder diagram of Freedom Square upgrading process. The pointed arrows of the stakeholders represent the intensity of their roles during the upgrading process. Stakeholders with a red arrow moved away from the process over time while stakeholders with a green arrow had increased involvement.

However, later in the phase, when government involvement increased for the pilot implementation of the FLTS, Delgado et al. 2020 [5] also mentioned that they treated the community as beneficiaries instead of partners in development. The active stakeholders of the later phase focused more on mass service provision and less on community participation, resulting in complaints and raised inhabitants' expectations. This impacted the sense of responsibility of the inhabitants and led them to the backseat, where they could make the authorities liable for all the decisions and developments. This directly impacted the current operations of the management committees as well. One interviewed committee leader, while discussing the lack of community efforts and responses, mentioned her frustration by saying, 'But the things in the beginning were very good, we finished together, the trenches, we did it together. Every time there was a meeting, they were there, but I don't know what happened with the people'.

The second outcome of changing stakeholder relationships affected the women participants. According to the interviewed experts, women have been excluded along the way, and their involvement decreased towards the end of the upgrading process. As was visible during the progression of the upgrading process, the involvement of women decreased near the end. There are multiple reasons behind this decreasing involvement of women, one of which is the change in stakeholder relationships. As NHAG and SDFN were less directly involved and the government organizations led the process, more men became dominant. This might have resulted from men's political ambitions rising due to the opportunities of networking with government organizations, or the absence of organizations inviting women and led by women. Tapiwa Maruza from NHAG confirmed this assumption, stating, 'In terms of leadership, it was very interesting to see more female leadership and females driving the process at the beginning. But then towards the end, when things were becoming a lot more formal, it became very male, more men representing the process than before'. A long-term impact of this participant shift is visible in the current disagreements between the inhabitants and the municipality regarding payment of the land. Participants were informed throughout the upgrading process of a minimum price to purchase the municipality land to be due upon certificate handover. However, currently, the community is not paying the municipality, which has been fueled by politicians and dragged further by the male inhabitants. The inhabitants not receiving their original deed until full payment risks the future of the current inhabitants.

3.4. Challenges of Participation for Women

During the nine-year (2012–2021) upgrading process, women from Freedom Square joined and withdrew from time to time to balance their other responsibilities. Eva (fictitious name), 35, a female married inhabitant, was living in Freedom Square with her husband, mother-in-law, and three children. Her involvement in the upgrading process started in 2014 with attending the planning studios with NUST students after the community achieved the first milestone. She wanted her family to own a plot and did not have a formal income back then, which encouraged and facilitated her participation. During the semi-structured interview, Eva mentioned that, in the beginning, there was no street in the settlement. So, it was difficult for them to reach their place when they came in a taxi if they had a lot of things to carry. Her husband was working outside the settlement, so he could not participate in the process even if he wanted to. On the days of community meetings, she would wake up earlier than usual to make food for the family, feed her children, and leave them with her mother-in-law so she could go attend the meetings. After a period of unpaid participation and a lack of visible results, her husband started restraining her participation. She argued and continued until they received joint land tenure as a married couple. She also mentioned during the interview, 'yes, it's also easier to stop because of the complaints (from her family). Also, the work was in the hot sun and all those things, but we went on until the end'. Although not satisfied that they have to pay for the land and services even though they provided physical labor during the service connection installation phase, Eva is still happy to be able to take a taxi to her doorstep when needed.

All of the interviewed participants shared similar stories with varied levels of challenges faced during their participation. During the FGD for data validation, the participating women were presented with the challenges identified from the semi-structured interviews and were asked to choose three of the most crucial challenges they faced. Lack of payment for participation and time-consuming participatory processes were chosen as the main challenges women, especially single mothers, faced during the process. From unhelpful family members to non-supportive husbands forbidding their wives to attend the processes, the women also faced restraints within their families. Table 2 summarizes the challenges identified from their responses. The challenges are categorized as primary, secondary, and tertiary, corresponding to the number of women selecting these as their main challenges.

Table 2. Challenges identified by women inhabitants.

Level of Challenges	Challenges
Primary	Lack of payment for participation Single mothers lacking time Time-consuming participatory process
Secondary	Single mothers hiring help to excavate trenches for lack of time Alcoholism Final results of the process are unclear
Tertiary	Lack of babysitter Non-supportive family

Some women stopped participating while others continued with support from their female friends and neighbors. During an interview with Dr. Anna Muller from NHAG, she mentioned, 'we had one member that was so strong and fluent in understanding and explaining the steps that she ended up in Johannesburg at the presentation of the governor to make it work. And her husband blocked her from further involvement'. Additionally, another major challenge identified in the workshop and by the expert interview (Anna Muller) was alcoholism in the family. This situation often restricted women from being able to participate.

This was long before the upgrading process started showing tangible results. This is also one major reason women kept slowly falling out of the process. A lot of the participating women became pregnant, stepped away, and came back when possible, as the process was strenuous and required physical labor. This also prohibited elderly people from participating. Additionally, since this process required reading maps, speaking up, and voicing their needs, illiterate and shy women often stayed away for fear of embarrassment.

Women also faced challenges due to their economic condition during the participation since the upgrading process did not provide any form of income. Instead, it involved saving from their current income to pay for land, services, and a house. Spending a long time in a process lacking payment also made them face retribution from their families, who were questioning their motives and choices to participate.

As single mothers mentioned during their interviews, when in the upgrading process, the inhabitants were excavating trenches for service connections, and single mothers often had to hire help to excavate for them. Earning and taking care of houses and their children did not leave them any spare time or energy to participate in the process. Since this was a mandatory step that the community agreed to partake in, single mothers were forced to spend money that other inhabitants did not need to. Elderly inhabitants also faced a similar issue; however, they received a government pension and some younger inhabitants also helped them to overcome this challenge. It was also revealed during data collection that in some families, the children and grandchildren are all dependent on one elderly family member's pension.

Several women participants left the process to seek work as they were aware of the upcoming costs and went outside the settlement due to a lack of work opportunities nearby. They often moved temporarily to Windhoek in search of domestic work. Some interviewees mentioned that participating in the upgrading process did not provide any income, and they had to prioritize earning and saving money to build a house when the opportunity arose. The lack of job opportunities in Gobabis impacted their overall participation. Many married or cohabiting women who did not have a formal income depended on their partner's income and economic decisions. The social norm of the earning member controlling the household prohibited women from participating in some cases.

3.5. Output of Participation

It is possible that at the initial phase of the upgrading process, land tenure and the necessity of water were chosen as the highest priorities because the majority of the participants were women from the settlements. Their suffering due to the lack of water connections made this their priority. This resulted in water connections in all houses, along with sanitary connections. Planning processes acknowledging women's point of view focus on the household needs of the inhabitants spending more time inside the settlements. The block planning of Freedom Square has agricultural plots for shared farming, business plots for shared informal businesses, and open spaces for playing and gathering. This could be the result of women's input, where they wanted to have gathering spaces and play areas close to their households and agricultural and business plots within the blocks so they could contribute or sell informally without having to go further away from their households. This plan to decentralize common facilities and put them within the range of each resident supports the women staying inside the settlement but performing different household duties.

Although the process was driven by women, it seems like ownership at the end of the process was often transferred to men. The reason behind this shift is the persisting notion of men being the head of the family. During his interview for this research, Robson Mazambani from GIZ mentioned that, during the final 'Head of Household' survey supported by GIZ, the heads were decided by several factors: the owner of the existing structure, an earning member, or simply, the socially hierarchical head. Participating women chose their husbands as 'Head of Households' and handed off land ownership because it was a socially acceptable decision. However, the parameters chosen, such as 'owner of the structure' or wage earner, gave men an advantage; as the formal earner of the family, men usually paid for the structures and therefore owned them. The unpaid work provided by women was disregarded as a parameter. Nevertheless, single mothers own their structures themselves, and couples married by law (contrary to the couples married in the community) were also recommended joint tenures. In cases of married couples in the community or cohabiting relationships, it is single ownership, leaving the other partner with no security of tenure—previously discussed as one of the aspects overlooked by the FLTA.

Moreover, it is notable that when land ownership was becoming more of a possibility during the upgrading process, men started involving themselves more. During the semi-structured interviews with the three male inhabitants, they indicated towards their concerns in regard to payment for the land, whereas the interviewed women stated more about the issues of lack of services. Additionally, men's involvement did not bring gender-mainstreamed participation in this case but resulted in less involvement of women due to men's dominance and disregarded gender-balanced outputs. As discussed earlier, this involvement resulted in setting non-gender-sensitive parameters for the previously discussed survey.

Many residents prioritize building their homes over paying for the land due to a lack of job opportunities or unclear information about payment terms. However, until the payments are completed, they do not receive their original land tenure certificates, and this raises concerns about inheritance, specifically whether their children will be able to safely inherit the land. This was highlighted by women residents during semi-structured interviews.

However, initial goals such as access to basic services were only partially met because of the government's incapability to provide electricity or cooking fuel even to tenured lands. The service connections for drinking water and sanitation have reached every household but the lack of economic development opportunities has refrained the inhabitants from investing in a water meter or building a toilet. Therefore, in most households, these services are still out of reach.

Nonetheless, now the women feel safer because they do not have to move around with their children and have land to settle down. This primary assurance helps them move forward in daily life. They can extend their shacks and run an informal business from within. Previously, businesses were often victims of vandalism but they are not anymore. Ownership has opened certain possibilities they can now build upon if they want to.

3.6. Impact of Participation

Previously shy to speak in gatherings, let alone make decisions, the women of Freedom Square are now more outspoken and aware of their citizen rights by participating in the upgrading process as a group. Freedom Square has management committees in each block where the elected leaders are gender-balanced and representative of each tribe. Some participant women were elected as community leaders and some others are now in important roles within SDFN. However, it is also notable that this percentage is not proportionate to the number of participating women.

In discussions with the community leaders during the data collection, they mentioned their frustration regarding the current management situation. Since the committee leaders have no authority to change their settlements' conditions, their leadership has become a symbol of tokenism to themselves and the rest of the inhabitants. Nonetheless, existing women leaders feel more accountable for their blocks due to their long term direct involvement in this process, and at least one woman per block is active in maintaining communication between inhabitants and the local authority. The interviewed women, who were not committee leaders, mentioned that they feel more comfortable sharing their issues regarding the lack of services to a female committee leader.

It was noticeable during the site visit and interview in Gobabis that women are less visible in governance with actual authority. Even though they are part of the management committees, the social norms and stigmas surrounding the gender of leadership persist. During the interview, a group of male inhabitants of Freedom Square also attended to discuss the possibilities of help regarding not having to pay for the land. It was visible that men were now leading the decisions of inhabitants, and the rest of the inhabitants were following this because it also served their needs since they barely had any money left to pay for land after paying for the necessary services and saving for building a house. Robson Mazambani also highlighted that it is worth looking at whether women are making decisions in Freedom Square anymore. However, during the brief interaction with the community, while performing data collection, it was not visible as such. Women expressed their hopelessness about the situation, while men advocated for not paying.

During the upgrading process, other stakeholders informed the inhabitants that they could receive recommendation letters for their participation, which might lead to more work opportunities. The inhabitants of Freedom Square, being the pilot in implementing the FLTA, continue to help other communities by sharing their experiences. Committee leaders, mostly women, are often invited by NHAG and SDFN to travel to other settlements. Some inhabitants also had temporary jobs with the municipality during the re-enumeration process, and some received jobs in NGOs and international organizations after they participated in the upgrading. However, this number is quite low.

By sharing opinions, participating in public gatherings, and public speaking to share knowledge, the women of Freedom Square have built or increased their social and technical capacities. They can negotiate with the municipalities and committee leaders for their needs and voice their frustrations. However, their visibility is significantly lower now than at the beginning of this process. The participating women are not visible as final decision makers or politicians; that arena is still navigated by men who stepped in later in the process that women had started. Additionally, according to Delgado et al. [5], the participants have been empowered with the technical skills they acquired, social organizing, engagement with authorities, and understanding of urban development processes, which is a valuable result of every co-creation process and could have been put to better use in Freedom Square for the longer-term betterment of the inhabitants.

During interviews, the women inhabitants mentioned not completely understanding the costs associated with their development and role. In the case of Freedom Square, NHAG and SDFN had disseminated information in earlier stages in detail but, often due to a lack of time and required emergency actions, the pre-process information dissemination failed to continue throughout the process. During the relocation and service installation, initially, active stakeholders were less concerned with information dissemination and more with distributing labor. The participating women inhabitants did not completely understand the cost of being formalized and that the urban area's formal inhabitants pay lifelong service charges for each service they receive, such as water, waste management, or electricity.

4. Discussion

4.1. Results Synthesis

The research findings confirm that women were key to initiating this process and leading the process in the beginning. Therefore, women's engagement was not a by-product of the process but drove it. However, the findings also indicate that their pattern of involvement changed due to several factors as the tenure certificates were about to be received.

Stakeholders involved in the upgrading process directly influenced women's engagement over the process's lifespan. NHAG and SDFN created an environment for women to engage and lead; therefore, their participation advanced the process and benefited them personally. However, the government's centralized approach reduced women's active engagement after relocating to their plots. Additionally, the increased involvement of men just before and after the relocation further reduced women's participation. This trend appears common in processes that lack a gender-mainstreaming agenda and gender-sensitive actions. The lack tends to overburden the actively engaged participants while others enjoy the fruit of their labor. On the contrary, completely excluding inhabitants' participation tends to create situations that are not in alignment with the inhabitants' needs. A sustainable approach requires coherence of goals from stakeholders coupled with coherent participation of the inhabitants. Although a multi-stakeholder engagement approach is suggested to localize SDG 11 in the African context [17], these research findings synthesize that this engagement, if not aligned, can adversely affect the participatory process and lengthen the process.

The findings also indicate that the participating women faced certain social and economic challenges within their households and settlements. These challenges often went unnoticed for two reasons. First, the lack of a gender-mainstreaming agenda did not prompt other stakeholders to identify the challenges faced by the women driving the process. Second, the upgrading process relied heavily on the time and labor of women, which led stakeholders to encourage their participation without addressing their additional burdens. The women's economic struggles were also overlooked due to the absence of a holistic plan for the economic development of the women inhabitants, most of whom lacked formal income or any income opportunities. While NHAG and SDFN introduced the women to savings programs and the possibility of loans for building permanent houses, they did not address the lack of income to lead to said savings.

The research shows that the primary motivation for women was to settle in their lands, which has been achieved by land occupation. However, their need for services has only been partially fulfilled, and since these services entail ongoing costs, women without economic solvency, especially single mothers, cannot afford them. Moreover, original land tenure certificates have not been provided to the inhabitants due to the remaining land payment installments. Currently, the inhabitants occupy the land with photocopies of the certificates. Not having an original formal document might hinder them from enjoying the full benefits of land ownership. The women inhabitants of Freedom Square are largely unaware of these legal issues due to a lack of legal knowledge or information.

Finally, the findings provide evidence that the women participating in Freedom Square who have been elected as leaders of block management committees have no actual power to ensure proper management. As a result, this leadership position burdens them instead of empowering them. Notably, women leading the upgrading process did not have the opportunity to be part of the governance later. The local authority is predominantly composed of men in decision-making positions. This is a continuation of the traditional practice where men continue being in decision-making positions while women even with their active participation are not considered in these roles. This ties this discussion back to the setbacks arising due to the lack of an agenda for mainstream gender.

According to the discussion at the beginning of this article, in Namibia, women's civil status had been a criterion in determining their status and land ownership. As also mentioned, FLTA does not set any criteria for choosing the 'Head of Households', leaving this decision to the applying family. Social norms heavily impact these decisions, which help strengthen their connection with the traditional communal laws. The assumptions of the consequences of the lack of explanation regarding terminologies such as the 'Head of Household' are visible in the Freedom Square process, where even the non-participating or uninterested male participants own land. The lack of direction on the situation after any separation makes it hard to assume the future situation of the women inhabitants who actively engaged in the process to make it happen. Such situations require clear descriptions of terms and lawmakers' farsighted decision-making.

Additionally, the findings highlight the lengthy process and women dropping out due to various challenges. Women's biological needs and social responsibilities are major considerations when designing activities for women. This reinforces the importance of gender mainstreaming with evidence from the case study. In a gender-mainstreamed system, the necessity of the participating women would have been weighed along with the necessities of the process itself, thus creating a balanced output and having a more positive impact on the participating women than it has now.

4.2. Conclusions

Analyzing the case of Freedom Square reveals that for women to be included in designed spaces, they must participate in the process and voice their needs. The urban planning and development standards are developed with average men's standards over time and often fail to be inclusive towards the needs of others who do not identify as such, physically or socially. Women, blending into the 'others' category, are excluded from decisions and planning, and the outcome often does not suit their needs. Alternatively, in some cases, women are categorized as secondary users of the space with certain facilities provided targeting them, again from the point of view of men, most of the time without consultation. Relaying the previously shared examples of walking several miles daily for water or relieving themselves in open spaces, women's active engagement ensures a better outcome by avoiding these unfortunate incidents and making their lives easier. Unfortunately for women, in a society with a persisting gender gap, the change often happens when they advocate for it themselves. Although this overburdens the pioneers of the process, hopefully, in a better context in the future, inclusivity regarding basic services will not need to be ensured only through active participation.

However, on a positive note, the process also translates into a bottom-up planning process for creating inclusive spaces. Unlike top-down planning, where the authorities decide the needs of communities, women residing in those communities have more opportunities to contribute as users of bottom-up planning processes. However, for this process to follow through, it is important that the women's needs are not only conveyed but fulfilled and the results are evaluated and monitored. As seen in the case study, the lack of monitoring in Freedom Square has resulted in the women not receiving full participation benefits. This also connects their participation to Arnstein's [19] ladder, where in a well-designed participatory process, citizens achieve control of their development. In the case of Freedom Square, the research indicates that the participating women have not been able to reach the final steps of the ladder. The post-upgrading situation presents that the citizens, or in this case, the inhabitants, do not have delegated power or have not achieved citizen control in their true sense. They had been in partnership with the stakeholders during the upgrading process but the current situation does not indicate that they are in partnership anymore.

Annan-Aggrey, Bandauko, and Arku [17] highlighted multi-stakeholder engagement in development processes as an opportunity to create inclusive cities in the African context. However, this case study research can add to such multi-stakeholder engagement recommendations, highlighting the necessity of common goals among stakeholders. The lack of a common framework among stakeholders might lead to the exclusion of certain active groups from the community. Gender mainstreaming, for example, depends on conducting a process by inviting all genders and allowing the inhabitants to understand the necessity of mainstreaming. If not understood, the challenges women face in their households during their participation will continue to burden them. In between social norms, practices, and inclusive planning processes, the inhabitants, regardless of their gender, need to acknowledge the necessity and reasoning of these actions so that they can accommodate them within their lives. When provided with the much-required support from their households, women can continue contributing in various ways, especially in creating their settlements. This calls for door-to-door advocacy, highlighting the challenges women face when they participate in something in addition to their social responsibilities and how reducing these challenges goes a long way in creating inclusive cities.

5. Recommendations

The research concludes by presenting recommendations based on the analysis of Freedom Square. The recommendations can be divided into three major segments, which shall be simultaneously improved to mainstream gender and localize SDG 11 in the process.

5.1. Policy Recommendations

According to the discussion in the previous section, the Flexible Land Tenure Act is sufficient for rapid land delivery but can be upgraded to serve an integrated purpose by taking gender mainstreaming into consideration and clarifying gender terminologies in terms of land ownership. This will add to Namibia's law reforms, which now aim to separate law from women's civil status. Land ownership requirements for couples shall consider civil statuses specific to the common practices of the culture, and especially acknowledge the ones that disadvantage women such as not addressing co-habiting partners in joint titling. Although aimed at a swift delivery, the implications of the process last longer and all aspects of ownership including gender, civil status, etc., shall be carefully considered and not overlooked due to the limited capacity of stakeholders or to streamline a process.

Additionally, bottom-up development policies must acknowledge the term 'inclusivity'. This will prevent women's needs from being overlooked if they do not physically involve themselves throughout the process. An inclusive gender policy tailored to the local cultural nuance can ensure that the challenges faced in Freedom Square do not continue to be the case going forward. This, however, does not mean reducing the involvement of the community to eliminate challenges but to encourage involvement in a holistic development plan with long-term outcomes. Although in the Namibian context, the stakeholders act under financial duress and time limitations, a roadmap of inclusive community involvement regardless of gender shall aid in speeding up the process, as opposed to overburdening certain residents and slowing down the process in the long run. The gender quotas in community leadership are one example of inclusive policy in Freedom Square.

Finally, the decentralized development policies shall clearly be implemented so that political parties in power do not override the local authorities and hamper an ongoing development process. SDG localization is a long-term process that requires the constant engagement of relevant stakeholders. For example, independent and local monitoring committees in each municipality can contribute to tracking the progress of localization in the area.

5.2. Development Process Recommendations

Inclusive development processes must be assessed promptly. The strategic planning of actions with periodical assessment and post-process monitoring and evaluation is required to successfully conduct an inclusive process and also for the process to make a lasting impact, thus localizing a sustainable goal. Such a framework could be developed by following the Freedom Square upgrading process milestones as important stages of assessment. As an example, at the end of the pre-planning phase, data can be collected regarding gendered participation and strategies can be adjusted to gender mainstream the process in the next phase, at the end of which, audits can be carried out to evaluate the challenges faced by participants in order to mitigate them prior to execution. For further implementation, gender mainstreaming, and inclusive process agendas can be re-assessed and adjusted prior to each phase, such as pre-planning, planning, implementation, and management phases.

Also, in Freedom Square, at the initial stage, the NGOs conducted door-to-door advocacy to explain the process to the inhabitants and encouraged them to join. However, the male inhabitants refrained due to their skeptic outlook towards the outcome. This experience can be utilized in the next processes and the advocacy to join processes can be tailored to resolve the confusions that arise within the inhabitants, such as having FGDs with each gender to address their specific concerns. Even if there is one family member joining the process, the understanding of the whole family is crucial in supporting the partner and the process as a whole.

Also, for participatory processes, the local authorities play a crucial role in continuing the interaction with the community. Thus, sharing responsibilities with the community and, in turn, empowering them will help the community to move forward and take more responsibility and control over their development. Certain powers need to be handed over to the community post-bottom-up development process, utilizing the capacity developed by the community during the process and relieving certain responsibilities of the local authorities, which, in most cases, are overburdened. As discussed previously in the article, titles without responsibilities have induced the attitude of withdrawal within the elected community leaders of Freedom Square, which can be attempted to be revived with engagement in the management of the settlement. As an example, training or team-building sessions between the community leaders and the local authorities can help bring them on an equal footing with a similar goal of the community's betterment in mind.

5.3. Social Awareness Recommendations

Gender mainstreaming requires policies and social awareness to establish its importance to the participants of any process. Stakeholders of the process play a crucial role in raising awareness in the participating community. In addition to their advocacy for participation, they can map out the possible future benefits of gender mainstreaming and how it requires cooperation that should start from the household scale. It is also important that these awareness-raising actions are shaped according to the community's social norms. For Freedom Square, actionable recommendations include organizing community forums promoting the re-involvement of all elected community leaders in the community's decision-making, and such forums shall also be joined by the local authority representatives to ensure trust.

Finally, this research concludes by stating that, for women to be included in all processes, a culture of inclusiveness needs to be raised from within the household. Inclusive decision-making regarding everyday decisions, considering the needs of all family members, can create a firm root for the community to conduct inclusive processes in their neighborhood. Religious leaders, community leaders, and the elderly can play an effective role in conveying these messages in a culturally sensitive way.

Author Contributions: Conceptualization, T.A., J.K., A.L. and M.S.; methodology, T.A., A.L. and M.S.; software, T.A.; validation, T.A. and J.K.; formal analysis, T.A., A.L. and M.S.; resources, A.L.; data curation, T.A.; writing—original draft preparation, T.A.; writing—review and editing, T.A., J.K., A.L. and M.S.; visualization, T.A.; supervision, A.L. and M.S.; project administration, A.L. and M.S.; funding acquisition, A.L. and M.S. All authors have read and agreed to the published version of the manuscript.

Funding: This research was conducted under the umbrella of the SDGs GoGlocal! project as a trilateral partnership between the University of Stuttgart, Ain Shams University, and Namibia University of Science and Technology, co-funded by the German Academic Exchange Service (DAAD) and German Federal Ministry for Economic Cooperation and Development (BMZ).

Institutional Review Board Statement: This study was conducted in accordance with the Declaration of Helsinki, and the protocol was approved by the Ethics Committee of Namibia University of Science and Technology on 24 January 2025.

Data Availability Statement: The raw data supporting the conclusions of this article will be made available by the authors on request.

Acknowledgments: The authors would like to express gratitude to Namibia Housing Action Group (NHAG), Slum Dwellers Federation of Namibia (SDFN), Deutsche Gesellschaft für Internationale Zusammenarbeit (GIZ), and the inhabitants of Freedom Square for their support in fieldwork in Namibia.

Conflicts of Interest: The authors declare no conflicts of interest. The funders had no role in the design of the study; in the collection, analyses, or interpretation of data; in the writing of the manuscript; or in the decision to publish the results.

Appendix A

Expert interview questions (Sample):

1. In your experience, what has motivated women to participate in the land tenure securing process of Freedom Square, Gobabis?

2. In your experience, what kind of socio-economic challenges women in Gobabis has faced that has an impact on their ability to secure land tenure there?

3. Also, what are the setbacks that affect them during the participatory planning process that hinder their active participation throughout the process? (Example: as married, single, widowed women)

4. In your knowledge, how far has women been involved during the land tenure securing process in Freedom Square, Gobabis?

5. How has participation in the land tenure securing process positively mitigated the social challenges women face in Gobabis?

6. How would you evaluate gender- sensitivity of the outcomes of the land tenure securing process?

Appendix B

Inhabitant interview questions (Sample):

Conducted in English through an Interpreter

1. What is her name?
2. How old is she?
3. Who does she live with, like her family?
4. Do you have children?
5. And who has the land?
6. And they are from Gobabis? Like from when?
7. Do they have land anywhere else?
8. And what did they do when they came here in 1998?
9. What did they do?
10. And who paid for it?
11. Now who is earning?
12. Are they paying for the land?
13. Do they have water or other services?
14. Why don't they apply for water?
15. So when the land process started, who informed her?
16. So when was this?
17. Okay, so they came and told her what?
18. And who were in the committees?
19. Deena is here?
20. She was active?
21. Was she part of the process?
22. Was the new area far from where she was living?
23. Was anyone else in her family also doing this work?
24. How long were they involved in the process?
25. Beginning of what?
26. So it was 2015 and afterwards?
27. So who would she leave her kids with when she went to the process?
28. Did she know from before that they had to pay for the land?
29. How did she feel when she found out?
30. She asked who?
31. And what did they say?
32. And what happens if they cannot pay? Do their children pay?
33. Do they give you a receipt if you pay?
34. So what is her next plan with her house?
35. And is she part of any savings group?
36. What does she mean by that?
37. Was it Shack Dwellers or other groups?
38. Did she also lose her money?
39. For how long she saved?
40. Are there any committees in her block?
41. Are shack dwellers members the same as committee members?
42. So they don't trust them anymore?
43. So no one of the seven people are working at the moment?
44. What does she want them to do?
45. What does she want the committee to do?
46. She goes outside for the toilet?
47. Is it safe?

48. But there were community toilets before?
49. Has having the land made her situation better than before?
50. Why does she think there were more women participating in the process?

Appendix C

Interview analysis codes:
Upgrading process timeline
Planning with community
Executing with community
Maintaining land & services
Planned settlement
Basic services and infrastructure
Payments of land and services
Management committees
Flexible Land Tenure Act
Initiating settlement
Planning with community
Executing with community
Participation motivation
Participation outcome
Perception of participation
Challenges of participation
Challenges of services
Economic challenges
Financial activities
Savings groups
Government stakeholders
Civil society stakeholders
Shack Dwellers Federation Namibia
NGO stakeholders
Private sector stakeholders
International stakeholders
Academic sector stakeholders
Stakeholder relationships
Civil status
Land ownership
Leadership
Social role
Gender perspective

References

1. Weber, B.; Mendelsohn, J. *Informal Settlements in Namibia: Their Nature and Growth: Exploring Ways to make Namibian Urban Development More Socially Just and Inclusive*; Development Workshop Namibia: Windhoek, Namibia, 2017.
2. UN-Habitat. Adequate Housing and Slum Upgrading. 2018. Available online: https://unhabitat.org/sites/default/files/2020/06/indicator_11.1.1_training_module_adequate_housing_and_slum_upgrading.pdf (accessed on 17 September 2024).
3. United Nations (UN). Universal Declaration of Human Rights (UDHR). 2015. Available online: https://www.un.org/en/udhrbook/pdf/udhr_booklet_en_web.pdf (accessed on 17 September 2024).
4. Ulrich, J.; Meurers, D. The Land Delivery Process in Namibia A Legal Analysis of the Different Stages from Possession to Freehold Title. GIZ. 2015. Available online: http://the-eis.com/elibrary/sites/default/files/downloads/literature/The%20land%20delivery%20process%20in%20Namibia.pdf (accessed on 17 September 2024).

5. Delgado, G.; Muller, A.; Mabakeng, R.; Namupala, M. Co-producing land for housing through informal settlement upgrading: Lessons from a Namibian municipality. *Environ. Urban.* **2020**, *32*, 175–194. [CrossRef]
6. UNICEF. Gender Equality: Glossary of Terms and Concepts. 2017. Available online: https://www.unicef.org/rosa/media/1761/file/Genderglossarytermsandconcepts.pdf (accessed on 17 September 2024).
7. Office of the Special Advisor on Gender Issues and Advancement of Women (OSAGI). Gender Mainstreaming: Strategy for Promoting Gender Equality. 2001. Available online: https://www.un.org/womenwatch/osagi/pdf/factsheet1.pdf (accessed on 17 September 2024).
8. World Economic Forum. Global Gender Gap Report 2024. 2024. Available online: https://www.weforum.org/publications/ (accessed on 17 September 2024).
9. Shack Dwellers Federation of Namibia (SDFN). Community Land Information Program (Clip): Profile of Informal Settlements in Namibia. 2009. Available online: https://sdinet.org/wp-content/uploads/2015/10/NAMclip_-1.pdf (accessed on 17 September 2024).
10. UN-Habitat; GIZ; Gender CC. Gender and Urban Climate Policy Gender-Sensitive Policies Make a Difference. 2015. Available online: https://unhabitat.org/gender-and-urban-climate-policy (accessed on 17 September 2024).
11. Chigbu, U.E.; Enemark, S. Land governance and gender in support of the Global Agenda 2030. In *Land Governance and Gender: The Tenure-Gender Nexus in Land Management and Land Policy*, 1st ed.; CABI: London, UK, 2022; pp. 35–50. [CrossRef]
12. Chigbu, U.; Uwayezu, E.; Katjiua, M. Customary and conventional land administration in evolution—Potential for a new land administration system in Sub-Saharan Africa. In Proceedings of the Conference on Land Policy in Africa (CLPA), Kigali, Rwanda, 1–4 November 2021.
13. Chigbu, U.; Paradza, G.; Dachaga, W. Differentiations in women's land tenure experiences: Implications for women's land access and tenure security in sub-Saharan Africa. *Land* **2019**, *8*, 22. [CrossRef]
14. Chigbu, U.E. Negotiating land rights to redress land wrongs: Women in Africa's land reforms. In *Rethinking Land Reform in Africa New Ideas, Opportunities and Challenges*; African Development Bank: Abidjan, Côte d'Ivoire, 2020; pp. 156–167.
15. Ministry of Land Reform (MLR). *Guide to Namibia's Flexible Land Tenure Act, 2012*; Deutsche Gesellschaft für Internationale Zusammenarbeit (GIZ) and the Land, Environment and Development Project of the Legal Assistance Centre (LAC); Ministry of Land Reform: Windhoek, Namibia, 2016.
16. Buyana, K.; Walubwa, J.J.A.; Mukwaya, P.; Sseviiri, H.; Byarugaba, D.; Nakyagaba, G.N. Global Norms, African Contexts: A Framework for Localizing SDGs in Cities. In *Localizing the SDGs in African Cities*; Croese, S., Parnell, S., Eds.; Springer International Publishing: Cham, Switzerland, 2022; pp. 31–45. [CrossRef]
17. Annan-Aggrey, E.; Bandauko, E.; Arku, G. Localizing the Sustainable Development Goals in Africa: Implementation challenges and opportunities. *Commonw. J. Local Gov.* **2021**, 4–23. [CrossRef]
18. Namibia Statistics Agency (NSA). Namibia 2011 Population and Housing Census Main Report. 2011. Available online: https://cms.my.na/assets/documents/p19dmn58guram30ttun89rdrp1.pdf (accessed on 17 September 2024).
19. Arnstein, S. A ladder of community participation. *J. Am. Inst. Plan.* **1969**, *35*, 216–224. [CrossRef]

land

MDPI

Article

Participatory Land Planning, Community Land Trusts, and Managed Retreat: Transforming Informality and Building Resilience to Flood Risk in Puerto Rico's Caño Martín Peña

Ivis García * and Leslie Martínez-Román

Department of Landscape Architecture & Urban Planning, Texas A&M University, College Station, TX 77843, USA; lmartinezroman@tamu.edu
* Correspondence: ivis.garcia@tamu.edu

Abstract: El Caño Martín Peña (CMP), a cluster of eight historically informal neighborhoods in San Juan, Puerto Rico, faces long-standing challenges of inadequate infrastructure, persistent flooding, and displacement pressures. In response, the G8 collective—a coalition representing these communities—and the Martín Peña Canal Community Land Trust (CLT) establishment have developed a participatory planning model prioritizing resilience and housing security. This unique and exemplary model of a participatory planning case study, grounded in document analysis and participant observation, explores how the G8 and CLT have driven transformative improvements in CMP, addressing informality, mitigating flood risks, and preventing displacement. Findings highlight how community-driven governance fosters trust and enhances long-term sustainability, offering valuable insights for equitable urban development and spatial justice.

Keywords: participatory land planning; community land trusts; managed retreat; resilience to flood risk; urban informality; climate adaptation; equitable development; land tenure security

1. Introduction

Urban planning has often prioritized efficiency over equity, neglecting marginalized communities. Participatory planning, rooted in advocacy and communication, empowers residents in decision-making, aligning with Lefebvre's "Right to the City" and Harvey's spatial justice [1–3]. In climate-vulnerable areas, it ensures that managed retreat fosters resilience rather than displacement. This study examines three entities in the Martín Peña Canal, San Juan, Puerto Rico, that exemplify participatory land planning: (1) Proyecto ENLACE, a public corporation, (2) G8, a coalition of eight canal-side communities, and (3) the Martín Peña Canal Community Land Trust (CLT).

Martín Peña Canal is a 3.75-mile-long tidal channel located in the center of San Juan, Puerto Rico with a total of eight informal communities. Since the late 1930s, about 5000 people have settled in this flood-prone mangrove habitat to live near employment opportunities [4]. The 1928 San Felipe and the 1932 San Ciprián hurricanes accelerated the movement of those displaced to move into this area [5,6]. Many of its original settlers moved from rural communities—displaced from agricultural work, such as on coffee and sugar cane plantations, and in search of new manufacturing, industrial and construction work opportunities [6].

During Puerto Rico's Operation Bootstrap industrialization program in the 1940s, low-income residents built informal housing near jobs, transit, and commerce [7]. In addition, because of urban renewal projects, many living in Old San Juan, Condado, Milla de Oro, and other areas of Santurce in San Juan were displaced by the eminent domain and ended up settling near the Caño [6].

In response to the Canal's environmental degradation, San Juan designated the eight surrounding communities as a Special Planning District in 2002, proposing dredging the Caño. As one resident asked, 'But for whom?' [8]. Immediately people were concerned that a new infrastructure project that made the community better could result in their displacement, given they are in an area where there are real estate interests. To address this issue, the community leaders in each of these communities figured that they needed to create a collective, and they started to meet in 2002 [9]. In 2004, these eight communities formed Proyecto ENLACE, a public corporation ensuring that the project addressed local needs, including flooding, poor sanitation, and health risks [8]. The G8 oversees the implementation of the district's comprehensive development and land use plan developed by ENLACE [8].

After Hurricane María in 2017, 1200 homes lost their roofs and 100 homes were destroyed [10]. In 2022, Joe Biden assigned $163 million to the Corps of Engineers for dredging, which is the "removal of sediments and debris from the bottom of lakes, rivers, harbors, and other water bodies" [11]. The work started in 2023, as shown in Figure 1. A total of 1000 families needed to be relocated because they lived along the right-of-way where the dredging and managed retreat initiatives were being planned [12].

Figure 1. Housing Units Cleared for Martín Peña Canal Dredging Project. A sign posted outside the acquired house informs residents that construction in this area is prohibited and indicates the planned future width of the channel. Authors' source.

Using $46 million of Community Development Block Grant-Mitigation (CDBG-MIT) provided by the U.S. Department of Housing and Urban Development (HUD), people were given the choice of whether they would like to find another home inside within the eight communities or if they would like to move to a different location with a housing relocation voucher [13]. ENLACE came to this conclusion through a community-driven process that ensured fair treatment and sought to minimize displacement-related risks as a main goal [14]. One of the reasons for having to relocate is that the Government of Puerto Rico does not allow the elevation of homes in the floodplain [15]. However, the way this community managed its own relocation process has been more than 20 years in the making, since residents understood that they needed to organize and come up with a plan that reflected what they wanted to achieve as a collective.

This article is about how they arrived at the participatory model that won the 2016 United Nations World Habitat Award and allows a just relocation, along with life improvement for those who stay [16]. Our aim is that other researchers and decision-makers can learn from this unique and exemplary case study and implement some of the lessons learned. This case study will be helpful to those working with informal settlements in both the Global South and North, where relocation, along with massive infrastructure projects, is used to solve environmental, social and spatial justice issues, making communities more sustainable and resilient in the future.

Using document analysis (Appendix A) and participant observation, this study evaluates how these initiatives have improved housing stability, enhanced infrastructure, and facilitated socio-economic development. Furthermore, it highlights the resilience and ingenuity of the communities in navigating political and environmental challenges to achieve sustainable urban regeneration. This article aligns closely with several of the United Nations Sustainable Development Goals (SDGs), particularly those that focus on inclusive, sustainable cities, and climate resilience [17]. Figure 2 is a flowchart illustrating the structure of participatory planning along the Martín Peña Canal.

Figure 2. Flowchart illustrating the structure of participatory planning for the Martín Peña Canal.

2. Literature Review

2.1. Non-Participatory Models

Rational planning, a historically significant urban planning framework, emphasizes systematic, data-driven approaches to address urban challenges [18]. Rooted in modernist ideals, rational planning prioritizes efficiency, logic, and evidence in decision-making. This approach relies on expert-driven, top-down strategies rather than participatory models [19]. This approach follows five steps—goal setting, evaluating alternatives, selecting a solution, implementation, and assessment—all still used today, with greater public participation [20]. Perhaps the only difference today is that we incorporate this participation in each one of these steps. While rational planning has shaped cities, its lack of public input has often harmed marginalized communities. Haussmann's redesign of Paris and Robert Moses' highway projects in New York optimized urban infrastructure but were criticized for prioritizing efficiency over social equity [21,22]. In the context of El Caño Martín

Peña, rational planning was evident during the mid-20th century, when the urban renewal policies of the Government of San Juan, Puerto Rico, were implemented to modernize Puerto Rico's areas, which were deemed as slums [23]. This approach focuses on clearing substandard housing, often equated with informal housing, to make way for formalized development projects [24]. Many of the new projects created were public housing units for low-income residents [25]. The problem is that the Urban Renewal Program, which was instituted by Housing and Urban Development (HUD) in 1949, demolished more affordable housing units than those that, in the end, were created [26].

Although the program at first meant to carry out a one-to-one replacement, it is estimated that only one in every four units was replaced nationwide during the duration of the program (between 1949 and 1973). This means that approximately 600,000 housing units were demolished, resulting in the displacement of around 2,000,000 residents. These efforts ended up displacing thousands of families and exacerbated existing inequalities, as many residents relocated on their own without adequate support. Unlike many participatory planning initiatives that struggle to gain long-term institutional support, this case stands out for its successful integration into formal governance structures, ensuring its sustainability beyond initial grassroots mobilization. Many residents ended up in Canal communities after being in public housing, as people did not like their original environment (citing high crime) and wanted to have single-family homes, instead of living in what they felt were crowded apartments that were not their own [27]. Even when they lived in informal communities, these Canal communities saw themselves as homeowners—they valued having a place they could call their own.

Rational planning interventions often reflected a technocratic approach, prioritizing infrastructure development and land use optimization from the planners', and not the communities', perspective [28]. It is, then, no surprise that the approach failed to account for the socioeconomic realities of the residents. In the end, government-led urban renewal efforts, including slum clearance programs sponsored by the U.S. Federal Government and policies such as the Housing Act of 1949 (Title I), displaced thousands of families [29], leaving those who remained with inadequate infrastructure and sanitation systems.

Rational planning focuses on achieving the "greatest good for the greatest number", rooted in utilitarianism, frequently prioritizing urban functionality over the well-being of minority groups [30]. Planning is a practical field that necessitates pragmatics as opposed to utopianism and, because of this, it is highly susceptible to the ideas of utilitarians [31]. Driven by the pragmatism required to have a home, people placed trash and debris in the wetlands to create buildable land, which contributed to significant environmental degradation. The government also played its role in promoting and filling the wetlands [32]. The technical solutions proposed at the time, such as partial dredging or infrastructure upgrades, often lacked meaningful community input, rendering them unsustainable and inequitable.

The limitations of rational planning became apparent, as these top-down strategies failed to address the "wicked problems" facing El Caño, such as entrenched poverty, informal housing, and ecological degradation [33]. These challenges required more nuanced and inclusive approaches that moved beyond the purely technical. In the late 1990s and early 2000s, the community-led efforts spearheaded by the G8 coalition and the establishment of ENLACE and the CLT marked a shift toward participatory planning [34]. By involving residents in decision-making and integrating local knowledge, these initiatives countered the shortcomings of earlier rational planning models.

2.2. Participatory Models

Contrary to rational planning, participatory planning emphasizes the active involvement of communities in shaping their environments and the act of empowering marginalized groups to influence decisions that directly affect their lives [35]. This approach roots itself in the democratic ideal that those who live in a community are best positioned to identify its needs and co-develop sustainable and equitable solutions [36]. Participatory planning integrates principles of diversity, equity, inclusion (DEI), and collaboration—making it a vital tool for addressing structural inequities and fostering long-term community resilience [26].

Paul Davidoff's advocacy planning, introduced in the 1960s, challenged traditional technocratic planning models by emphasizing the idea of pluralism as opposed to that of the public interest [37]. Public interest presupposes one single public, but pluralism explains that there are multiple publics or stakeholders [38]. Each public could organize and advocate for what they want. The role of the planner in this context is to help the community write a plan that reflects what they would like to advocate for. Davidoff argued that, by using this approach, planners can secure the representation of diverse community interests. Davidoff's theory of advocacy planning could be used for any group or type of advocacy. For example, a planner could represent the needs of business owners and create a plan for them and another planner could represent the interest of homeless individuals in a different plan. This means that, in the same neighborhood, competing interests and competing plans might be found.

However, Paul Davidoff practiced his own advocacy planning to ensure that marginalized groups had a voice in urban planning processes [39]. This is why a figure like Norman Krumholz conflated the idea of advocacy planning with equity planning. However, equity planning, the term developed by Krumholz, was unique in the sense that the city government prioritized marginalized groups in decision-making, as well as in the outcome of urban development [40]. Still, Davidoff's approach laid the groundwork for equity planning, which Norman Krumholz later advanced to explicitly redistribute resources and power toward low-income and underserved populations [37,40].

In the context of El Caño Martín Peña, advocacy planning is reflected in the work of community organizations, like the G8 coalition, which represents the voices of the eight neighborhoods surrounding the channel, through the general idea of mutual aid [16]. By uniting diverse stakeholders, the coalition has amplified community concerns, particularly regarding displacement risks and infrastructure deficits, ensuring that the planning process aligns with the residents' aspirations for equity and justice [8,41].

Equity planning, as championed by Krumholz in the 1970s, extends advocacy planning (which calls for planners to pick their side and to create plans, not with government necessarily, but as advocates working alongside community stakeholders) by focusing on the redistribution of resources to rectify systemic inequalities from the perspective of the planners working for the city. Communicative planning emerged as a response to fragmented societies, emphasizing the importance of dialogue, mutual understanding, and consensus in urban planning. Patsy Healey and John Forester, leading figures in this paradigm, argue that planners should act as facilitators, who mediate between diverse stakeholders, building trust and fostering collaborative solutions [18,19]. Judith Innes further elaborates on this idea through her research on consensus-building and collaborative rationality, which stresses the value of inclusive dialogue in resolving complex urban challenges [42–44].

The participatory processes for the Martín Peña Canal embody both communicative planning and transactional planning. In transformative planning (now identified as insurgent), which was developed by Friedman in the 1990s, through sustained community engagement and mutual learning between groups of groups, such as residents, planners, and policymakers, the community has developed a shared understanding of (and, we might argue, vision for) sustainable urban regeneration. The inclusive approach has not only enhanced the quality of plans but also strengthened the community's capacity to advocate for itself, creating a model of collaborative governance [8]. Miraftab further develops this idea through her research, articulating insurgent planning as a radical planning practice, which argues the importance of rethinking participation [45].

In the global south, particularly in the case of Sao Paulo in Brazil, where participatory planning has been institutionalized, the disadvantaged groups depend on a constant struggle, both within and outside institutional levels, to pressure the government into action [46]. In other words, citizen participation is not enough to ensure that common understanding and agreement leads to action. The same situation has been observed in the case of the G8 coalition and El Caño Martin Pena, where the residents have devoted a considerable amount of time and energy during the last two decades to pressuring the authorities, in order to uphold to the comprehensive plan [47–49]. John Friedmans' radical planning theory calls for a shift in power from centralized authorities to grassroots movements [50]. The literature on the different types of participatory planning underscores the transformative potential of community-centered approaches in addressing systemic urban challenges (Table 1). These frameworks challenge the limitations of traditional top-down, rational planning models by emphasizing inclusion, equity, and the value of local knowledge.

Table 1. Comparison between Non-Participatory (Rational) Planning and Participatory Planning Models. Source: Created by authors.

Category	Non-Participatory (Rational) Planning	Participatory Planning
Core Principles	Efficiency, logic, technical expertise; focus on data-driven, top-down approaches.	Empowering communities, inclusive dialogue, and collaboration; focusing on local knowledge and equitable outcomes.
Decision-Making Approach	Centralized, led by urban planners or technocrats.	Decentralized, led by community stakeholders with planners as facilitators.
Goals	Modernize infrastructure, optimize land use, and improve urban functionality.	Address systemic inequalities, prevent displacement, and foster long-term community resilience.
Process	Step-based: (1) Set goals, (2) Evaluate alternatives, (3) Select an optimal solution, (4) Implement, (5) Assess outcomes.	Ongoing: Workshops, focus groups, participatory events, and collaboration between stakeholders.
Outcomes in El Caño Martín Peña	Urban renewal efforts displaced thousands of residents and degraded the channel through environmental mismanagement; limited community input led to inequitable outcomes.	G8 coalition, Fideicomiso de la Tierra (CLT) and Proyecto ENLACE ensured housing stability, environmental restoration, and long-term governance through inclusive processes.
Key Figures and Theories	Rooted in modernist ideals; influenced by figures like Hausman (Champs-Elysees Neighborhood in Paris) and Robert Moses (New York Highways).	Informed by theories such as advocacy planning (Davidoff), equity planning (Krumholz), communicative planning (Healey, Forester), and transactive and radical planning (Friedmann).
Environmental Impact	Significant degradation of wetlands and ecosystems; technical solutions failed to address root causes of flooding and contamination.	Ecological restoration through community-led efforts, including mangrove replanting, dredging, and green infrastructure projects.

Table 1. *Cont.*

Category	Non-Participatory (Rational) Planning	Participatory Planning
Housing Impacts	Slum clearance destroyed more affordable housing than it created, exacerbating displacement and marginalization.	CLT decoupled land ownership from property value, securing tenure for 1500 families and preventing gentrification [16].
Long-Term Sustainability	Unsustainable due to lack of community buy-in and socio-economic considerations.	Sustainable due to empowered governance structures, equity-driven outcomes, and ongoing community involvement.

2.3. *Spatial Justice*

Spatial justice is a critical theory concept from the Marxian perspective in urban planning that examines how access to space, resources, and opportunities is distributed across different social groups, based in class. French philosopher Henri Lefebvre's *The Production of Space* (1974) argues that space is socially produced and contested and often reinforces existing inequalities between the working class and the owner class [51]. Lefebvre's notion of the "right to the city" asserts that those who contribute to manufacturing production as workers should have equal agency in shaping the urban environments they inhabit. This perspective is echoed by David Harvey (2008), who emphasizes that urban planning should ensure that marginalized communities have equitable access to housing, infrastructure, and public services [3].

Additionally, scholars like Robert Bullard (2005) have emphasized the intersection of environmental justice, spatial justice, and hazards—showing that marginalized communities often bear the brunt of environmental degradation. The environmental restoration efforts in El Caño Martín Peña, including dredging the canal and reintroducing green infrastructure, are essential components of spatial justice, as they mitigate the long-standing environmental burdens disproportionately affecting residents, like flooding and threats from hurricane events. Similarly, *After Shocks of Disaster* (2019) argues that colonial legacies continue to shape spatial inequalities, particularly in the Global South [52]. Puerto Rico's colonial history and its continued struggle for equitable development are reflected in the challenges faced by the communities along the canal, making the participatory planning model a vital counterbalance these historical injustices.

3. Methods

This study employs a qualitative case study approach to answer the following questions: (1) What are the socio-economic and environmental challenges faced by marginalized communities in vulnerable urban areas? (2) How have participatory planning models helped address these challenges and empowered communities to achieve sustainable development? Table 2 outlines the methodological framework for the Martín Peña Canal case study, emphasizing the research characteristics and how they align with the goals of this study. As Yin explains, the research questions should be both descriptive and explanatory. In this case, we focus on identifying the socioeconomic and environmental challenges faced by the Martín Peña Canal and exploring how participatory planning has addressed these issues to promote sustainable development [53]. This study adopts a single case study approach to provide an in-depth analysis of a contemporary phenomenon, specifically the transformation of Martín Peña Canal from a marginalized urban area to a model of community-led planning. Stake [54] emphasizes the importance of the single case study in exploring the unique attributes and complexities of a specific instance, particularly when the case provides critical insights into broader theoretical questions.

Table 2. Case study characteristics and their application to Martín Peña Canal research. Source: Authors.

Case Study Characteristic	Description	How It Is Reflected in This Research
Research Question	"What" (descriptive) and "How" (explanatory) research questions [53].	What are the socio-economic and environmental challenges faced by El Caño Martín Peña? How have participatory planning models helped address these challenges and empowered the community to achieve sustainable development?
Time/Space	A qualitative single-case study approach focusing on a contemporary phenomenon [53].	The study focuses on El Caño Martín Peña, a marginalized urban area in San Juan, Puerto Rico, examining events and interventions from the 1990s to the present day, including the formation of the G8 coalition and the establishment of the Community Land Trust (CLT).
Approach	Concluding a single case study, with a focus on depth and contextual understanding [53,54].	Martín Peña Canal is selected as a critical case to explore the impacts of participatory planning, offering insights into how community-led models can transform marginalized urban spaces globally.
Sources of Evidence	Multiple sources of evidence, with data converging in a triangulating fashion [53].	Evidence is gathered from participant observation and conservations with stakeholders (e.g., G8 members, residents, policymakers), document analysis (e.g., project reports, legal frameworks), and secondary sources (e.g., academic articles and news reports).
Limitations	Can produce generalizable knowledge but findings might lack scientific rigor, be prone to researcher bias, or be difficult to replicate [55].	The findings are context-specific but offer transferable lessons for other urban areas. Researcher bias is mitigated through triangulation, and the focus on a single case may limit generalizability but provide rich, in-depth insights.

Using Yin's framework, the Caño Martín Peña (CMP) case presents a unique and exemplary model of participatory planning, offering critical insights into the intersection between land tenure security, community-driven governance, and environmental resilience. Unlike many participatory planning initiatives that are limited in scope or duration, CMP stands out as one of the few large-scale, community-managed land trusts within an informal settlement, demonstrating long-term land tenure security as a resilience strategy. The participatory governance model institutionalized through Proyecto ENLACE, the G8 coalition, and the Community Land Trust (CLT) has successfully embedded community engagement into policy-making processes, ensuring that decision-making remains in the hands of residents rather than external stakeholders. This continuity beyond short-term interventions sets CMP apart from conventional participatory planning efforts. Additionally, CMP's model is particularly noteworthy because it bridges environmental restoration, housing security, and climate adaptation within an equity-driven framework, illustrating how integrated participatory planning can simultaneously address social, economic, and environmental vulnerabilities. Given its holistic approach and its success in preventing displacement while enhancing resilience, the CMP case provides a valuable precedent for other marginalized communities facing similar challenges.

Multiple sources of evidence, including document analysis (Appendix A) and participant observations, ensure triangulation and credibility. While the findings may be context-specific, they offer valuable lessons for similar urban contexts globally. Recognizing the limitations of the single-case study method, such as potential researcher bias and challenges to generalizability, this research mitigates these issues through data triangulation and situating the findings within broader urban planning discourses. The table highlights how this methodological approach provides a robust foundation for exploring the transformative potential of participatory planning in urban regeneration.

Flyvbjerg argues that single case studies can produce generalizable knowledge when they are strategically chosen as critical, extreme, or paradigmatic examples [55]. This methodology is particularly effective in examining sociopolitical and environmental phenomena, as it enables researchers to analyze interconnected processes, stakeholder dynamics, and context-specific outcomes [53,54]. The methodology is well-suited for the study of El Caño Martín Peña, where a comprehensive understanding of the participatory planning processes and their socio-economic and ecological impacts requires a focused, context-driven analysis.

3.1. Participant Observation and Conversations

Author 2 served as an AmeriCorps VISTA member at Proyecto ENLACE Martín Peña Canal (July 2016–November 2017), advancing participatory planning efforts in housing rights, disaster relief, and community engagement. She developed workshops, facilitated the Community Reallocation Committee, co-managed Raíces del Caño, and coordinated Universidad del Barrio. Through participant observation, she gained firsthand insights into housing, land tenure, and environmental risks. Informal conversations with residents, planners, and policymakers enriched her understanding of grassroots urban regeneration. In March 2023, Author 1 conducted a formal interview on post-María recovery, supported by IRB approval from the University of Utah (00155027).

3.2. Analysis

We analyzed the reports using a systematic approach of content extraction, interpretation, and synthesis. First, we identified key themes, like "land trust", "participatory planning", "housing", and "equity", focusing on executive summaries, methodologies, and conclusions. We then contextualized the findings within urban planning frameworks, prioritizing data related to Puerto Rico and El Caño. Finally, we synthesized the information into literature reviews, tables, and summaries, ensuring accuracy through citations and highlighting any gaps. This method provided a clear and relevant analysis for our research.

3.3. Limitations

While the case study approach provides an in-depth understanding, its findings may not be directly generalized to other contexts. Additionally, reliance on qualitative data introduces subjectivity, particularly in interpreting participant perspectives. Efforts were made to mitigate these limitations by triangulating data sources and maintaining transparency in the analysis process.

4. Case Study Findings

The questions that this case study seeks to answer are as follows: (1) What are the socio-economic and environmental challenges faced by marginalized communities in vulnerable urban areas? (2) How have participatory planning models helped address these challenges and empowered communities to achieve sustainable development? The sections titled Location, History, and Demographics answer the first question, which is descriptive. Meanwhile, the second question is answered by the final two sections.

4.1. Location

The history of Martín Peña Canal reveals a trajectory of struggle, resilience, and innovative solutions to longstanding socio-environmental challenges. El Caño Martín Peña, a tidal channel spanning 3.75 miles, is a vital yet historically neglected part of San Juan, Puerto Rico [56]. It is surrounded by eight neighborhoods collectively known as the Martín Peña Special Planning District: Barrio Obrero Oeste, Barrio Obrero, San Ciprián, Parada 27, Buena Vista Santurce, Las Monjas, Buena Vista Hato Rey, and Israel–Bitumul (Figure 3).

Figure 3. Map showcasing San Juan, Puerto Rico, including the Martín Peña Channel and the eight surrounding communities. Source: Authors.

Martín Peña Canal links the San José Lagoon to the San Juan Bay Estuary. The San Juan Estuary, with lagoons, wetlands, and mangroves, is a vital part of Puerto Rico's water–land ecosystem. This area is home to several marine species, such as manatees, which are a highly engendered species in Puerto Rico but, due to urbanization, their populations have been affected. The channel is in the center of the San Juan metropolitan area and is near the central business district and other important transportation nodes, including seaports and airports. This location highlights how it is well positioned for urban links and economic development [32,41]. This contradiction between a strategic location and social exclusion demonstrates Martín Peña Canal as both a neglected area and a project for change from below in terms of planning and governance [41].

4.2. History

Figure 4 shows a timeline for Caño Martín Peña. These communities, established during the 20th century due to waves of rural–urban migration, reflect a pattern of filling in land consisting of debris, trash, and dirt to construct homes along mangrove wetlands, spurred by limited access to formal housing and infrastructure [56,57]. It was believed that the mosquitos that gave malaria to people were associated with the stagnant waters of the Caño so, as the government demolished areas due to urban renewal, it would use the debris and trash to cover the waters of the Caño [6].

During this time, marginalization began, because disease was associated with the areas [6]. Along the Caño, as in many places around the world, the water was used for multiple purposes—fishing, bathing, drinking, and even defecating—leading to severe environmental degradation and public health challenges [6]. These informal settlements grew rapidly, and by the 1950s much of the wetlands were filled in. Figure 5 shows a 1936 aerial photograph of the Martín Peña Canal, showcasing extensive mangrove swamps, approximately 250 feet wide, that once lined the channel. In contrast, an aerial view from the 2000s reveals that nearly all the mangrove swamps have been filled in, drastically narrowing the canal. This expansion resulted in dense neighborhoods but also caused significant environmental degradation, including the obstruction of the tidal channel and severe flooding issues.

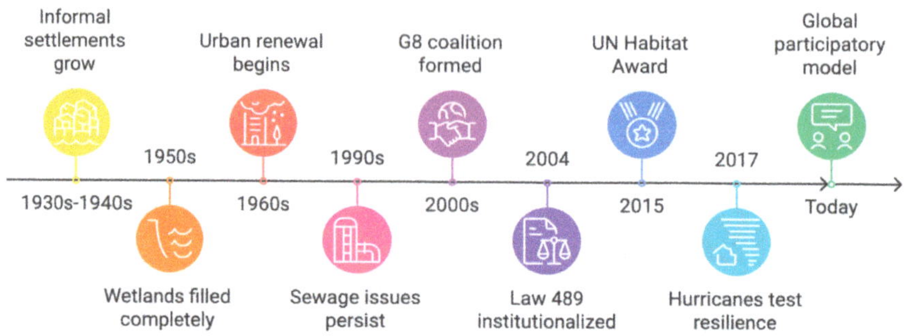

Figure 4. Community Timeline. Source: Image created by authors using napkin ai.

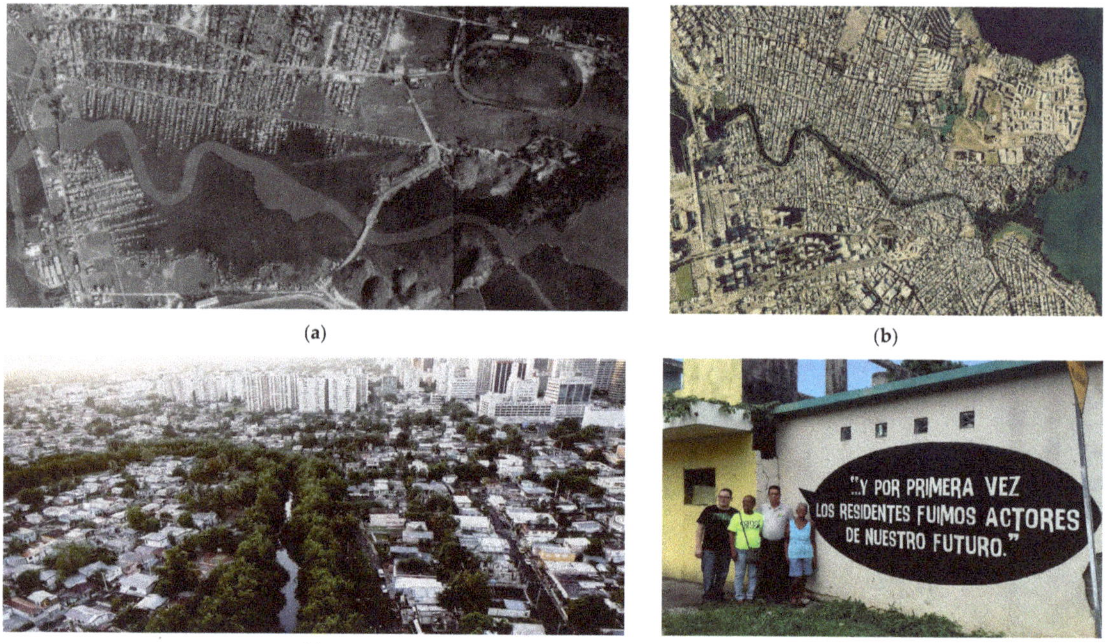

(a) (b) (c) (d)

Figure 5. Transformation of the Martín Peña Canal: From Extensive Mangroves in 1936 to a Narrowed and Filled Canal in the 2000s. (**a**) A 1936 aerial photograph of the Martín Peña Canal reveals the extensive mangrove swamps about 250 feet wide that once bordered the waterway during that period.; (**b**) An aerial view of the Martín Peña Canal from the 2000s reveals that nearly all the mangrove swamps have been filled in, resulting in a significantly narrower canal; (**c**) Aerial view of the communities surrounding the Caño Martín Peña shows how close the community is to the main economic center; (**d**) Community leaders shown in a mural that reads, "And for the first time we, the residents, were the leaders of our future". Source: Proyecto ENLACE.

Using a highly rational planning model, where the planner as a technical agent tries to solve problems with little or tokenistic public input, by the 1960s the urban renewal programs led by the government aimed at erasing slum areas and transferring thousands of people to new residential zones [58]. However, most of the remaining neighborhoods were left with poor sewage systems, thus compounding the social and environmental problems [59]. Such conditions were extant until the 1990s, with untreated water sewage leading to the extremely poor water quality of the channel and its surroundings [23].

The challenges that beset the communities of Martín Peña Canal were addressed in the late 1990s and early 2000s, which proved to be a significant shift for the eight neighborhoods surrounding the channel [23]. In 2001, the government of San Juan partnered with the U.S. Army Corps of Engineers to launch an ambitious project aimed at dredging the canal and restoring its water flow [4]. At first, residents were scared of this infrastructure project, as they thought it would result in their displacement. There was a lack of trust in the government. In 2002, residents started to organize and come up with solutions [9]. They started to group together and, finally, in 2004, formed the G8 coalition made up of eight neighborhoods surrounding the channel [9].

The Puerto Rican government institutionalized these efforts in 2004 with the passage of Law 489, which formalized the ENLACE Project Corporation to manage environmental restoration and community development initiatives [60]. At the same time, this decade also saw significant strides in participatory planning, where residents directly had a say in infrastructure improvements, housing policies, and social services.

In a way, we can say that these participatory practices proved to be helpful when disaster struck, like when Hurricanes Irma and María in 2017 devastated much of Puerto Rico. Despite widespread damage, the community's resilience, bolstered by the CLT and participatory governance structures, allowed for more coordinated and effective recovery efforts compared to other areas in San Juan and Puerto Rico. These achievements earned international recognition, including the prestigious United Nations World Habitat Award in 2016 [16].

4.3. Demographics

The area is home to approximately 10,997 residents [61]. There has been a significant decrease in the population living in this area since 2010. The population used to be approximately between 15,693 and 18,103 residents [62]. This population decrease may be linked to a variety of factors and is not limited to the demographics presented in Table 3. For instance, community leaders have reported issues related to governmental discrimination and constant stigmatization of the community surrounding the Caño [6]. In addition to this, the Caño Martín Peña's communities have been living through a constant state of disaster that threatens the quality of life of thousands of residents. Table 3 shows a comparison of demographics for Caño Martín Peña, San Juan, and Puerto Rico. The median household's income in Caño Martín Peña is $17,817, compared to San Juan's $26,981, and Puerto Rico's $25,096. The neighborhoods bordering the Caño are among the most economically disadvantaged in Puerto Rico, with over 61% of households living below the poverty line and facing high unemployment rates of 22.8% Contrary to other key demographics, in the Caño 33.69% of people have only obtained a High School Diploma compared to 21.44% and 27.70% in San Juan and Puerto Rico. The median age is 48.7 and over 2673 people are over 65 years old in our community of interest. However, according to the American Community Survey, people over 25 years old with a High School Diploma living under the poverty rate is 68.6% higher when compared to the 55% and 46% in San Juan and Puerto Rico [63].

There is a total of 6946 housing units in Caño Martín Peña. The total of owners living in the Caño is 2619, while renters are approximately 2563. Before the establishment of the Fideicomiso de la Tierra (CLT), most of the residents lacked a clear title, compared to approximately 55% of households in Puerto Rico. Currently, 2000 families (around 16.51%) share a collective clear title [62,64]. The median home value of those with a mortgage is $116,400 compared to $181,200 and $149,200 in San Juan and Puerto Rico. Meanwhile, the median home value of those without a mortgage is lower than those with a mortgage ($100,400) in our community of interest. The median home value in the surrounding area is higher by $61,500 in San Juan. However, in the case of Puerto Rico, the median home value is $110,900, representing only a slight difference between the Caño and Puerto Rico as a whole. The higher home values between the Caño and the surrounding areas could represent the threat of displacement. Unfortunately, surrounding informal communities that used to share similar characteristics to the Caño were forcibly displaced to prioritize higher-cost housing initiatives for middle and upper-income families [65].

Table 3. Key Demographics. Source: Created by authors.

Demographic Indicator	Caño Martín Peña's Communities	San Juan	Puerto Rico	Source/Year
Population	10,997	342,259	3,285,874	(El Nuevo Día, 2023); U.S. Census Bureau (2020)
Median Household Income	$17,817	$26,981	$25,096	American Community Survey 5-Year Estimates (2023)
Poverty Rate	61.1%	36.1%	38.2%	American Community Survey 5-Year Estimates U.S. (2023)
Unemployment Rate	22.8%	14.5%	12.1%	American Community Survey 5-Year Estimates U.S. (2023)
Median Age	48.7	45.1	45.2	U.S. Census Bureau (2020)
Percent Over 65 Years Old	24.3%	22.8%	22.3%	U.S. Census Bureau (2020)
Life Expectancy	N/A	N/A	80.86 years	United Nations—World Population Prospects (2024)
Educational Attainment (High School or Equivalent)	3325	61,160	750,934	American Community Survey 5-Year Estimates U.S. (2023)
Housing Units	6946	193,244	1,598,159	U.S. Census Bureau (2020)
Renters	2563	75,020	399,879	U.S. Census Bureau (2020)
Owners	2619	82,333	940,655	U.S. Census Bureau (2020)
Median Home Value with Mortgage (owner-occupied units)	$116,400	$181,200	$149,200	American Community Survey 5-Year Estimates U.S. (2023)
Median Home Value without Mortgage (owner-occupied units)	$100,400	$161,900	$110,900	American Community Survey 5-Year Estimates (2023)
Percentage of Households with Access to Basic Services (access to potable water, sanitation, electricity, etc.)	50% do not have sanitary service	N/A	250,000 people in Puerto Rico without basic access to safe drinking water and sanitation	El Nuevo Día, (2023); Dig Deep and US Water Alliance, (2019); Proyecto ENLACE/Local Surveys
Percentage of Population Living in Informal Housing	100% in the past but now 16.51% is part of the CLT	N/A	Approx. 55%	Proyecto ENLACE/Local Surveys; U.S. Census Bureau (2020); Puerto Rico Builders Association (2022)

Some of the communities still lack infrastructure for basic services, such as sewer and stormwater systems. This has resulted in widespread issues, including raw sewage discharge, frequent flooding, and the accumulation of debris, which has severely degraded the ecosystem of the tidal channel [8,32,66].

4.4. Before and After Community Organizing

Table 4 shows a comparative analysis of conditions before and after community organizing in El Caño Martín Peña. Before the formation of Proyecto ENLACE, G8, and the CLT, these informal settlements were more acutely characterized by overcrowding and precarious living conditions [4]. The physical and environmental challenges facing these communities have been compounded by systemic neglect. Historically viewed as a barrier to urban development rather than an integral part of San Juan, the Caño and its adjacent neighborhoods were excluded from formal urban planning processes. This marginalization left the residents to confront not only the environmental consequences of unplanned urban growth but also the social stigmas associated with informal settlements [8,41,57].

Table 4. Comparative Analysis of Conditions Before and After Community Organizing in El Caño Martín Peña.

Category	1990 (Before G8)	2010 (After G8 Formation)	2024 (Current)
Land Tenure	Informal; high risk of displacement	Regularized for 1500 families (CLT)	1000 families who needed to move (inside or outside of the CLT, to allow for dredging) continue to prevent displacement
Housing Condition	Precarious; informal structures	Improved housing with repairs and upgrades	Further improvements, long-term stability
Flooding Frequency	Frequent (70% of areas affected)	Reduced through initial infrastructure	Significantly reduced due to dredging
Sewage Infrastructure	Poor; untreated sewage flows into the channel	Initial upgrades, some sanitation facilities	Expanded sewer system and improved sanitation
Water Quality	Severely degraded, highly polluted	Moderate improvements in post-participatory projects	Improved but challenges persist
Community Engagement	Minimal or token participation	G8 established; 700+ participatory events	Continued active governance by G8
Environmental Condition	Mangroves severely degraded	Restoration efforts started	Replanted mangroves and restored biodiversity
Disaster Preparedness	Minimal	Framework established post-G8 formation	Tested successfully during hurricanes

The area is susceptible to flooding as, when the area is blocked, it is filled with debris. When there are storms or hurricanes, like Hurricane Maria in 2017, this results in additional debris, blockage and flooding. This causes water to enter people's homes, exposing them to health complications, such as mold, which causes respiratory issues, gastrointestinal diseases related to water contamination, and mosquito-borne diseases.

Despite these adversities, the residents of Martín Peña Canal have demonstrated remarkable resilience and solidarity. Their collective efforts have given rise to innovative initiatives, such as the Proyecto ENLACE and the CLT, which aim to address long-standing issues of land tenure insecurity, inadequate infrastructure, and environmental degradation [23].

These community-driven organizations have fostered a participatory approach to urban planning, enabling residents to advocate for their rights and lead transformative changes in their neighborhoods [8,41,64]. The Caño Martín Peña Comprehensive Infrastructure Master Plan (CIMP) was developed in 2004 to meet the challenges of flooding and other environmental and social justice issues [67]. The plan is to carry out infrastructure redevelopment, improve water quality, and create new housing and climate adaptation measures.

This context of persistent marginalization, coupled with grassroots resilience, provides the backdrop for understanding how participatory land planning initiatives have evolved to empower the community and address systemic inequities in El Caño Martín Peña. The challenges faced and the solutions developed in this community serve as a compelling case for exploring equitable and inclusive urban regeneration practices [8,32,64].

4.5. Transformative Strategies

The transformation of Martín Peña Canal from a marginalized and environmentally degraded area into a globally recognized model for community-led resilience underscores the tangible impact of participatory planning. This section critically examines the specific mechanisms through which community-driven strategies have contributed to urban resilience, focusing on housing stability, governance, environmental restoration, socio-economic development, policy advocacy, and managed retreat.

Unlike top-down, technocratic interventions, which often neglect local contexts, participatory planning in Martín Peña Canal have ensured that resilience-building efforts were grounded in lived experiences and community priorities. By integrating resident-led decision-making with technical expertise, these initiatives addressed both immediate vulnerabilities and long-term sustainability. Resilience in this context is not merely about infrastructural improvements but about fortifying social, economic, and political structures to withstand future environmental and socio-economic shocks.

One of the key factors in this transformation was the Community Land Trust (CLT), which prevented displacement while securing affordable, climate-adaptive housing solutions for residents. By decoupling land ownership from property value, the CLT stabilized the community against speculative real estate pressures, a critical factor in long-term resilience. This model not only ensured tenure security but also established a governance framework that enabled proactive environmental risk management.

Furthermore, participatory planning facilitated infrastructure resilience by incorporating local knowledge into flood mitigation strategies. Unlike previous flood control projects that displaced vulnerable populations, the dredging and sanitation improvements in Martín Peña Canal were co-designed with residents, ensuring equitable access to resources and minimizing harm. These efforts highlight how participatory planning fosters context-sensitive climate adaptation rather than externally imposed solutions.

Beyond physical infrastructure, resilience was reinforced through capacity-building programs and governance structures, such as the G8 coalition. These initiatives empowered residents with political agency, allowing them to advocate for policy protections, secure funding, and influence urban development decisions. This political empowerment is a fundamental aspect of resilience, as it enables communities to continuously adapt to evolving socio-environmental conditions rather than relying on static interventions.

Policy advocacy also played a crucial role in institutionalizing resilience within the legal and regulatory frameworks of Puerto Rico. Through sustained community mobilization, residents successfully lobbied for legislative protections that reinforced land tenure security and environmental conservation efforts. This illustrates how participatory planning not only produces site-specific solutions but also reshapes broader urban governance structures to support long-term resilience.

Finally, managed retreat in Martín Peña Canal differed significantly from conventional relocation strategies, which often result in social fragmentation. Instead of forced displacement, residents were actively involved in relocation planning, ensuring that those affected by environmental risks had viable alternatives within the community. This participatory approach mitigated the psychological and economic disruptions commonly associated with displacement, reinforcing social cohesion as a key dimension of resilience.

5. Discussion

The research questions posed in this study aimed to answer the following questions: (1) What are the socio-economic and environmental challenges faced by marginalized communities in vulnerable urban areas? (2) How have participatory planning models helped address these challenges and empowered communities to achieve sustainable development? This discussion synthesizes the findings considering the research questions and the literature review, emphasizing the insights gained from this case study about public participation as a tool for addressing historical spatial/social inequities, mitigating environmental risks, and fostering long-term resilience.

5.1. Addressing Socioeconomic and Environmental Challenges

The challenges in Martín Peña Canal were multifaceted, stemming from decades of marginalization, environmental degradation, and socioeconomic vulnerabilities. Residents lacked access to basic services, such as potable water, sewage infrastructure, and flood protection, reinforcing cycles of poverty and vulnerability. Historically excluded from formal planning processes, the community faced precarious living conditions, frequent flooding, poor sanitation, and health risks exacerbated by climate-related hazards.

The introduction of participatory planning models transformed this context by enabling residents to address these challenges collaboratively. Through community-driven initiatives, like Proyecto ENLACE, G8, and the CLT, long-standing issues, such as land tenure insecurity, inadequate infrastructure, and environmental risks, were tackled comprehensively. What sets this case apart is its holistic integration of land tenure security, environmental justice, and participatory governance, demonstrating a rare example of a community-led planning model successfully institutionalized within formal policy structures. Unlike other CLTs, which often operate at smaller scales or face political resistance, Martín Peña Canal's model has gained legal and governmental support, ensuring its sustainability over time.

The CLT's innovative approach to land management emerged as a pivotal solution to housing instability. By regularizing land tenure for 2000 low-income families, the CLT not only secured housing rights but also established a governance framework that prevented speculative displacement. This ensured that land value increases resulting from infrastructural improvements benefitted the community rather than external investors. These outcomes align with advocacy and social justice planning principles, which prioritize addressing systemic and spatial inequities through community involvement and resource redistribution. Scientific evidence supports these improvements:

- Land Tenure Stability: Post-implementation, displacement rates significantly declined, with over 80% of families retaining their homes within the CLT framework.
- Flood Mitigation: Prior to intervention, 70% of the area experienced recurrent flooding. Infrastructure projects, including dredging and stormwater management, have since reduced flood incidents.
- Economic Empowerment: Community-led vocational training and micro-enterprises have bolstered employment opportunities, reducing economic vulnerability [12,68].

5.2. Advancing Participatory Planning and Governance

The participatory governance model implemented by the G8 coalition exemplifies the principles of communicative and equity planning. Instead of relying on purely technical expertise, residents became active agents in decision-making, co-developing policies and strategies alongside planners and policymakers. Over 700 participatory events empowered residents to influence decisions, ensuring that infrastructure projects and policies aligned with community needs.

This model not only fostered a sense of ownership but also strengthened social cohesion, as residents united around shared goals. A key strength of this approach was its inclusivity; women, youth, and historically marginalized groups actively participated in leadership roles, breaking traditional barriers to political engagement. The inclusion of women in leadership roles within the G8 was particularly significant, highlighting how participatory planning can amplify marginalized voices and prioritize equity.

In contrast to the technocratic approaches of rational planning, which often excluded local perspectives, the collaborative governance structures in Martín Peña Canal demonstrated how inclusive dialogue and consensus-building can achieve sustainable outcomes and spatial justice. The case underscores the value of empowering communities as active agents in urban regeneration, bridging the gap between technical expertise and local knowledge.

5.3. Insights into Public Participation from the Literature

The literature review highlighted the limitations of rational planning and the strengths of participatory models. The transition in Martín Peña Canal's urban governance—from top-down renewal efforts to grassroots-driven initiatives—aligns with critiques of non-participatory frameworks and right-to-the-city movements. Several key planning theories are reflected in this case:

- Advocacy Planning (Davidoff, 1965): The G8 coalition acted as an advocacy group amplifying residents' concerns about displacement and environmental risks.
- Equity Planning (Krumholz, 1982): The CLT directly redistributed land and housing resources to ensure long-term affordability.
- Communicative Planning (Forester, 1989): Sustained engagement with local knowledge strengthened planning outcomes.
- Radical Planning (Friedmann, 1987): The grassroots activism behind the CLT exemplifies bottom-up planning that directly challenges systemic inequities.

This case study offers a tangible demonstration of how these planning theories can transition from academic discourse into institutionalized urban policy, providing valuable lessons for global urban planning.

5.4. Broader Implications for Managed Retreat and Sustainable Development

The managed retreat efforts in Martín Peña Canal provide a replicable framework for integrating public participation into urban resilience planning. Key takeaways include:

- Transparent communication: Open dialogue with residents mitigated concerns and fostered trust.
- Community involvement: Residents played an active role in shaping the relocation and infrastructure process.
- Equitable resource allocation: The prioritization of local needs ensured fair distribution of resources and minimized forced displacement.

Despite challenges, such as securing funding and navigating bureaucratic complexities, the participatory nature of this initiative fostered resilience among stakeholders. Unlike conventional relocation programs that often result in social fragmentation, this model-maintained community cohesion and strengthened governance structures.

This case study contributes to global discussions on sustainable urban development by demonstrating how formalizing participatory governance within legal and policy frameworks can ensure long-term community stability despite external pressures. Planners and policymakers should prioritize inclusive engagement, recognizing the expertise and agency of local communities in shaping their urban futures.

6. Conclusions

The case of Martín Peña Canal illustrates the transformative potential of participatory planning in addressing socio-economic vulnerabilities, environmental degradation, systemic marginalization and spatial inequalities. By centering community voices and fostering collaborative governance structures, initiatives like Proyecto ENLACE, G8, and the Community Land Trust have empowered residents to take control of their urban future. These efforts have not only regularized land tenure and improved housing stability but also catalyzed infrastructure and environmental improvements that enhance quality of life. The resilience demonstrated by the community underscores the importance of grassroots leadership and sustained public engagement in achieving equitable urban regeneration, offering a replicable model for participatory governance in other vulnerable urban areas.

This study reinforces the critical role of participatory planning as an antidote to the limitations of traditional, top-down approaches. The inclusion of residents in decision-making processes exemplifies how advocacy and communicative planning principles can bridge the gap between technical expertise and local knowledge. By addressing issues such as displacement, gentrification, and environmental risks through inclusive frameworks, the initiatives in Martín Peña Canal align with equity-focused planning theories and demonstrate how these principles can be operationalized. What makes this case particularly compelling is that it not only mitigates displacement but also establishes a legally protected model of land tenure that can resist external market pressures—something rarely achieved at this scale. The success of these efforts provides a compelling argument for planners and policymakers to prioritize participatory frameworks, especially in marginalized communities facing complex challenges.

Looking forward, the insights gained from this case study have broader implications for urban planning and policy. As climate change and urbanization continue to intensify socio-environmental challenges globally, the lessons from Martín Peña Canal offer a roadmap for integrating resilience, sustainability, and equity into planning practices. Planners must adopt approaches that center the lived experiences of communities, fostering trust and co-ownership in the planning process. More importantly, this case demonstrates that participatory planning is not just an idealistic approach but a viable, institutionalized strategy that can be replicated in other contexts facing similar socio-environmental pressures. By doing so, they can not only address immediate challenges but also build the foundations for inclusive and sustainable urban futures. This case serves as a powerful testament to the ability of participatory planning to achieve not just community engagement, but actual policy transformation, setting a precedent for other cities grappling with urban inequities. This case study reinforces that meaningful participation is not just about engaging residents in consultations but about embedding community leadership in governance structures.

Author Contributions: Conceptualization, I.G. and L.M.-R.; methodology, I.G. and L.M.-R.; software, I.G.; validation, I.G. and L.M.-R.; formal analysis, I.G. and L.M.-R.; investigation, I.G. and L.M.-R.; resources, I.G. and L.M.-R.; data curation, I.G. and L.M.-R.; writing—original draft preparation, I.G.; writing—review and editing, L.M.-R.; visualization, I.G. and L.M.-R.; supervision, I.G.; project administration, I.G. and L.M.-R.; funding acquisition, I.G. All authors have read and agreed to the published version of the manuscript.

Funding: This research received no external funding.

Institutional Review Board Statement: IRBS approved by the University of Utah 00155027.

Data Availability Statement: No new data were created. Data sharing is not applicable to this article.

Acknowledgments: We thank two anonymous colleagues who were available to clarify concerns about the socioeconomic data available to the Caño.

Conflicts of Interest: The authors declare no conflicts of interest.

Abbreviations

The following abbreviations are used in this manuscript:

CLT	Community Land Trust
CMP	Caño Martín Peña
CDBG-DR	Community Development Block Grant—Disaster Recovery
CDBG-MIT	Community Development Block Grant—Mitigation
G8	G8 Initiative or G8 Coalition
HUD	U.S. Department of Housing and Urban Development
SDGs	United Nations Sustainable Development Goals

Appendix A. Document Analysis

Archival documents, policies reports, organizational records, and published research were reviewed to gather information about the historical context, goals, and strategies of ENLACE, G8, and the CLT. The following 40 texts were read and analyzed: plans (2), laws (2), dissertation or thesis (2), journal articles (6), website articles (6), magazines (1), newsletters (1), books (1), reports (2), videos (14) and websites (3). Notable references included analyses of similar urban planning models and case studies from other regions. Some of the articles and reports collected include:

Plans

1. Comprehensive Infrastructure Master Plan for Caño Martín Peña: Comprehensive Development and Land Use Plan. 2017.
2. OLIN. Caño Martín Peña Comprehensive Infrastructure Master Plan. Corporación del Proyecto ENLACE del Caño Martín Peña, 2023. Available at: https://worldlandscapearchitect.com/cano-martin-pena-comprehensive-infrastructure-master-plan-olin/ (accessed on 7 December 2024).

Laws

3. Act 32 (2009).
4. Legal frameworks such as Puerto Rico Law 489 (2004).

Dissertation or Thesis

5. Fuentes Viera, Kaira Thais. The Role of Community-Based Organizations in Natural Disaster Recovery and Community Resilience: A Case Study of El Grupo de Las Ocho Comunidades Aledañas al Caño Martín Peña, San Juan, PR. PhD diss., State University of New York, College of Environmental Science and Forestry, 2021.
6. Vega Ramirez, Gustavo Andrés. 2022. "Improving Urban Resiliency through a Framework of Equitable Development in El Caño Martin Peña". M.Arch., United States—Ohio: University of Cincinnati. https://www.proquest.com/docview/2688171216/abstract/5AAE3570B21C429CPQ/1 (accessed on 7 December 2024).

Journal Articles

7. Algoed, Line, and María E. Hernández Torrales. "Communities as Global Actors in Counterhegemonic Policy Mobilities: Lessons from the Community Land Trust Movement". International Journal of the Commons 18, no. 1 (2024): 53–65. DOI: 10.5334/ijc.1217.
8. Algoed, Line, and María E. Hernández Torrales. "The Land is Ours: Vulnerabilization and Resistance in Informal Settlements in Puerto Rico: Lessons from the Caño Martín Peña Community Land Trust". Radical Housing Journal 1, no. 1 (2019): 29–47.
9. Algoed, Line, María E. Hernández Torrales, and L. Rodríguez Del Valle. "El Fideicomiso de la Tierra del Caño Martín Peña: Instrumento Notable de Regularización de Suelo en Asentamientos Informales". Working Paper. Cambridge: Lincoln Institute of Land Policy, 2018.
10. Hernández Torrales, María E. "The Caño Martín Peña Community Land Trust: Corollary of a Model of Community Involvement". Progress, Revista del Colegio de Abogados de Puerto Rico, 68, no. 4 (2007): 794–817.
11. Lorenzo Pérez, M.A., and S. Contreras. "Understanding the Impacts of Managed Retreat and Resettlement on Informal Communities". Natural Hazards Center Mitigation Matters Research Report Series, Report 8. Natural Hazards Center, University of Colorado Boulder, 2022. Available at: https://hazards.colorado.edu/mitigation-matters-report/understanding-the-impacts-of-managed-retreat-and-resettlement-on-informal-communities (accessed on 7 December 2024).
12. Veronesi, Mariangela, Line Algoed, and María E. Hernández Torrales. 2022. "Community-Led Development and Collective Land Tenure for Environmental Justice: The Case of the Caño Martín Peña Community Land Trust, Puerto Rico". International Journal of Urban Sustainable Development 14 (1): 388–97. https://doi.org/10.1080/19463138.2022.2096616.

Website Articles

13. Brodine, Maria. "Proyecto ENLACE del Caño Martín Peña: Restoring an Ecosystem and Building Resilient Communities in Puerto Rico". Urban Waters Learning Network, 22 December 2017. Available at: https://urbanwaterslearningnetwork.org/ (accessed on 7 December 2024).
14. Brooklyn Rail. "Caño Martín Peña: A Case Study in Community Action". The Brooklyn Rail, 2022.
15. Cruz, Lucy. Grupo de las Ocho Comunidades Aledañas al Caño Martín Peña (G8). "Women's Leadership in Community Development". Interview with Lucy Cruz, President of G8, 26 March 2021. Women Build Community. Available at: https://www.womenbuildcommunity.org/ (accessed on 7 December 2024).
16. Proyecto ENLACE del Caño Martín Peña. "Restoring an Ecosystem and Building Resilient Communities in Puerto Rico". Urban Waters Learning Network, 2017.
17. Urbanet. "By the People, For the People: Social and Environmental Revitalization of the Caño Martín Peña, Puerto Rico". Urbanet, 2022.
18. Zárate, Lorena. "By the People, For the People: Social and Environmental Revitalization of the Caño Martín Peña, Puerto Rico". Urbanet, 22 August 2018. Available at: https://www.urbanet.info/cano-martin-pena/ (accessed on 7 December 2024).

Magazines

19. Shelterforce. "El Caño Vive, La Lucha Sigue! Community-Controlled Land in Puerto Rico". Shelterforce, 2017.

Newsletters

20. Vila Biaggi, Ingrid. Proyecto ENLACE del Caño Martín Peña: Restoring an Ecosystem and Building Resilient Communities in Puerto Rico (Urban Waters Learning Network, December 2017), https://www.urbanwaterslearningnetwork.org (accessed on 7 December 2024).

Books

21. Fuller Marvel, Linda. Listen to What They Say. Planning and Community Development in Puerto Rico. San Juan: La Editorial, Universidad de Puerto Rico, 2008.

Reports

22. PBS and J. Environmental Justice Study: Caño Martín Peña Ecosystem Restoration Project, San Juan, Puerto Rico. Prepared for Estudios Técnicos, Inc. Document No. 110036, February 2011.
23. Vila Biaggi, Ingrid M. Caño Martín Peña Urban Waters Location Work Plan. Urban Water Ambassador, 31 October 2019.

Videos

24. Aileen se deja sentir como lider comunitaria en el Caño Martín Peña, Sagrado Corazón, 2019.
25. APA Awards 2009: Peña Special Planning District, American Planning Association, 2009.
26. Caño Martín Peña: A Blueprint for Better, AIA National, 2018.
27. Construyendo La Utopía. Proyecto Enlace del Caño Martín Peña, 2023.
28. El Caño se Transoforma, Proyecto Enlace, 2012.
29. El Espejo del Caño Martín Peña: Qué es, Cómo Nació, y La Historia Detrás. El Nuevo Día, 2023.
30. El Fidecomiso de Caño Martín Peña, Proyecto Enlace, 2019.
31. Huertos del Caño Martín Peña. Proyecto Enlace, 2022.
32. Noticentro, La Tierra Es Nuestra, 2013.
33. Orientation Caño 3.7, Proyecto Enlace, 2023.
34. Pérez Hicks, Ryan, et. al. Documental Agua Mala (Caño Martín Peña), Proyecto Enlace del Caño Martín Peña, 2010.
35. Prensa Comunitaria PR 2009. Barrio TV 605 Caño Martín Peña, 2009.
36. Una Comunidad que Camina. Jonny Walker, 2019.
37. Vivir en el Caño Martín Peña, Primera Hora, 2009.

Websites

38. Proyecto ENLACE del Caño Martín Peña. Martín Peña Channel ENLACE Project: Comprehensive Development and Environmental Justice for Marginalized Communities. 2011.
39. Proyecto ENLACE Martín Peña Canal Website. http://www.martinpena.org/ (accessed on 21 December 2024).
40. World Habitat Awards—Community Land Trust. https://www.world-habitat.org/ (accessed on 7 December 2024).

Table A1. Critical Analysis of Document Contributions.

Document Type	Reference	Contribution to the Work
Plans	Comprehensive Infrastructure Master Plan for Caño Martín Peña (2017) (1)	Provided foundational frameworks for land use and infrastructure development, guiding participatory planning efforts.
	OLIN. Caño Martín Peña Comprehensive Infrastructure Master Plan (2023) (2)	Updated infrastructure plans incorporating resilience and environmental sustainability.
Laws	Act 32 (2009) (3)	Established policy frameworks influencing participatory planning and community development.
	Puerto Rico Law 489 (2004) (4)	Enabled the legal formation of the Community Land Trust, ensuring long-term housing stability.
Dissertation/Thesis	Fuentes Viera (2021) (5)	Examined the role of community-based organizations in disaster recovery.
	Vega Ramirez (2022) (6)	Provided an urban resiliency framework applicable to participatory planning.
Journal Articles	Algoed and Hernández Torrales (2024) (7)	Discussed counterhegemonic policy mobilities and lessons from the Community Land Trust movement.
	Algoed and Hernández Torrales (2019) (8)	Analyzed vulnerability and resistance in informal settlements within the CLT framework.
	Algoed et al. (2018) (9)	Assessed the CLT as a notable instrument for regularizing informal settlements.
	Hernández Torrales (2007) (10)	Outlined the role of community involvement in land tenure security.
	Lorenzo Pérez and Contreras (2022) (11)	Investigated managed retreat and resettlement impacts on informal communities.
	Veronesi et al. (2022) (12)	Linked collective land tenure to environmental justice and community-led development.
Website Articles	Brodine (2017) (13)	Documented ecosystem restoration and community resilience in Puerto Rico.
	Brooklyn Rail (2022) (14)	Featured case studies on community action in urban renewal.
	Cruz (2021) (15)	Highlighted women's leadership in community-driven development.
	Proyecto ENLACE (2017) (16)	Provided project updates and engagement initiatives.
	Urbanet (2022) (17)	Analyzed social and environmental revitalization strategies.
	Zárate (2018) (18)	Showcased participatory urban planning examples.
Magazines	Shelterforce (2017) (19)	Covered community-controlled land practices and sustainable urbanization.
Newsletters	Vila Biaggi (2017) (20)	Reported on the restoration efforts and policy advocacy at Caño Martín Peña.
Books	Fuller Marvel (2008) (21)	Explored historical perspectives on planning and development in Puerto Rico.
Reports	PBS and J (2011) (22)	Conducted an environmental justice study relevant to ecosystem restoration.
	Vila Biaggi (2019) (23)	Developed an urban water location work plan emphasizing resilience.
Videos	APA Awards (2009) (25)	Recognized participatory planning initiatives.
	Caño Martín Peña: A Blueprint for Better (2018) (26)	Visualized the infrastructure transformation and urban regeneration process.
	Construyendo La Utopía (2023) (27)	Showcased community-driven participatory planning successes.
	El Espejo del Caño Martín Peña (2023) (29)	Chronicled the history and evolution of the community-led initiative.
	Pérez Hicks et al. (2010) (34)	Documented the environmental struggles in Agua Mala.
	Una Comunidad que Camina (2019) (36)	Highlighted resident-led efforts in reclaiming urban space.
Websites	Proyecto ENLACE Martín Peña Canal Website (39)	Provided ongoing documentation of participatory planning projects.
	World Habitat Awards (2023) (40)	Recognized Caño Martín Peña's community-led planning as a global model.

Summary: These archival documents, reports, and academic sources provided a robust foundation for understanding participatory land planning in the Caño Martín Peña region. Each reference contributed uniquely to contextualizing historical challenges, evaluating legal frameworks, analyzing case studies, and recognizing the successes and shortcomings of participatory planning approaches.

References

1. Davidoff, P. Advocacy and Pluralism in Planning. *J. Am. Inst. Plan.* **1965**, *31*, 331–338. [CrossRef]
2. Innes, J.E. Information in Communicative Planning. *J. Am. Plan. Assoc.* **1998**, *64*, 52–63. [CrossRef]
3. Harvey, D. The Right to the City. In *Rebel Cities: From the Right to the City to the Urban Revolution*; Verso Trade: London, UK, 2013; ISBN 978-1-78168-074-2.
4. Bachmann, L. An Informal Settlement in Puerto Rico Has Become the World's First Favela Community Land Trust—RioOnWatch 2017. Available online: https://rioonwatch.org/?p=36081 (accessed on 1 February 2025).
5. Proyecto Enlace del Caño Martín Peña, dir. Documental: Agua Mala (Caño Martín Peña). 2013. Available online: https://www.youtube.com/watch?v=rPPbFM-Rvok (accessed on 1 February 2025).
6. El Nuevo Dia, dir. El Espejo del Caño Martín Peña: Qué es, Cómo Nació y la Historia Detrás de su Gran Lucha. 2022. Available online: https://www.elnuevodia.com/noticias/locales/videos/el-espejo-del-cano-martin-pena-que-es-como-nacio-y-la-historia-detras-de-su-gran-lucha-276038/ (accessed on 1 February 2025).
7. Toro-Morn, M.; García, I. *Puerto Ricans in Illinois*; Southern Illinois University Press: Carbondale, IL, USA, 2024.
8. Brodine, M. Proyecto ENLACE del Caño Martin Peña: Restoring an Ecosystem and Building Resilient Communities in Puerto Rico. Urban Waters Learning Network. 2017. Available online: https://urbanwaterslearningnetwork.org/resources/proyecto-enlace-del-cano-martin-pena-restoring-ecosystem-building-resilient-communities-puerto-rico/ (accessed on 1 February 2025).
9. PrensaComunitariaPR, dir. Barrio Tv 605 Caño Martin Peña Seg 2 de 2. 2009. Available online: https://www.youtube.com/watch?v=bFb2OzKAjak (accessed on 1 February 2025).
10. AIANational, dir. Caño Martín Peña: A Blueprint for Better. 2018. Available online: https://www.youtube.com/watch?v=mo7umxuCZPY (accessed on 1 February 2025).
11. National Oceanic and Atmospheric Administration. What Is Dredging? Available online: https://oceanservice.noaa.gov/facts/dredging.html (accessed on 30 December 2024).
12. U.S. Environmental Protection Agency. Ongoing Projects in Caño Martín Peña. Available online: https://www.epa.gov/urbanwaterspartners/ongoing-projects-cano-martin-pena (accessed on 30 December 2024).
13. Notiséis WIPR, dir. Gobernador Anuncia Firma de Acuerdo Millonario con Comunidades del Caño Martín Peña. 2023. Available online: https://hi-in.facebook.com/notiseiswiprtv/videos/gobernador-anuncia-firma-de-acuerdo-millonario-con-comunidades-del-ca%C3%B1o-mart%C3%ADn-p/622644879283772/ (accessed on 1 February 2025).
14. Lorenzo Pérez, M.A.; Contreras, S. *Understanding the Impacts of Managed Retreat and Resettlement on Informal Communities*; Natural Hazards Center Mitigation Matters Research Report Series; Natural Hazard Center, University of Colorado Boulder: Boulder, CO, USA, 2022.
15. Puerto Rico Department of Housing "Action Plan Amendment 8 (Substantial Amendment)". 2022. Available online: https://recuperacion.pr.gov/en/download/8th-amendment-to-the-cdbg-dr-action-plan-nonsubstantial-effective-on-february-25-2022/ (accessed on 1 February 2025).
16. Veronesi, M.; Algoed, L.; Hernández Torrales, M.E. Community-Led Development and Collective Land Tenure for Environmental Justice: The Case of the Caño Martín Peña Community Land Trust, Puerto Rico. *Int. J. Urban Sustain. Dev.* **2022**, *14*, 388–397. [CrossRef]
17. United Nations. *The Sustainable Development Goals Report 2024*; United Nations: New York, NY, USA, 2024.
18. Healey, P. 'Rational Method' as a Mode of Policy Formation and Implementation in Land-Use Policy. Available online: https://journals.sagepub.com/doi/abs/10.1068/b100019 (accessed on 30 December 2024).
19. Friedmann, J. *Planning in the Public Domain: From Knowledge to Action*; Princeton University Press: Princeton, NJ, USA, 1987; ISBN 9780691022680.
20. Glasson, J.; Marshall, T. *Regional Planning*, 2nd ed.; Routledge: London, UK, 2007.
21. Harvey, D. *Paris, Capital of Modernity*, 1st ed.; Routledge: New York, NY, USA, 2005.
22. García, I. Community Participation as a Tool for Conservation Planning and Historic Preservation: The Case of "Community As A Campus" (CAAC). *J. Hous. Built Environ.* **2018**, *33*, 519–537. [CrossRef]
23. Vega Ramírez, G.A. Improving Urban Resiliency Through a Framework of Equitable Development in "El Caño Martín Peña". Ph.D. Thesis, University of Cincinnati, Cincinnati, OH, USA, 2022.
24. Stuart, D.G. Rational Urban Planning: Problems and Prospects. *Urban Aff. Q.* **1969**, *5*, 151–182. [CrossRef]

25. Zipp, S. The Roots and Routes of Urban Renewal. *J. Urban Hist.* **2013**, *39*, 366–391. [CrossRef]
26. Garcia, I.; Garfinkel-Castro, A.; Deidre, P. *Planning With Diverse Communities*; American Planning Association: Chicago, IL, USA, 2019.
27. Primera Hora. Vivir en el Caño Martín Peña—YouTube. 2009. Available online: https://www.youtube.com/watch?v=88MwRRKwf0Y&sttick=0 (accessed on 1 February 2025).
28. Eraydin, A. "Resilience Thinking" for Planning. In *Resilience Thinking in Urban Planning*; Eraydin, A., Taşan-Kok, T., Eds.; Springer: Dordrecht, The Netherlands, 2013; pp. 17–37, ISBN 978-94-007-5476-8.
29. Hunt, D.B. Public Housing in Urban America. In *Oxford Research Encyclopedia of American History*; Oxford University Press: Oxford, UK, 2018; ISBN 978-0-19-932917-5.
30. Guehlstorf, N.P. The Utilitarian Assessment. In *The Political Theories of Risk Analysis*; Guehlstorf, N.P., Ed.; Springer: Dordrecht, The Netherlands, 2004; pp. 45–75, ISBN 978-1-4020-2882-3.
31. Hoch, C. Utopia, Scenario and Plan: A Pragmatic Integration. *Plan. Theory* **2016**, *15*, 6–22. [CrossRef]
32. Nagy, C. El Caño Vive, La Lucha Sigue! Community-Controlled Land in Puerto Rico. Available online: https://shelterforce.org/2017/03/27/el-cano-vive-la-lucha-sigue-community-controlled-land-in-puerto-rico/ (accessed on 30 December 2024).
33. Rittel, H.W.J.; Webber, M.M. Dilemmas in a General Theory of Planning. *Policy Sci* **1973**, *4*, 155–169. [CrossRef]
34. Fuentes Viera, K.T. The Role of Community-Based Organizations in Natural Disaster Recovery and Community Resilience: A Case Study of El Grupo de Las Ocho Comunidades Aledañas al Caño Martín Peña, San Juan, PR. Ph.D. Thesis, State University of New York, College of Environmental Science and Forestry, Albany, NY, USA, 2021.
35. Healey, P.; Booher, D.E.; Torfing, J.; Sørensen, E.; Ng, M.K.; Peterson, P.; Albrechts, L. Civic Engagement, Spatial Planning and Democracy as a Way of Life Civic Engagement and the Quality of Urban Places Enhancing Effective and Democratic Governance through Empowered Participation: Some Critical Reflections One Humble Journey towards Planning for a More Sustainable Hong Kong: A Need to Institutionalise Civic Engagement Civic Engagement and Urban Reform in Brazil Setting the Scene. *Plan. Theory Pract.* **2008**, *9*, 379–414. [CrossRef]
36. Szetey, K.; Moallemi, E.; Ashton, E.; Butcher, M.; Sprunt, B.; Bryan, B. Participatory Planning for Local Sustainability Guided by the Sustainable Development Goals. *Ecol. Soc.* **2021**, *26*, 16. [CrossRef]
37. Davidson, M.E. Alternative Planning Models and the Boston Model Cities Planning Experience. Ph.D. Thesis, Brandeis University, Waltham, MA, USA, 1975.
38. Mattila, H. Can Collaborative Planning Go beyond Locally Focused Notions of the "Public Interest"? The Potential of Habermas' Concept of "Generalizable Interest" in Pluralist and Trans-Scalar Planning Discourses. *Plan. Theory* **2016**, *15*, 344–365. [CrossRef]
39. Clavel, P. The Evolution of Advocacy Planning. *J. Am. Plan. Assoc.* **1994**, *60*, 146–149. [CrossRef]
40. Krumholz, N. *Making Equity Planning Work: Leadership in the Public Sector*, 3rd ed.; Temple University Press: Philadelphia, PA, USA, 1990.
41. Women's Leadership in Community Development. Grupo de las Ocho Comunidades Aledañas al Caño Martín Peña (G8). 26 March 2021. Available online: https://www.womenbuildcommunity.org/case/grupo-de-las-ocho-comunidades-aledanas-al-cano-martin-pena-g8/ (accessed on 1 February 2025).
42. Healey, P. *Collaborative Planning: Shaping Places in Fragmented Societies*; Palgrave: London, UK, 1997.
43. Innes, J.E.; Booher, D.E. *Planning with Complexity: An Introduction to Collaborative Rationality for Public Policy*, 2nd ed.; Routledge: London, UK, 2018; ISBN 978-1-315-14794-9.
44. Forester, J. *Planning in the Face of Power*; University of California Press: Berkeley, LA, USA, 1989.
45. Insurgent Planning: Situating Radical Planning in the Global South—Faranak Miraftab. 2009. Available online: https://journals-sagepub-com.srv-proxy1.library.tamu.edu/doi/abs/10.1177/1473095208099297 (accessed on 3 February 2025).
46. Caldeira, T.; Holston, J. Participatory Urban Planning in Brazil. *Urban Stud. Sage Publ. Ltd.* **2015**, *52*, 2001–2017. [CrossRef]
47. Vocero, L.M.Q. EL Residentes del Caño se Sienten Discriminados Sin Dragado. Available online: https://www.elvocero.com/gobierno/politica/residentes-del-ca-o-se-sienten-discriminados-sin-dragado/article_8652b35e-8aa0-11e8-87f4-1756fa261fe1.html (accessed on 3 February 2025).
48. Protesta de Residentes del Caño Martín Peña Tras Decisión de Tribunal Apelativo Federal. Available online: https://www.primerahora.com/noticias/policia-tribunales/notas/protesta-de-residentes-del-cano-martin-pena-tras-decision-de-tribunal-apelativo-federal/ (accessed on 3 February 2025).
49. Torres, J. Comunidades del Caño Martín Peña: "Tenemos Derecho al Servicio de Energía Eléctrica". Available online: https://www.politicaaccesible.com/post/comunidades-del-cano-martin-pena-tenemos-derecho-al-servicio-de-energia-electrica (accessed on 3 February 2025).
50. Friedmann, J. *Insurgencies: Essays in Planning Theory*; Routledge: London, UK, 2011; ISBN 978-0-203-83211-0.
51. Lefebvre, H. *The Production of Space*, 1st ed.; Wiley-Blackwell: Malden, MA, USA, 1992; ISBN 978-0-631-18177-4.
52. Bonilla, Y.; LeBrón, M. Aftershocks of Disaster: Puerto Rico Before and After the Storm. Haymarket Books. 2019. Available online: https://focmedia.org/2020/12/aftershocks-of-disaster-puerto-rico-before-and-after-the-storm/ (accessed on 1 February 2025).

53. Yin, R.K. *Case Study Research: Design and Methods*, 1st ed.; Sage Publications, Inc.: Thousands Oaks, CA, USA, 1994.
54. Stake, R.E. *The Art of Case Study Research*; Sage Publications, Inc.: Los Angeles, CA, USA, 1995.
55. Flyvbjerg, B. Five Misunderstandings About Case-Study Research. *Qual. Inq.* **2006**, *12*, 219–245. [CrossRef]
56. Urban Waters Federal Partnership. *Urban Waters Federal Partnership, Urban Waters Progress Report: Understanding the Value of Partnerships*; U.S. Environmental Protection Agency: Washington, DC, USA, 2021.
57. Zárate, L. By the People, For the People: Social and Environmental Revitalisation of the Caño Martín Peña, Puerto Rico. Available online: https://www.urbanet.info/cano-martin-pena/ (accessed on 1 December 2024).
58. Fuller Marvel, L. *Listen to What They Say. Planning and Community Development in Puerto Rico*; La Editorial, Universidad de Puerto Rico: San Juan, PR, USA, 2008.
59. García-López, G.A. The Multiple Layers of Environmental Injustice in Contexts of (Un)Natural Disasters: The Case of Puerto Rico Post-Hurricane Maria. *Environ. Justice* **2018**, *11*, 101–108. [CrossRef]
60. Algoed, L.; Hernández Torrales, M. The Land Is Ours. Vulnerabilization and Resistance in Informal Settlements in Puerto Rico: Lessons from the Caño Martín Peña Community Land Trust. *RHJ* **2019**, *1*, 29–47. [CrossRef]
61. U.S. Census Bureau. 2020 Census. Census.Gov. Available online: https://www.census.gov/2020census (accessed on 1 February 2025).
62. Algoed, L.; Hernández Torrales, M.; Rodríguez Del Valle, L. *El Fideicomiso de la Tierra del Caño Martín Peña: Instrumento Notable de Regularización de Suelo en Asentamientos Informales: Documento de Trabajo WP18LA1SP*; Lincoln Institute of Land Policy: Cambridge, MA, USA, 2018.
63. U.S. Census Bureau. Aerican Community Survey American Community Survey 5 Year Estimates (Data Set). 2023. Available online: https://www.census.gov/data.html (accessed on 1 February 2025).
64. Fideicomiso de la Tierra–Fideicomiso de la Tierra del Caño Martín Peña. (n.d.). Available online: https://fideicomisomartinpena.org/ (accessed on 1 February 2025).
65. Quiñones Sierra, W. Ensayo Fotográfico. In *Trabajo Comunitario y Descolonización*; Fundación Francisco Manrique Cabrera: Hato Rey, PR, USA, 2012; pp. 263–274, ISBN 978-1-881748-86-1.
66. Estudios Técnicos, Inc. *Environmental Justice Study: Caño Martín Peña Ecosystem Restoration Project*; Estudios Técnicos, Inc.: San Juan, PR, USA, 2011.
67. Popowsky, R.; Du, S. Drawing Out Climate Action: The Role of Graphic Representation in Climate-Centered Landscape Architectural Practice. In *Representing Landscapes: Visualizing Climate Action*; Routledge: London, UK, 2024; ISBN 978-1-003-40479-8.
68. US EPA. Achievements and Accomplishments in Caño Martín Peña. Available online: https://19january2017snapshot.epa.gov/urbanwaterspartners/achievements-and-accomplishments-cano-martin-pena (accessed on 3 February 2025).

land

MDPI

Article

Participatory Planning and Gamification: Insights from Hungary

Katinka Tóbiás and Lajos Boros *

Department of Human Geography, University of Szeged, Egyetem u. 2., H-6722 Szeged, Hungary; tobiaska@outlook.hu
* Correspondence: borosl@geo.u-szeged.hu

Abstract: Involving citizens in public decision making has become more and more important recently. However, activating citizens is challenging in urban planning, especially in post-socialist countries, such as Hungary, where civil activity is weak, and citizens' attitudes are characterized by distrust towards decision-makers. The gamification of planning processes aims to address this issue and support a more democratic planning process. Gamification is the application of game-like elements (e.g., rewards, storytelling, feedback, competition, etc.) in non-game contexts, thus transforming them into more engaging and enjoyable activities. This study aims to present how gamification is used in the Hungarian urban planning processes and the obstacles to using gamified practices. The research is based on semi-structured interviews with urban planning experts in major Hungarian cities. Our results show that although decision-makers are open to using participatory approaches and gamification techniques, no clear guidelines and principles support these aspirations. The lack of trust and the apathy of citizens makes participatory planning a challenging task. Furthermore, the lack of resources is an obstacle and motivation at the same time when innovative planning methods are used.

Keywords: gamification; urban planning; Hungary; participation; citizen engagement; post-socialism; interviews

1. Introduction

Gamification strategically integrates game design elements in non-game contexts to boost motivation and engagement [1]. In recent years, particularly over the last decade, gamification has gained significant traction in participatory planning. Notably, studies by Gordon and Baldwin-Philippi [2], Aguiar-Castillo et al. [3], Thiel and Lehner [4], and Mandujano et al. [5] highlight the growing importance and effectiveness of these techniques in fostering community involvement and enthusiasm in urban planning projects.

The adoption of gamification in urban planning reflects a shift in approach, moving away from traditional methods of public engagement to more innovative and interactive practices. By incorporating elements such as point scoring, competition, and rewards, planners are finding new ways to capture the interest of citizens, encouraging them to participate actively in the decision-making processes that shape their communities [6,7]. This evolution in engagement strategies signifies a broader recognition of the value of citizen input and collaboration, ultimately contributing to more inclusive and responsive urban environments.

Integrating game elements into presenting serious societal issues can considerably enhance citizen motivation and engagement. By transforming complex topics into more accessible and interactive experiences, game-based approaches can facilitate greater participation among community members [8]. This is particularly important in the context

https://doi.org/10.3390/land14030573

of participatory urban planning, where the persistent problem of citizen demotivation drives the need for innovation in methodologies. Addressing this challenge is essential for fostering meaningful involvement and ensuring that urban development reflects the needs and aspirations of the community. Furthermore, gamification is also a learning process for all the parties involved. Thus, the effects of such initiatives go beyond opinion polls or contributions to the planning and implementing of particular programs or projects [9].

The pressing challenge we encounter on a global scale is the widespread belief among citizens that their ideas and requests are often ignored or undervalued by local governments. This pervasive sentiment not only diminishes civic engagement but contributes to a substantial erosion of trust in public institutions. Many individuals arrive at this conclusion based on their negative experiences with local government interactions, where their voices may have gone unheard or unacknowledged. As a result, this has created a significant disconnect between residents, governmental bodies, and various stakeholders. This gap poses profound implications for community collaboration, effective governance, and the overall health of democratic processes [10].

In recent decades, Hungary's urban planning field has undergone substantial transformation, reflecting evolving societal needs and priorities. This transformation has been marked by a progressive shift towards participatory approaches, prioritizing citizen engagement in urban planning. These methodologies aim to empower individuals and communities by inviting their active participation in decision making, ensuring that diverse perspectives and local knowledge are integrated into planning outcomes. This trend not only fosters greater transparency and accountability in urban development but also enhances the overall effectiveness of urban policies by aligning them more closely with the needs and aspirations of the population [11,12]. Nonetheless, the previously mentioned gap continues to be challenging in this context. However, incorporating gamification as an innovative approach in participatory practices could bring the actors closer together and create a connection between them.

This paper aims to deliver an overview of the current landscape of urban planning in Hungary, focusing specifically on the participatory practices employed within this context. The paper seeks answers to three questions. Our first question was as follows: what is the current state of participatory urban planning in Hungary? The second question was the following: how is gamification used in Hungarian participatory urban planning? Finally, our third question is, what are the obstacles to gamified urban planning in Hungary? To answer these questions, we conducted in-depth interviews with professionals in the field and government officials involved in urban planning processes. The goal was to explore how these participatory approaches shape urban development, informing future policymaking and providing insights into the effectiveness and challenges of public participation in the Hungarian urban planning framework. The paper is structured as follows: first, we review the literature on participatory planning and gamification. In the following sub-section, we highlight the key features of the Hungarian context. The next section describes the process of data collection. Then, we present our findings within the last two sections, and we interpret our results in the light of the body of previous research and highlight the theoretical and practical implications (Figure 1).

To gain a comprehensive understanding of contemporary urban planning trends through participatory methods, it is essential to provide a succinct theoretical background that includes general attributes and those specific to Hungary.

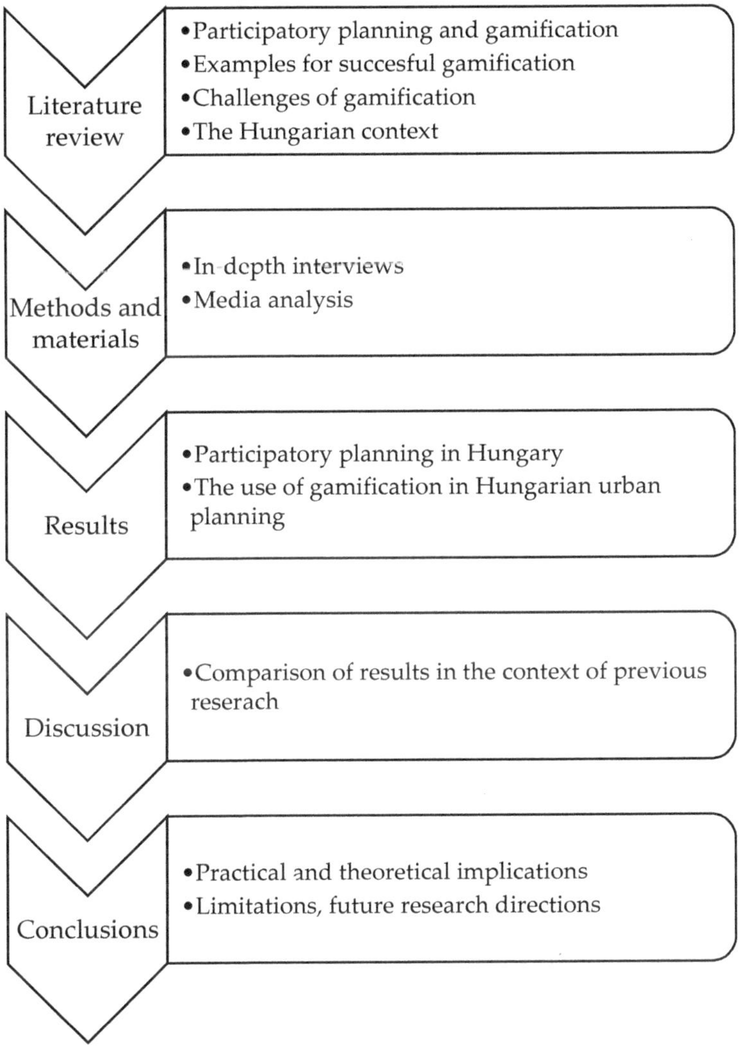

Figure 1. The structure of the paper. Elaborated by the authors.

2. Literature Review

Urban planning has increasingly been scrutinized for its propensity to perpetuate the existing inequalities and adhere to the established trajectories. Considering this criticism, there has been a growing emphasis on the necessity for planning processes that are more responsive, adaptable, and participatory. This evolution has resulted in various theoretical frameworks designed to address these challenges, including collaborative planning, transition management, co-creation, and design thinking [13].

Collaborative planning, influenced by Habermasian concepts of ideal speech, emphasizes the importance of stakeholder engagement and decision making in complex situations. Stakeholders with varying views, power dynamics, and levels of knowledge need to participate in deliberative processes to reach a consensus and minimize conflicts, as illustrated by the "Stuttgart 21" project case where the lack of communication and transparency and the lack of deliberation between stakeholders led to financial damage and delay to the project, along with protests [7].

However, an alternative perspective in the literature argues that stakeholder engagement should not be seen merely as a moral obligation but as a key component of a firm's strategic objectives [14]. Freeman's stakeholder analysis suggests that strategic decision making should consider the interests of a wide range of stakeholders, not just those with direct economic influence. Including "powerless" stakeholders in the analysis phase can provide valuable insights into the values and priorities that ultimately impact the final decision [15]. Furthermore, secondary stakeholders who advocate for primary stakeholders may play a vital role in synthesizing diverse viewpoints.

While the transparency and inclusivity of collaborative planning are essential factors, the strategic significance of stakeholder engagement must also be considered. This ensures that decision-making processes remain aligned with the organization's broader objectives rather than being overly narrow. Despite its merits, collaborative planning has been criticized for its focus on consensus-building, which may not always be practical or desirable in addressing deeply contested urban issues. A balanced approach that integrates stakeholder engagement's moral and strategic aspects may lead to more effective and sustainable urban planning outcomes.

Transitions management, on the other hand, takes a complexity-informed, normative approach to sustainable futures, emphasizing experimentation and adaptive pathways. This approach recognizes the inherent complexity of urban systems and the need for flexible, iterative planning processes. Design thinking, derived from the design disciplines, strongly emphasizes stakeholder engagement and collaborative decision making. Unlike collaborative planning, design thinking encourages a more experimental and iterative approach, focusing on prototyping and testing solutions [13].

While these three approaches share a common goal of fostering more participatory and transformative planning processes, they differ significantly in their theoretical foundations, methodologies, and priorities [13,16].

The increasing popularity of these participatory approaches, particularly in urban sustainability transitions, has led to a growing interest in gamified participatory planning [17]. The integration of gamification in urban planning has become an increasingly prominent focus in recent years. This trend is driven by the desire of municipalities and urban development professionals to find innovative strategies for engaging citizens in the planning process. By incorporating game-like elements, such as point scoring, competition, and interactive platforms, urban planners aim to enhance community involvement and ensure that residents have a voice in decisions that impact their neighborhoods [8,18,19]. Furthermore, gamification can improve decision making by simulating various urban development scenarios, allowing citizens to visualize the potential outcomes of different planning choices. This fosters a deeper understanding of urban issues and encourages collaboration among community members. Ultimately, the goal is to promote more sustainable and livable communities by aligning development initiatives with the needs and preferences of the residents they serve [18,20,21]. Various studies highlighted that gamification as a democratic innovation can serve as an empowering tool for citizens [22,23]. Gamification enables citizens to imagine and experiment with social innovations, creating and testing ideas for a more sustainable or democratic future [24].

The global integration of gamification in urban planning has gained momentum due to a growing recognition of its significant potential to enhance public engagement and participation. By incorporating game-like elements into planning processes, municipalities can foster an environment that encourages community involvement, allowing residents to actively contribute their insights and preferences [19,23–26].

Additionally, gamification serves to improve data collection by transforming complex information into more accessible and engaging formats, thereby encouraging a wider range of participants to provide valuable input. This innovative approach not only aids in gathering diverse perspectives but also improves the quality of data collected [21,27,28]. However, the flow of information is not one-directional: there is a mutual learning process and the co-creation of new knowledge [29]. Also, design alternatives can be tested using gamification, as Ehab and Heath [30] demonstrate through the example of urban design projects in London, where virtual reality was used as a co-design tool.

Furthermore, gamification facilitates collaborative decision making by creating interactive platforms that enable stakeholders, including citizens, planners, and policymakers, to work together more effectively. Urban planning can become more inclusive and adaptive through these collaborative efforts, leading to more sustainable and successful outcomes [31,32].

2.1. Successful Gamification

Various case studies that have employed gamified methodologies to enhance citizen engagement highlight the increasing importance of gamification in urban planning [33]. For instance, Kazhamiakin et al. [34] investigated using a service-based gamification framework in Rovereto, Italy, to encourage sustainable urban mobility. The study emphasizes gamification's potential to promote voluntary behavioral changes among citizens, a vital aspect of smart city initiatives.

Similarly, Gabrielli et al. [35] describe the design and implementation of gamified systems in several European cities, including Trento, Milan, Helsinki, and Barcelona, to involve citizens in sharing information about parking spaces and to encourage sustainable commuting habits. Kazhamiakin et al. [36] present a report on a successful gamified system in Eindhoven, Netherlands, which was developed to motivate employees to choose sustainable commuting options. Another noteworthy example is creating a gamified mobile application called WasteApp as a part of an EU-funded project. This application aims to promote recycling behavior among citizens by incorporating game design elements, and the study evaluates various factors influencing the intention to use this gamified solution [3].

In Sweden, the innovative game Cities: Skylines was harnessed through dynamic workshops to aid in planning the vibrant central area of Norra Djurgårdstaden. This engaging tool transformed the urban planning process into an interactive experience, allowing residents to immerse themselves in the creative design of their community. Through this collaborative platform, individuals were empowered to articulate their visions and suggestions, shaping their neighborhood's future in a meaningful and impactful way [37].

While the potential benefits of gamification in urban planning are clear, it is also essential to consider critical perspectives. A viewpoint paper titled "Cities and the Politics of Gamification" underscores the need for a deeper understanding of the cultural and power dynamics involved in the gamification processes within urban contexts [38].

2.2. Challenges of Gamifying

However, the application of gamification in urban planning does come with challenges. Souza et al. [39] offer a critical perspective on the political and cultural implications of gamification in cities, warning against the risk of "gamipulation"—where game design elements can be used to influence user behavior. Nonetheless, the overall evidence from the case studies suggests that gamified urban planning practices can serve as a valuable tool for fostering citizen engagement, promoting sustainable behaviors, and enhancing the livability of cities. At the same time, gamification can also have adverse effects: it can compromise the quality of citizen participation or enhance competition instead of collaboration. In addition, gamified techniques are more likely to engage certain demographic groups (mostly younger generations) [33].

Cultural contexts play a crucial role in adapting new techniques and approaches to urban planning. In the post-socialist region, civil activity was weak as a legacy of the totalitarian era. It resulted in low trust in public institutions, weak community engagement, and a lack of social cohesion. As a result, a certain "participatory apathy" was a widespread feature of these societies [40–42]. However, the urban planning landscape in Hungary has undergone significant transformation in recent decades, characterized by an increasing emphasis on participatory approaches that seek to engage citizens actively in the decision-making process [11,12,43,44]. This shift has emerged from the recognition that traditional top-down planning models frequently fail to address local communities' diverse needs and perspectives adequately. Consequently, a more inclusive and collaborative methodology is essential for creating sustainable and livable urban environments. Furthermore, participation became crucial due to the lack of financial resources at the local level. The centralized political system decreased local authorities' revenue; thus, mobilizing non-financial resources through increasing citizen engagement is crucial.

2.3. *The Potential of Applying Gamification in Hungary*

The transition toward enhanced civic engagement in urban development has been an evolutionary process, reflecting a broader societal movement towards democratic and participatory governance models in post-communist Hungary. This transformation has been influenced by several factors, including the decentralization of political power after the regime change, the growth of civil society organizations, and an increasing acknowledgment of the significance of public participation in shaping urban spaces [45]. Various initiatives and mechanisms have been established to promote greater citizen involvement in the planning and decision-making processes. These efforts aim to ensure that the needs and aspirations of local communities are more effectively represented in the development of their urban environments [46,47].

The evolution of participatory planning practices in Hungary can be linked to the post-communist transition. This transformative period commenced in the late 20th century as the country embarked on the journey towards a more democratic framework. This transition was characterized by significant sociopolitical changes, which prompted a reevaluation of government structures and public policy approaches [11]. During this crucial time, there was an increasing recognition among policymakers and urban planners of the essential role that civic engagement plays in shaping urban environments. Stakeholders began to understand that involving citizens in the planning process enhances the legitimacy of decisions and ensures that developmental plans reflect the needs and desires of the community [48,49].

In response to this growing awareness, various initiatives and programs were launched to foster greater public participation in urban planning and decision making. These efforts included community workshops, public forums, and collaborative planning sessions, all designed to encourage dialog and input from citizens. Such initiatives aimed to create more inclusive and transparent governance structures, ultimately contributing to Hungary's cities' and communities' long-term sustainability and vitality. This shift towards participatory planning has played a pivotal role in shaping contemporary urban development in the country [50,51]. However, practitioners have often experienced citizen mobilization not being easy due to cultural features and the lack of know-how (best practices, skills, approaches, etc.) [52]. As the examples presented above, gamification can overcome these challenges.

Gamification has emerged as a practical approach to bolstering participatory planning practices in Hungary. By integrating play, competition, and reward components, gamification can make the planning process more engaging and accessible to a broader spectrum of stakeholders, including those historically marginalized from such processes [53,54]. Gamification can increase citizen engagement and involvement in urban planning by making the process more interactive, fun, and rewarding. It can incentivize participation, encourage collaboration, enhance the flow of information between various actors, and foster a sense of community ownership over the development of cities. Using game-like elements such as points, leaderboards, and badges, gamification can motivate citizens to contribute ideas, provide feedback, and actively participate in the planning and decision-making processes that shape their local environment [55–58].

After 2010, the Hungarian government's recentralization processes transformed the urban planning environment. This process has meant centralizing financial resources and forcing municipalities to be more effective and use as many resources as possible [59]. Hungary's changing urban planning landscape suggests that incorporating gamification and other participatory methods will be essential for shaping the future of the country's cities. Implementing these approaches can create a more inclusive and collaborative planning process. This empowers citizens to engage in the development of their communities actively, ensuring that the resulting plans better represent the needs and aspirations of residents [54]. At the same time, it is worth highlighting that the authoritarian tendencies within the political sphere have strengthened Euroskeptic attitudes and generated a certain level of mistrust against "external" ideas—such as gamification.

The literature highlights two main characteristics: first, participatory methods have recently gained attention internationally and in Hungary; second, while gamified methods are not yet widely used, their effectiveness has been demonstrated. Based on the interviews conducted with professionals, these findings provided the foundation for our research.

3. Materials and Methods

First, we identified the scope of this study, which is the presence of gamification in participatory urban planning practices. After that, we defined the national context for our paper. In Hungary, this is a little-researched topic, which made it necessary to deal with it. Above all, the Central European context of the study could contribute to the research of the neighboring countries as well. According to the literature, participatory methods and gamification can be studied through various methods. The experiences of locals are often analyzed through questionnaire surveys, while experts' viewpoints can be revealed both through surveys and interviews (e.g., [18,60]). Case study analyses (e.g., presenting the resources used or analyzing the results of the actions) are also used to reveal the effects of gamification (e.g., [5,48]). Studies also analyzed the available or work-in-progress applications to present the tools of civic engagement (e.g., [3,22]). Testing games and analyzing recorded game sessions were also used to understand how gamification affects decision making (e.g., [19,25,61]). Since our study is exploratory research and aims to analyze the general tendencies of participatory planning and gamification in Hungary, acquiring information from experts was the most important task.

Our initial attempt was to conduct an online questionnaire; however, it was unsuccessful. The failure occurred due to insignificant responses preventing us from concluding. Consequently, we have determined that in-depth interviews are needed instead to uncover the current state of participatory urban planning and its relation to gamification in Hungary. According to Tuza and Kovács [60], this method enables a deeper and more nuanced understanding of the processes, revealing the perceptions, attitudes, and opinions of actors. In addition, we also conducted focused media research to identify the discourses in relation to participatory budgeting, which is one of the most often used participatory tools in Hungarian planning practice. The aim of the media analysis was to supplement the information acquired through in-depth interviews.

For our research, we conducted semi-structured interviews with government officials and professionals from organizations implementing participatory and/or gamified methods in urban planning. Interviews provide opportunities to reveal the experiences of professionals not just regarding their work but also the broader processes (e.g., the general status of gamified urban planning in Hungary, the results, and challenges arising in other municipalities). We selected interviewees who are professionals based on the criteria that they have work experience in participation and are currently employed in their field. After choosing the first interviewee, their insight led us to use the snowball sampling technique and contact the following partners (Figure 2).

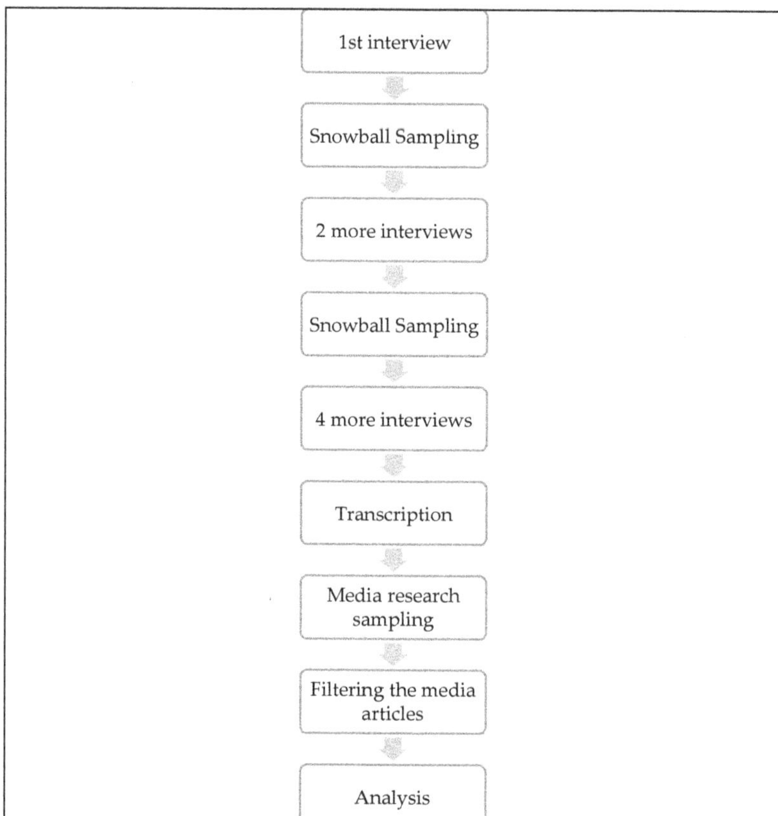

Figure 2. Process of interviewing. Edited by the authors.

These interviews were tailored to each interviewee and focused on three main areas: their work experience related to participation, their understanding of gamification and gamified methods, and their likelihood of using gamified approaches. Additionally, the discussions explored the potential usefulness of a collection of gamification best practices. In most cases, the discussions raised additional questions not covered in the guide, allowing for a more thorough exploration of the topic. The interviews were recorded and transcribed; then, these transcriptions were analyzed. Following the analysis, we formulated the study's theoretical and practical implications (Figure 3).

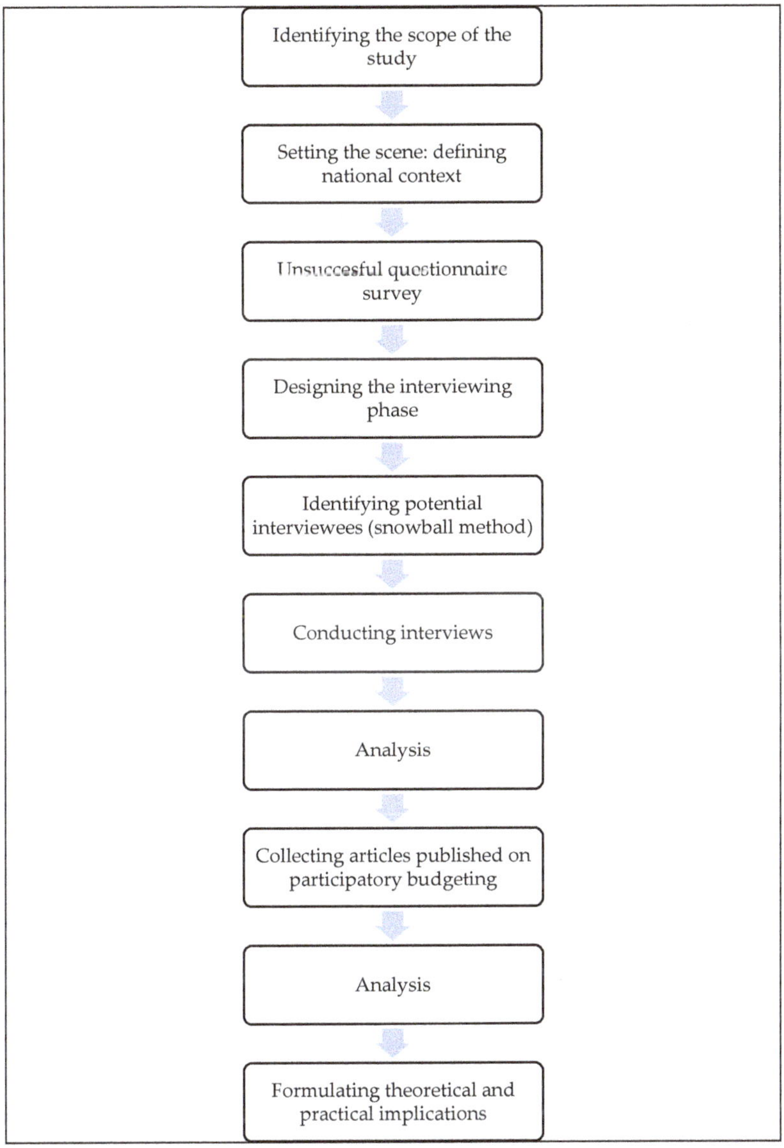

Figure 3. The course of the research. Edited by Lajos Boros.

During the study period, from September 2024 to December 2024, we conducted seven interviews with an average length of 35 min, where the minimum time was 25 min, and the maximum time was 45 min. The interviewees were informed about the research, and we assured them of complete anonymity.

Although it is uncommon for such a small sample size to be significant, it is essential to note that both gamification and public participation are relatively recent phenomena in Hungarian urban planning practices. As a result, there is a limited number of potential interviewees. However, the interviewees possessed valuable insights into the functioning of municipalities at various settlement levels (Table 1). Several analyses deal with the issues of small interview samples, highlighting the usability of research with few participants. We used "purposive sampling" [62], selecting interview partners with extensive knowledge of their city and the whole Hungarian planning scene. Thus, they also provided general information beyond describing the processes within their municipalities. According to the analyses (e.g., [63–65]) focusing on the appropriate sample sizes in interview research, the scope and the approach of the study play an important role. Our research aims to uncover the general trends and experiences regarding participatory planning and the role of gamification in it. This scope requires experts on the field, whose number is limited—in a sense, the sampling has a resource constraint [66]. On the other hand, some studies point out that in many interview research, the first interviews generate most of the data [67]. Thus, Guest et al. [68] state that 6–12 interviews can be enough for research in the case of a homogenous sample. Our group of interview partners was homogenous because they all represented the professional planning side of the processes. Thus, although the sample sizes can consistently be increased, our sample still provides insights into the analyzed topic.

Table 1. Interviewees by occupation and workplace. Edited by Katinka Tóbiás.

Interviewee	Profession	Workplace
Interviewee No. 1.	project lead	non-profit organization
Interviewee No. 2.	municipality officer	local government, major city
Interviewee No. 3.	municipal representative	local government, major city
Interviewee No. 4.	community organizer	local government, major city
Interviewee No. 5.	program leader	climate policy institute
Interviewee No. 6.	municipal representative	local government, major city
Interviewee No. 7.	consultant	non-profit organization

Our main challenges during the interviews included reaching out to individuals and coordinating suitable discussion times. Additionally, since most interviews were online, connectivity issues occasionally disrupted conversations.

To supplement the results of the interviews, media research was also conducted (Figure 4). We collected articles that focus on participatory budgeting published in online media outlets. We focused on this topic because, according to the interviews, this is the most well-known and most often used participatory method among Hungarian municipalities. The following keywords were used (in Hungarian): participatory budget interview, participatory budget evaluation, and participatory budget opinion. Our search returned 412 results. Since our aim was to identify the related discourses, we only analyzed those articles that presented interviews with decision-makers. Thus, articles describing the method or the background of participatory budgeting, presenting the calls for participation, or describing the submitted ideas or results of the voting were left out of the analysis. Altogether, 31 articles remained in the sample after the filtering. The low number of articles after filtering demonstrates the novelty of participatory methods and highlights the lack of knowledge on the topic within society. The articles made it possible to gain more information on the attitudes of decision-makers and to supplement the results of the in-depth interviews.

Figure 4. The process of media analysis. Elaborated by the authors.

We conducted a qualitative analysis of the articles focusing on the following topics: How are participation and participatory budgets presented? What are the expected benefits of participation? What kinds of results and critiques are mentioned?

4. Results

The main results of the interviews are presented in this chapter, focusing on the attitudes and experiences towards gamification and participatory planning. According to the literature, participatory methods are becoming increasingly common in urban planning, and Hungary is no exception. Our results support this statement since all the interviewees were familiar with participation and had previously attended participatory urban planning events. They unanimously agreed that these methods are becoming more mainstream:

> "*. . .participation is becoming very fashionable recently, and we like to use this word participation, and we can also insist on every process that it is participative. . .*" (community organizer, major city)

According to the media analysis, one of the key reasons behind this process is that decision-makers aim to bridge the widening gap between policymaking and local societies:

> "*Everywhere in the world we experience the professionalization of policymaking. This separates locals, decision makers and politicians. That is why citizen assemblies are initiated [. . .] to discuss issues in various topics and formulate policy proposals. Participatory budgeting is about helping the creation and implementation of ideas that are invisible for officials but contribute to the well-being of residents.*" [69]

However, some have argued that this process is much slower than it appears (and should be):

> "*Based on our experience, it is starting to develop very slowly in Hungary, so that local governments start using it.*" (program leader, climate policy institute)

The contrasting perceptions of participation in Hungary stem from the diverse viewpoints held by various stakeholders involved in urban planning. Each actor, whether they are government officials, community members, or urban planners, approaches the idea of participation from unique angles, influenced by their experiences, interests, and objectives. This variance shapes the overall dynamics of how urban development initiatives are conceived and implemented, highlighting the complexity of collaboration in the urban environment.

Another intriguing finding from the interviews was a prevalent skepticism regarding the government and its stance on various issues. This skepticism starkly contrasts the growing adoption of participatory methods, which suggest a push for more inclusive and engaged governance. The tension between the public's distrust and the idea of collaborative decision making highlights a complex relationship between citizens and their government:

> "*. . .basically, a municipality has long-term visions, and it's good if it involves the local population, so this in itself is point zero, that we don't really expect municipalities to want or be able to fulfill this role.*" (community organizer, major city)

This issue is closely related to the broader challenge of balancing the interests of the public with those of local government authorities. Finding common ground is crucial, as it involves addressing the concerns and needs of the community while ensuring that government policies and decisions align with their objectives:

> "*Therefore, we always say that one of the most difficult tasks of working in local government, but also the most beautiful challenge, is to somehow reconcile the very diverse needs and opinions of the population and, with that, to find compromises that obviously represent the majority opinion.*" (municipal representative, major city)

This challenge may stem from or intersect with a common communication issue and a lack of transparency. The interviewees all agree that communication is essential, yet it is a long process of building up the public's trust:

> "*...actually, the operation of the local government should be made participatory, which has a lot of not so spectacular parts [...] so what is not that you can come to the general assembly or that its materials are public, but that the local government actively informs you about what is happening.*" (municipality officer, major city)

> "*and let's say the people who come to the forum experience a very honest communication, where we don't use such political election texts, but we really tell you what the situation is and what could be done and what the difficulties are and what we want, but it may not be possible to do it and then facilitated in a workshop-like manner, they can actually say what their needs are.*" (municipality officer, major city)

On the other hand, there are some cases where the local government tries to apply all possible forms of communication to reach more people:

> "*...we appealed to the population through all kinds of channels: the municipality's website, Facebook page, representatives' Facebook pages, local press, print media, TV.*" (municipal representative, major city)

> "*We use a lot of different channels, newsletters, website, Facebook, Instagram, and then the offline posters, there are biger info points with small signs, and mini-info points that are posted on the fences of all such public areas.*" (community organizer, major city)

However, they often struggle to find effective communication methods and the necessary human and financial resources:

> "*So that we have very few communication channels. Actually, mostly Facebook, which is a joke somewhere.*"— (municipality officer, major city)

> "*They don't always have an effective communication channel with the population, but even if they did, the other big obstacle is obviously money. So that they are usually forced to the minimum required.*" (program leader, climate policy institute)

The lack of resources endangers the sustainability of participatory methods, for example, in Budapest or one of the major cities, Győr, where due to the austerity policies of the national government, the local authorities face difficulties in providing the required funds for the participatory budget.

> "*...It is a difficult situation, since due to the government cutbacks, the participatory budget can be initiated only as a pilot project with a much smaller budget.*" [70]

Having limited funds not only decreases the magnitude of the projects but sustains a low level of participation. According to some experiences, the more money is spent through participatory budgets, the higher the participation rate among locals [71]. Some have recognized that the effectiveness of participative initiatives largely depends on their scale. Engaging smaller communities, particularly at the neighborhood level, is generally more straightforward, as this requires fewer financial and human resources. In many cases, the focus of these initiatives revolves around specific aspects of the community, such as the renovation of public spaces, which directly impacts residents (Figure 5).

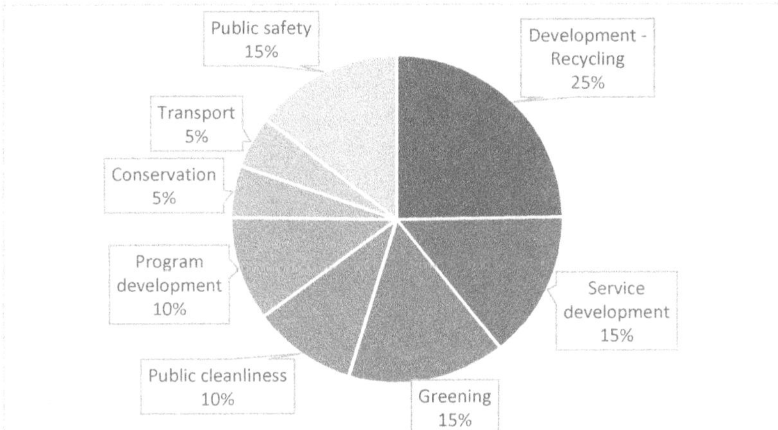

Figure 5. Distribution of votes of residents cast by idea groups. Edited by Katinka Tóbiás based on the document of the VIII. District of Budapest called "Numbers and the Józsefváros participation budget." Source: https://jozsefvaros.hu/downloads/2022/12/rk_adatbazisok_elemzese-pdf.pdf?ver=20221220074612, accessed on 27 February 2025.

Consequently, it is more likely that residents will participate when a development affects their area.

> *"But you also must see that these are primarily local things. So, the local community is the one really interested in the renovation of a public park or a playground."* (municipal representative, major city)

This also makes it easier to reach out to the public in different ways, such as the info signs or points mentioned above, as well as by sending letters or personal visits.

Based on the experiences of municipal representatives from major cities and given the scale at which they think, smaller settlements or towns may not require participative or gamified methods for community engagement. This is likely because the residents know each other well, and the mayor is familiar with the community. There may be little need to motivate people to communicate in small villages and towns, as they tend to engage naturally. However, this conclusion is merely an assumption based on the interviews and would require further research for validation.

Another key aspect of the interviews was motivation, specifically the methods used to encourage public participation in urban planning. This was highly important considering gamification—the main topic of the study—since it serves as a tool to increase motivation. When we analyzed the qualitative data, we found different ways the interviewees used or knew as a motivational tool that led to other perspectives on the matter:

> *"…perhaps this is the best way to encourage residents to give them the opportunity to express their opinions on as many channels as possible."* (municipal representative, major city)

> *"Perhaps the other incentive, which is again not a classic incentive, is to present the renovation processes where fellow residents participated in their opinions so that they can see that, yes, what your neighbor said has come true […] so perhaps we can also motivate residents to participate, if you see good examples of how other opinions have been given and the municipality has accepted them, then there is no danger, as you mentioned, that it will not happen here anyway, and why should I participate in this matter"*— (municipal representative, major city)

"Well, if we think about the project, that Foodwave, the food waste event, someone made the video and won an iPhone and an iPad. We took the whole team to Vienna on a study trip." (project lead, non-profit organization)

The media analysis supported these findings; for example, the most often emphasized aims of participatory budgets were to encourage public participation and to strengthen local democracy.

"We wanted to bring decision-making closer to the locals, thus they can be more familiar with the processes. This changes their perspective on the work done by the municipality, and finally it also increases their trust between decision-makers and locals." [72]

"We often describe the voting on the ideas as democracy project—since it is not enough to submit an idea, gaining support is also crucial. Collaboration is required to have the support of neighbors." [73]

The interviews indicate that idea submissions, competitions (i.e., the work of Budapest Tuning), and workshops are standard. While there are larger gamified projects, they are not clearly labeled as such and seem to appear unknowingly. Some examples of that are from the capital city Budapest, such as the Budapest13 Smart City mobile app, which helps connect the citizens of XIII. district with the municipality through mobile parking, mobile tickets, district news, and prize games (Figure 6).

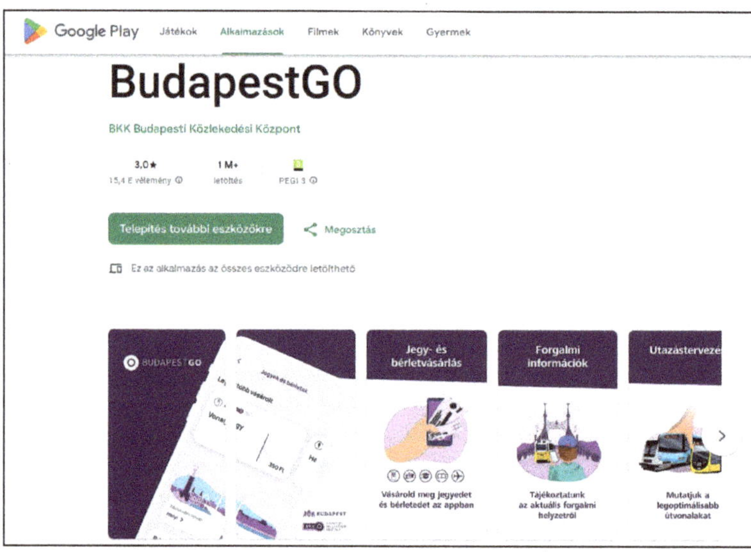

Figure 6. Budapest13 Smart City application on GooglePlay. Source: https://play.google.com/store/apps/details?id=hu.webvalto.bkkfutar, accessed on 7 March 2025.

Another notable example is the Foodwave Óbuda initiative, where participants were required to upload pictures or videos related to food waste and its reduction. As highlighted in the quote, participants could win prizes and go on a field trip.

In another instance, a drawing competition for children was held, with the winner's artwork being displayed on the mosaic of a newly renovated public square.

Another notable instance is the "Green Hackathon" (Figure 7). This event aimed to encourage teams to compete in creating the most sustainable and green ideas for their city. The winner received a prize, and their ideas were incorporated into the final conception.

Figure 7. Moment of the Green Hackathon. Source: https://www.delmagyar.hu/helyi-kozelet/2024/06/a-varos-zolditesevel-kapcsolatosan-rendeztek-otletversenyt-galeria, accessed on 29 January 2025.

Only one individual was initially unfamiliar with gamification but became open to it once they learned more about it. They recognized the potential in it regarding citizen mobilization:

> *"I think it is likely that this would increase the number of participants."*— (municipal representative, major city)

As the media analysis revealed, participatory actions are sometimes considered political actions, and their main aim is to increase the popularity of the local mayor. Other critics highlight that the implementation of projects is often delayed, and participation is quite low.

> *"Only 1.7% of the locals participated in the formulation of participatory planning in the 8th District of Budapest. There is a general disinterest towards such projects, thus they are just wasting the money."* [74]

Some argue that the ideas raised through participatory methods do not bring innovation to urban planning either:

> *"While the method itself is innovative, most of the ideas are quite traditional. The most supported proposals are related to children, dogs, bicycles or trees. In some cases, locals voted for creating sidewalks or pedestrian crossings."* [75]

On the other hand, the advocates of participatory methods emphasize that the methods require time: since locals are not accustomed to having real opportunities for participation, it is difficult to mobilize them.

> *"This is a slow democracy tool that is useful to change the apathy of people and their subordinate role in public matters"* [76]

There is some skepticism regarding the potential use of gamification to enhance motivation as well. Our observation suggests that this hesitance stems from a lack of familiarity with gamification in the context of urban planning, as most examples they have encountered are primarily focused on education. This limited exposure may contribute to their reluctance to explore gamification as a viable tool in urban planning initiatives. In summary, while they know gamification and may attempt to incorporate some games or playful elements into their events or citizen engagement efforts, they do not know how to implement it effectively.

Thus, the consensus among the participants was that they were likely to use gamification; however, a significant concern was the absence of established guidelines to help them apply this new method effectively. As a result, all the interviewees agreed that compiling adaptable best practices on gamification would be instrumental.

> "...*the one thing that is useful is if there is such a toolbox, and actually, it might not even be a toolbox, but somehow, so this gamification should obviously have such a definition background*..." —(project lead, non-profit organization)

To overcome the difficulties concerning the lack of know-how, civil organizations can play a significant role in helping local governments implement gamification. These organizations often use more gamified approaches and have valuable experience that can serve as an excellent starting point. This knowledge can be tailored to fit the specific needs of each municipality and its residents, and it can also be transferred to other localities in the form of best practices to adapt.

The interviews conducted for this study indicate that urban planners and municipal representatives generally demonstrate a willingness to explore the integration of gamified elements into their public engagement strategies. However, they currently face challenges regarding the practical knowledge and resources necessary to implement these elements effectively.

A significant finding from the interviews indicates that participants predominantly associate gamification with educational environments rather than with urban planning. This limited exposure to gamification within the community engagement framework may contribute to their reluctance to consider it an effective tool for enhancing public participation.

Despite initial skepticism, the interviewees acknowledged the potential advantages of gamification, including enhanced citizen engagement and increased participation in the planning process. Nevertheless, they expressed concerns regarding the lack of knowledge and adaptable best practices for implementing gamification in a manner that is tailored to the specific needs of their municipalities.

The interviews also underscored the significant potential of civil society organizations in supporting local governments with their gamification initiatives. These organizations typically have extensive experience with gamified methodologies and are well positioned to share their knowledge and expertise, thereby aiding in developing tailored solutions for specific municipalities.

The interviews demonstrate an increasing interest in applying gamification to enhance public participation in urban planning. However, the widespread implementation of such strategies is currently impeded by a lack of know-how and resources (both human and financial). Thus, the transfer and adaptation of the existing knowledge, combined with enhanced collaboration between local governments and civil society organizations, can address this deficiency and unlock the full capabilities of gamified approaches in fostering community engagement and promoting urban development.

5. Discussion

This research investigates the status of public participation in urban planning within Hungary, specifically examining the potential role of gamification in enhancing stakeholder engagement. The findings indicate that while there is a general openness among stakeholders to consider the application of gamification in urban planning initiatives, there exists a significant lack of familiarity and guidance regarding its practical implementation. This finding also aligns with the previous research (e.g., [7]).

The advancement of collaborative efforts among participants is considerably impeded by a pervasive atmosphere of mutual distrust, as revealed by the interview findings. This challenge is not solely confined to Hungary; the existing literature indicates that demotivation and distrust are widespread internationally [10]. Nevertheless, the literature suggests that gamification possesses the potential to create platforms that facilitate closer interactions among stakeholders, thus mitigating misunderstandings [31,32].

The current body of literature underscores the critical necessity for planning processes to be more responsive to emerging challenges and stakeholder inputs [13]. Furthermore, the insights derived from the recent interviews illuminate the importance of clear and effective communication in fostering meaningful participation among all the parties involved. This emphasizes the need for efficient communication channels to ensure that diverse perspectives are accounted for and integrated within the planning framework.

Using gamification to enhance public engagement and participation has gained increasing importance in Western Europe, the US, and Hungary [51]. However, due to the distinctive post-socialist context prevalent in Central European countries [11], the adoption of these progressive strategies remains in its early stages, as evidenced by the interviews conducted.

Research demonstrates that incorporating gamified elements into participatory processes can significantly enhance citizen engagement [3,34,36]. This assertion is corroborated by insights from interviews indicating that participants respond positively to the incentives and enjoyment provided by these elements, ultimately promoting more active and meaningful involvement in civic activities.

The interviewees identified a range of motivational techniques currently employed to stimulate public participation, including idea submissions, competitions, and workshops, which was already explained in the literature [50,51], indicating that applying such techniques cannot be considered a novelty. However, implementing more comprehensive gamified approaches appears to be limited, with only a few examples noted in the capital city of Budapest. Moreover, in many instances, these applications occur without the conscious recognition of their gamified characteristics. This suggests that many practitioners are utilizing gamification strategies to enhance engagement and motivation while remaining unaware of the foundational principles that underpin this methodology.

Our results align with previous findings that highlight the significance of financial resources [77]. Thus, the lack of funds is a motivation and an obstacle to gamifying urban planning processes simultaneously. However, the lack of funds endangers the existing or planned participatory projects, such as the cases of local participatory budgeting initiatives demonstrate.

Additionally, the interviews highlighted the potential role of civil society organizations in assisting local governments with their gamification efforts. These organizations often possess significant experience with gamified approaches and could share their expertise to help formulate tailored solutions for specific municipalities. Notably, the literature does not address the role of civil society organizations, leaving open the question of whether this is a uniquely Hungarian phenomenon or a gap that necessitates further examination in the literature.

The existing literature does not discuss the concept of a gamification handbook or formal guidelines; however, it documents the various challenges individuals may encounter [38]. Insights from interviews suggest that developing such a guide could effectively alleviate these challenges, presenting detailed best practices, the possibilities and challenges of adaptation, and the required human and financial resources.

Furthermore, the cases of gamification implemented in urban planning, as presented in this study, are primarily located in major municipalities [3,34,36], and the interviews were conducted with professionals from these larger cities. This indicates a significant need for gamified approaches, particularly in larger communities, due to the prevalence of social alienation. Our research also supports this statement.

Gamification is undoubtedly an innovative approach; however, like any new method, it has its drawbacks and is not a one-size-fits-all solution for the challenges found in urban environments (Figure 8). While applying gamified elements is significant, careful planning is essential. Without it, gamification can become a burden rather than a benefit.

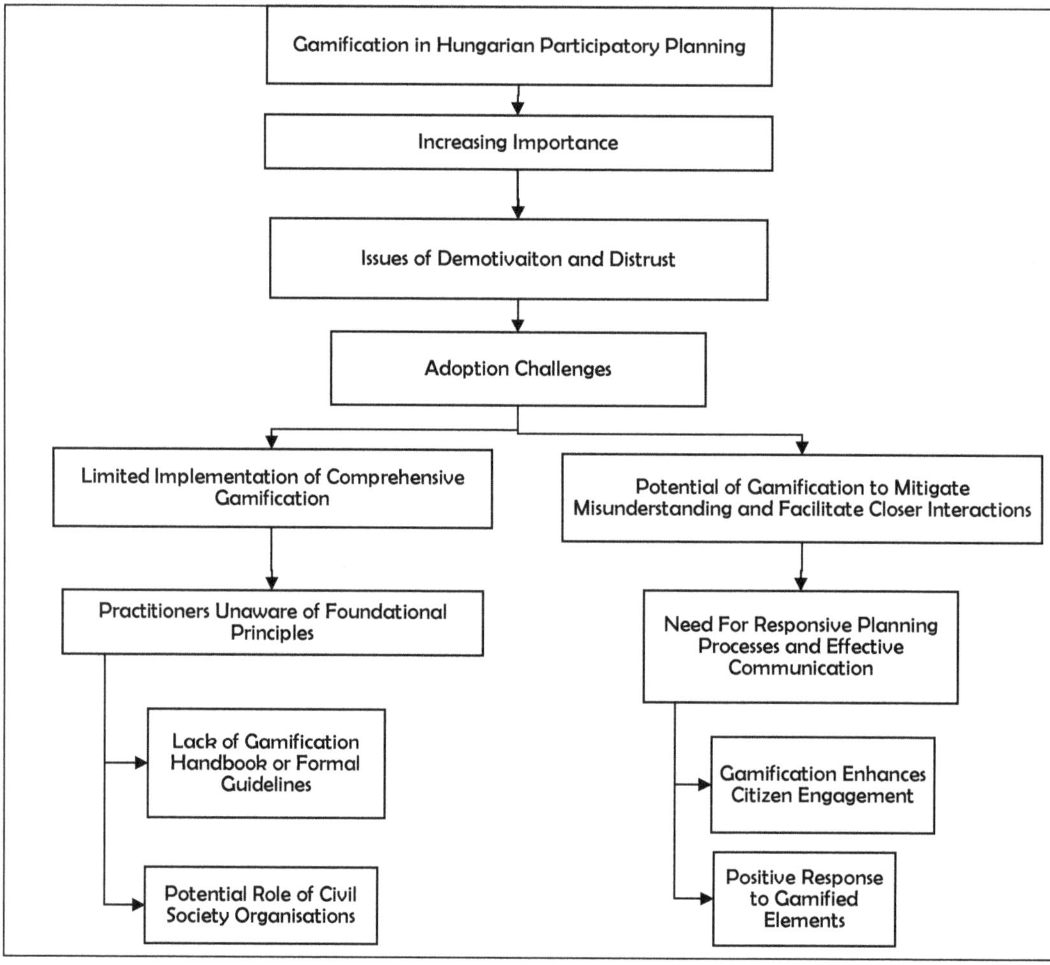

Figure 8. The social environment of gamified urban planning in Hungary. Source: elaborated by the authors.

The study demonstrated that participatory practice in Hungary is in its early stages. However, the experience gained and the obstacles identified can benefit other countries with a low level of participation. Therefore, while Hungary may not be the ideal model of citizen involvement overall, it provides valuable insights into pathfinding and identifying the challenges related to participation and gamification.

6. Conclusions

Our research had three questions to answer. The first focused on the current state of participatory urban planning in Hungary. The results show that participatory approaches are relatively novel in Hungary, and municipalities must deal with several issues when implementing them. Balancing between various interests and finding the appropriate channel for communication are among the key problems to address. Thus, the development of such approaches is slower than expected or desired. This situation makes the use of gamification quite challenging. With these circumstances, our second question aimed to understand the concern regarding using gamification techniques in Hungarian urban planning. According to our results, gamification is not an unfamiliar concept among decision-makers, and the attitudes toward gamification are positive. Mostly, the larger cities use gamification to enhance participation because social cohesion is usually stronger in smaller communities. The second question focused on the obstacles to using gamification in Hungarian urban planning. According to the interviews, there is a knowledge gap regarding the implementation of gamification. Thus, practical guidelines, toolkits, and exchanging experiences and ideas will be crucial in the future use of gamified planning. Furthermore, the lack of financial and human resources also hinders the use of participatory approaches and gamification techniques.

Since fostering participation is a relatively new and innovative methodology in Hungary, most municipalities are still in the early stages of trying to implement it. Unfortunately, many of them are unsure of how to proceed and struggle with finding the required financial and human resources. This creates a scenario where we must find a path forward, especially as we adopt methods that have been successful in several European (e.g., Germany [7], Sweden [37], Italy [3], Spain, and Finland [35]) countries. In Central Europe, particularly in Hungary, the post-socialist legacy poses unique challenges, such as the low level of civil activity, general distrust towards decision-makers, or hesitancy towards external ideas and solutions. It explains why there is hesitancy regarding gamification, as it is also a new and not widely understood concept. However, several municipalities are already open to new approaches and are attempting to implement participatory methods, including some gamified techniques. But there is still much to learn. Professionals and civil organizations that have successfully organized participatory and gamified events can be crucial in guiding these efforts. Their experience can help enhance methods, foster innovation, and strengthen public and local government relationships. Furthermore, involving civil organizations can also prevent the over-politicization of participatory actions. As the media analysis revealed, some of the critiques towards participatory methods consider these approaches parts of political campaigns. Civic organizations, acting as intermediaries, can distance actions from the political scene.

Our study, on the one hand, contributes to the understanding of the factors influencing gamification in urban planning, shedding light on the significance of post-socialist legacy. Distrust, skepticism, and the apathy of citizens are significant obstacles to urban planning in several post-socialist countries. Since participatory approaches are relatively new planning tools in these countries, municipalities often struggle to find appropriate methods and forms of communication. Enhancing citizen participation through gamification can contribute to more inclusive urban planning and overcome trust issues. By engaging locals, an exchange of knowledge can occur, which can make the processes not just more effective but more transparent as well. Gamification can motivate locals' involvement by expressing their opinions, sharing and generating ideas, or volunteering. This is particularly important when municipalities have minimal resources; thus, enhancing efficiency and providing adequate answers to local needs are complex tasks. Our results show that the lack of financial resources is a motivation and an obstacle for municipal participatory planning processes and gamification. On the other hand, the research provides practical insights regarding the necessity of adaptable know-how to manage the problems related to the lack of knowledge and skill sets. Involving civil organizations can bring expertise and build a bridge between decision-makers and citizens. The citizens' organizations are bottom-up initiatives that embody local aspirations. Thus, they are often more trusted and recognized by locals. This offers opportunities to overcome the difficulties arising from the erosion of trust between public institutions. Since local governments frequently struggle with financial and human resources, they cannot transfer existing knowledge—external actors, such as civil organizations, can support them to manage this problem. However, the centralization tendencies and politicization of the civil sphere in Hungary may limit the possibilities of such operations. Regarding communication channels, social networks and mobile phone applications are valuable tools. Still, it is essential to highlight that certain social groups can be omitted due to lacking digital skills or not being involved in social networks. Thus, despite applications and social networks seeming to be convenient and cheap means of citizen engagement, face-to-face interactions remain essential. Reaching out on Facebook or homepages is often a one-way communication and does not bring the expected results—it only creates an illusion of citizen dialog and participation. Furthermore, the transfer of know-how through handbooks or guidelines presenting best practices of participatory practices and gamification can contribute to successfully adapting these techniques, helping to overcome knowledge gaps. While compiling and analyzing best practices should be coordinated by national and/or regional actors (e.g., ministries, regional development agencies, etc.), the handbooks should be used mainly at the municipal level, where the best practices and recommendations can be tailored to local needs.

Like any piece of research, this study also has certain limitations. First, the analysis focuses on one country—thus, one should be careful with the generalization of the results. Furthermore, the results are based on interviews and media analysis, which provide valuable insight into the experiences and attitudes of decision-makers but do not provide a comprehensive overview of the topic. Future research should focus on comparing various countries to explore the effects of different social environments on gamification in urban planning. For example, the possibilities regarding the supporting role of civil organizations could be explored in such comparisons. In addition, other methods, such as surveys among residents or content analysis of policy documents, can also be used to gain a more detailed understanding of the issue. Different stakeholder groups can also be interviewed to reveal various points of view on the possibilities of gamification in public participation. Increasing the number of interviewees could incorporate more points of view into the research, thus providing a more nuanced overview of the processes and challenges of participatory urban planning and gamification. The effects of gamified planning techniques (e.g., increased participation and more informed decision-makers and citizens) could also be analyzed—this could provide valuable insights for decision-makers through benchmarking previous actions. Finally, longitudinal studies (e.g., continuing the media analysis and surveying users or decision-makers) are also needed to survey the development of gamification within urban planning.

Author Contributions: Conceptualization, K.T. and L.B.; methodology, K.T. and L.B.; formal analysis, K.T. and L.B.; investigation, K.T.; writing—original draft preparation, K.T. and L.B.; writing—review and editing, K.T. and L.B.; visualization, K.T.; supervision, L.B.; project administration, L.B.; funding acquisition, K.T. and L.B. All authors have read and agreed to the published version of the manuscript.

Funding: Project no TKP2021-NVA-09 has been implemented with the support provided by the Ministry of Culture and Innovation of Hungary from the National Research, Development and Innovation Fund, financed under the TKP2021-NVA funding scheme. It has been funded by the Research Fellowship Program (Code: 2024-2.1.1-EKÖP) of the Ministry of Culture and Innovation from the National Fund for Research, Development and Innovation.

Institutional Review Board Statement: Ethical review and approval were waived for this study due to the Act LXXVI of 2014 on Scientific Research, Development and Innovation (https://nkfih.gov.hu/palyazoknak/dokumentumtar/hatosagi-engedelyek, accessed on 31 January 2025).

Informed Consent Statement: Informed consent for participation was obtained from all the subjects involved in the study.

Data Availability Statement: The original contributions presented in this study are included in the article. Further inquiries can be directed to the corresponding authors.

Conflicts of Interest: The authors declare no conflicts of interest. The funders had no role in the design of the study; in the collection, analyses, or interpretation of data; in the writing of the manuscript; or in the decision to publish the results.

References

1. Deterding, S.; Dixon, D.; Khaled, R.; Nacke, L. From game design elements to gamefulness: Defining "gamification". In Proceedings of the 15th International Academic MindTrek Conference: Envisioning Future Media Environments, Tampere, Finland, 28–30 September 2011; pp. 9–15.
2. Gordon, E.; Baldwin-Philippi, J. Playful Civic Learning: Enabling Reflection and Lateral Trust in Game-based Public Participation. *Int. J. Commun.* **2014**, *8*, 759–768.
3. Aguiar-Castillo, L.; Rufo-Torres, J.; De Saa-Pérez, P.; Perez-Jimenez, R. How to Encourage Recycling Behaviour? The Case of WasteApp: A Gamified Mobile Application. *Sustainability* **2018**, *10*, 1544. [CrossRef]
4. Thiel, S.-K.; Lehner, U. Exploring the effects of game elements in m-participation. In Proceedings of the 2015 British HCI Conference, Lincoln, UK, 13–17 July 2015; pp. 65–73.
5. Guillen Mandujano, G.; Quist, J.; Hamari, J. Gamification of backcasting for sustainability: The development of the gameful backcasting framework (GAMEBACK). *J. Clean. Prod.* **2021**, *302*, 126609. [CrossRef]
6. Fox, N.; Campbell-arvai, V.; Lindquist, M.; Berkel, D.V.; Serrano-vergel, R. Gamifying Decision Support Systems to Promote Inclusive and Engaged Urban Resilience Planning. *Urban Plan.* **2022**, *7*, 239–252. [CrossRef]
7. Muehlhaus, S.L.; Eghtebas, C.; Seifert, N.; Schubert, G.; Petzold, F. Game.UP: Gamified Urban Planning Participation Enhancing Exploration, Motivation, and Interactions. *Int. J. Hum. Comput. Interact.* **2022**, *39*, 331–347. [CrossRef]
8. Thiel, S.-K.; Ertiö, T. Play It to Plan It? The Impact of Game Elements on Usage of a Urban Planning App. In *User Centric E-Government*; Integrated Series in Information Systems; Saeed, S., Ramayah, T., Mahmood, Z., Eds.; Springer International Publishing: Cham, Germany, 2018; pp. 203–229, ISBN 978-3-319-59441-5.
9. Barata, G.; Gama, S.; Jorge, J.; Gonçalves, D. Improving participation and learning with gamification. In Proceedings of the First International Conference on Gameful Design, Research, and Applications, Toronto, ON, Canada, 2–4 October 2013; pp. 10–17.
10. Cupps, D.S. Emerging Problems of Citizen Participation. *Public Adm. Rev.* **1977**, *37*, 478. [CrossRef]
11. Beluszky, P.; Timár, J. The Changing Political System and Urban Restructuring in Hungary. *Tijdschr. Voor Econ. En Soc. Geogr.* **1992**, *83*, 380–389. [CrossRef]
12. Istenič, S.P.; Kozina, J. Participatory Planning in a Post-socialist Urban Context: Experience from Five Cities in Central and Eastern Europe. In *Participatory Research and Planning in Practice*; The Urban Book Series; Nared, J., Bole, D., Eds.; Springer International Publishing: Cham, Germany, 2020; pp. 31–50, ISBN 978-3-030-28013-0.
13. Raynor, K.E.; Doyon, A.; Beer, T. Collaborative planning, transitions management and design thinking: Evaluating three participatory approaches to urban planning. *Aust. Plan.* **2018**, *54*, 215–224. [CrossRef]
14. Sautter, E.T.; Leisen, B. Managing stakeholders a Tourism Planning Model. *Ann. Tour. Res.* **1999**, *26*, 312–328. [CrossRef]
15. Goodpaster, K.E. Business Ethics and Stakeholder Analysis. *Bus. Ethics Q.* **1991**, *1*, 53–73. [CrossRef]
16. Peris, J.; Bosch, M. The paradox of planning for transformation: The case of the integrated sustainable urban development strategy in València (Spain). *Urban Transform.* **2020**, *2*, 7. [CrossRef]
17. Pasca, M.G.; Di Pietro, L.; Mugion, R.G. Is Gamification an Enabler of the Sustainable Urban Transition Process? In *Sustainable Urban Transitions*; Urban Sustainability; Allam, Z., Ed.; Springer Nature: Singapore, 2023; pp. 201–217, ISBN 978-981-99-2694-7.
18. Ampatzidou, C.; Gugerell, K.; Constantinescu, T.; Devisch, O.; Jauschneg, M.; Berger, M. All Work and No Play? Facilitating Serious Games and Gamified Applications in Participatory Urban Planning and Governance. *Urban Plan.* **2018**, *3*, 34–46. [CrossRef]
19. Özden, S.; Arslantürk, E.; Senem, M.; As, İ. Gamification in Urban Planning—Experiencing the Future City. *Archit. Plan. J.* **2023**, *28*, 530–547. [CrossRef]
20. Redondo, E.; Giménez, L.; Navarro, I.; Fonseca, D. "Gamification" for Teaching Collaborative Urban Design and Citizen Participation. In *Graphical Heritage*; Springer Series in Design and Innovation; Agustín-Hernández, L., Vallespín Muniesa, A., Fernández-Morales, A., Eds.; Springer International Publishing: Cham, Germany, 2020; Volume 7, pp. 431–441, ISBN 978-3-030-47986-2.
21. Akbar, P.N.G.; Maulidiyanti, M.; Wiwesa, N.R.; Auliya, A. The use of gamification for participatory smart city planning of Indonesia's new capital. *Int. J. Urban Sci.* **2024**, *29*, 1–22. [CrossRef]

22. Beutelspacher, L.; Mainka, A.; Siebenlist, T. Citizen Participation via Mobile Applications: A Case Study on Apps in Germany. In *Research Anthology on Citizen Engagement and Activism for Social Change*; Information Resources Management Association, Ed.; IGI Global: Hershey, PA, USA, 2022; pp. 988–996, ISBN 978-1-6684-3706-3.
23. Van Der Lubbe, L.M.; Gerritsen, C.; Klein, M.C.A.; Hindriks, K.V. Empowering vulnerable target groups with serious games and gamification. *Entertain. Comput.* **2021**, *38*, 100402. [CrossRef]
24. Milburn, C.; Wills, M. Citizens of the future: Science fiction and the games of citizen science. *Sci. Fict. Film Telev.* **2021**, *14*, 115–144. [CrossRef]
25. Poplin, A. Playful public participation in urban planning: A case study for online serious games. *Comput. Environ. Urban Syst.* **2012**, *36*, 195–206. [CrossRef]
26. Gordon, E.; Walter, S. 16. Meaningful inefficiencies : Resisting the logic of technological efficiency in the design of civic systems. In *The Playful Citizen*; Glas, R., Lammes, S., De Lange, M., Raessens, J., De Vries, I., Eds.; Amsterdam University Press: Amsterdam, The Netherlands, 2018; pp. 310–334, ISBN 978-90-485-3520-0.
27. Olszewski, R.; Turek, A.; Łączyński, M. Urban Gamification as a Source of Information for Spatial Data Analysis and Predictive Participatory Modelling of a City's Development. In Proceedings of the 5th International Conference on Data Management Technologies and Applications, Lisbon, Portugal, 24–26 July 2016; SCITEPRESS—Science and Technology Publications, pp. 176–181.
28. Fernández-Castrillo, C.; López-Varela Azcárate, A. Intermedial participatory culture: Ludification and gamification for sustainable development. *Arte Individuo Y Soc.* **2024**, *36*, 779–788. [CrossRef]
29. Ng, P.; Zhu, S.; Li, Y.; Van Ameijde, J. Digitally gamified co-creation: Enhancing community engagement in urban design through a participant-centric framework. *Des. Sci.* **2024**, *10*, e17. [CrossRef]
30. Ehab, A.; Heath, T. Exploring Immersive Co-Design: Comparing Human Interaction in Real and Virtual Elevated Urban Spaces in London. *Sustainability* **2023**, *15*, 9184. [CrossRef]
31. Foth, M.; Bajracharya, B.; Brown, R.; Hearn, G. The Second Life of urban planning? Using NeoGeography tools for community engagement. *J. Locat. Based Serv.* **2009**, *3*, 97–117. [CrossRef]
32. Gugerell, K.; Zuidema, C. Gaming for the energy transition. Experimenting and learning in co-designing a serious game prototype. *J. Clean. Prod.* **2017**, *169*, 105–116. [CrossRef]
33. Resek, C.L.; Siniscalchi, P.P.; Caetano, B.P.; Oliveira, K.W.R.; Paula, M.M.V.; Souza, J.M. Gamification of Citizen Participation: A Systematic Mapping. *IEEE Trans. Games* **2024**, *16*, 1–15. [CrossRef]
34. Kazhamiakin, R.; Marconi, A.; Martinelli, A.; Pistore, M.; Valetto, G. A gamification framework for the long-term engagement of smart citizens. In Proceedings of the 2016 IEEE International Smart Cities Conference (ISC2), Trento, Italy, 12–15 September 2016; pp. 1–7.
35. Gabrielli, S.; Forbes, P.; Jylhä, A.; Wells, S.; Sirén, M.; Hemminki, S.; Nurmi, P.; Maimone, R.; Masthoff, J.; Jacucci, G. Design challenges in motivating change for sustainable urban mobility. *Comput. Hum. Behav.* **2014**, *41*, 416–423. [CrossRef]
36. Kazhamiakin, R.; Marconi, A.; Perillo, M.; Pistore, M.; Valetto, G.; Piras, L.; Avesani, F.; Perri, N. Using gamification to incentivize sustainable urban mobility. In Proceedings of the 2015 IEEE First International Smart Cities Conference (ISC2), Guadalajara, Mexico, 25–28 October 2015; pp. 1–6.
37. Donelly, J.; PcGamer. How Cities: Skylines is Being Used to Build a Real-Life City District. 2016. Available online: https://www.pcgamer.com/how-cities-skylines-is-being-used-to-build-a-real-life-city-district/ (accessed on 11 February 2025).
38. Vanolo, A. Cities and the politics of gamification. *Cities* **2018**, *74*, 320–326. [CrossRef]
39. Souza, V.S.; Marques, S.R.B.D.V.; Veríssimo, M. How can gamification contribute to achieve SDGs?: Exploring the opportunities and challenges of ecogamification for tourism. *JHTT* **2020**, *11*, 255–276. [CrossRef]
40. Bron, A. The struggle for a new collective and democratic identity in post-communist Europe. In *Democratic Practices as Learning Opportunities*; Brill: Leiden, The Netherlands, 2007; pp. 113–124.
41. Ekiert, G.; Kubik, J. Civil Society in Postcommunist Europe: Oland in a Comparative Perspective. In *Civil Society Revisited: Lessons from Poland*; Berghahn Books: New York, NY, USA, 2017; pp. 39–62.
42. Foa, R.S.; Ekiert, G. The weakness of postcommunist civil society reassessed. *Eur. J. Political Res.* **2017**, *56*, 419–439. [CrossRef]
43. Baros, Z.; Dávid, L. Public participation in urban noise prevention: The case of a hungarian town. *Int. J. Sustain. Dev. Plan.* **2012**, *7*, 101–114. [CrossRef]
44. Bajmócy, Z. Participation in Urban Planning and the Post-Socialist Legacy. Revisiting Maier's Hypothesis Through the Case of Hungary. *Deturope Cent. Eur. J. Reg. Dev. Tour.* **2021**, *13*, 4–23. [CrossRef]
45. Szabó, T.; Szabó, B.; Kovács, Z. Polycentric urban development in post-socialist context: The case of the Budapest Metropolitan Region. *Hung. Geogr. Bull.* **2014**, *63*, 287–301. [CrossRef]
46. Mikecz, D. Civil Society, Social Movements and Political Participation in Hungary. In *Politics and Society in Hungary*; Bos, E., Lorenz, A., Eds.; Springer Fachmedien: Wiesbaden, Germany, 2023; pp. 121–136, ISBN 978-3-658-39825-5.

47. Kovács, É. The participatory budgeting and the collaborative governance movement at the Hungarian local government level. *Intersections* **2024**, *10*, 74–94. [CrossRef]
48. Dąbrowski, M.; Piskorek, K. The development of strategic spatial planning in Central and Eastern Europe: Between path dependence, European influence, and domestic politics. *Plan. Perspect.* **2018**, *33*, 571–589. [CrossRef]
49. Barati-Stec, I. Strategic Planning in Post-Communist Settings: The Example of Hungary. In *Strategic Planning in Local Communities;* Governance and Public Management; Hințea, C.E., Profiroiu, M.C., Țiclău, T.C., Eds.; Springer International Publishing: Cham, Germany, 2019; pp. 45–69, ISBN 978-3-030-03435-1.
50. Enyedi, G. *Public Participation in Socially Sustainable Urban Development*; UNESCO: Pécs, Hungary, 2004; p. 39.
51. Kőszeghy, L.; Csizmady, A. Hungarian Urban Planning and the Challenge of Integrated Planning: A Sociological Analysis. *Corvinus Reg. Stud.* **2019**, *4*, 61–69.
52. Tóbiás, K. A gamifikáció eszközei és alkalmazásuk lehetőségei a településfejlesztési folyamatokban. *Földrajzi Közlemények* **2024**, *147*, 343–359. [CrossRef]
53. Prilenska, V. Current Research Trends in Games for Public Participation in Planning. *Archit. Urban Plan.* **2019**, *15*, 113–121. [CrossRef]
54. Kurucz, O. Urban Gamification: Analyzing the Effects of UFLab's Tangible Gamified Tools in Four Hungarian Urban Public Participation Processes. *World Acad. Sci. Eng. Technol. Int. J. Humanit. Soc. Sci.* **2021**, *15*, 1.
55. Green City Times. 4 Ways Gamification Can Promote Sustainable Urban Planning. 2021. Available online: https://www.greencitytimes.com/4-ways-gamification-can-promote-sustainable-urban-planning/ (accessed on 31 January 2025).
56. Infinitivity Design Labs. The Power of Play: How Gamification Shapes Urban Development. 2021. Available online: https://infinitivitydesignlabs.com/the-power-of-play-how-gamification-shapes-urban-development/ (accessed on 31 January 2025).
57. ArchDaily. Gattupalli, A. What is Urban Gamification? 2022. Available online: https://www.archdaily.com/989548/what-is-urban-gamification (accessed on 31 January 2025).
58. Kim, C.; Nam, K.Y. Policy Puzzle Game: Making policy ideas feasible and acceptable in policy co-design. *CoDesign* **2022**, *18*, 448–465. [CrossRef]
59. Pálné Kovács, I. Governance Without Power? The Fight of the Hungarian Counties for Survival. In *Contemporary Trends in Local Governance*; Local and Urban Governance; Nunes Silva, C., Ed.; Springer International Publishing: Cham, Germany, 2020; pp. 45–65, ISBN 978-3-030-52515-6.
60. Tuza, B.; Kovács, Z. A részvételi demokrácia kihívásai a városfejlesztésben: Budapesti tapasztalatok. *Civ. Szle.* **2024**, *21*, 133–150. [CrossRef]
61. Ampatzidou, C.; Vervoort, J.; von Flittner, Z.F.; Vaajakallio, K. New Insights, New Rules: What Shapes the Iterative Design of an Urban Planning Game? *Urban Plan.* **2022**, *7*, 295–305. [CrossRef]
62. Bekele, W.B.; Ago, F.Y. Sample Size for Interview in Qualitative Research in Social Sciences: A Guide to Novice Researchers. *REPAM* **2022**, *4*, 42–50. [CrossRef]
63. Saunders, M.N.K.; Townsend, K. Reporting and Justifying the Number of Interview Participants in Organization and Workplace Research. *Br. J. Manag.* **2016**, *27*, 836–852. [CrossRef]
64. Crouch, M.; McKenzie, H. The logic of small samples in interview-based qualitative research. *Soc. Sci. Inf.* **2006**, *45*, 483–499. [CrossRef]
65. Marshall, B.; Cardon, P.; Poddar, A.; Fontenot, R. Does Sample Size Matter in Qualitative Research?: A Review of Qualitative Interviews in is Research. *J. Comput. Inf. Syst.* **2013**, *54*, 11–22. [CrossRef]
66. Lakens, D. Sample Size Justification. *Collabra Psychol.* **2022**, *8*, 33267. [CrossRef]
67. Young, D.S.; Casey, E.A. An Examination of the Sufficiency of Small Qualitative Samples. *Soc. Work Res.* **2019**, *43*, 53–58. [CrossRef]
68. Guest, G.; Bunce, A.; Johnson, L. How Many Interviews Are Enough? An Experiment with Data Saturation and Variability. *Field Methods* **2006**, *18*, 59–82. [CrossRef]
69. Budavári Önkormányzat. Szabad a Pálya, Ötletelhetnek a Budapestiek (Interjú—Várnegyed Újság). Available online: https://budavar.hu/szabad-a-palya-otletelhetnek-a-budapestiek-interju-a-varnegyed-ujsagbol/ (accessed on 27 February 2025).
70. Ugytudjuk! Dömötör, G. 1 Milliárdról 100 Millió Forintra Csökkent a Győri Közösségi Költségvetés. Available online: https://ugytudjuk.hu/cikk/2025-02-07_honnan-es-mire-lesz-penz-gyorben-2025-ben (accessed on 27 February 2025).
71. G7; Bucsky, P. Minél Több Pénzről Dönthetnek, Annál Többen Szavaznak a Közösségi Költségvetésekről. Available online: https://g7.hu/kozelet/20230729/minel-tobb-penzrol-donthetnek-annal-tobben-szavaznak-a-kozossegi-koltsegvetesekrol/ (accessed on 27 February 2025).
72. Pesti Srácok; Balogh, B. Valódi Választópolgári Részvétel—Interjú Szepesfalvy Annával, Budafok-Tétény Alpolgármesterével. Available online: https://pestisracok.hu/valodi-valasztopolgari-reszvetel-interju-szepesfalvy-annaval-budafok-teteny-alpolgarmesterevel/ (accessed on 27 February 2025).

73. Huszonkettő. Budafok-Tétény Még Élhetőbb, Biztonságosabb és Tisztább Lesz: Interjú Szepesfalvy Anna Alpolgármesterre. Available online: https://huszonketto.hu/budafok-teteny-meg-elhetobb-biztonsagosabb-es-tisztabb-lesz-interju-szepesfalvy-anna-alpolgarmesterrel/ (accessed on 27 February 2025).
74. Magyar Nemzet. Hatalmas blamázs a józsefvárosi részvételi költségvetés. Available online: https://magyarnemzet.hu/belfold/2023/07/hatalmas-blamazs-a-jozsefvarosi-reszveteli-koltsegvetes (accessed on 27 February 2025).
75. HVG. Nagy Sikertől a Kínos Bukásig—Megnéztük, Mi Lett a Közösségi Költségvetések Sorsa. Available online: https://hvg.hu/gazdasag/20231220_kozossegi_koltsegvetes (accessed on 27 February 2025).
76. Szabad Európa; Kerényi, G. Mit Adtak Nekünk az Ellenzéki Vezetésű Kerületi Önkormányzatok? 1. Rész. Available online: https://www.szabadeuropa.hu/a/mit-adtak-nekunk-az-ellenzeki-vezetesu-keruleti-onkormanyzatok-1-resz/32681705.html (accessed on 27 February 2025).
77. Sergeyeva, O.; Bogomiagkova, E.; Orekh, E.; Kolesnik, N. Gamification as a Trend in the Development of Civic and Political Participation. In *Electronic Governance and Open Society: Challenges in Eurasia*; Communications in Computer and Information Science; Chugunov, A., Khodachek, I., Misnikov, Y., Trutnev, D., Eds.; Springer International Publishing: Cham, Germany, 2020; Volume 1135, pp. 125–137, ISBN 978-3-030-39295-6.

land

MDPI

Article

(Re)designing the Rules: Collaborative Planning and Institutional Innovation in Schoolyard Transformations in Madrid

Manuel Alméstar [1,2,*] and Sara Romero-Muñoz [3]

1 Escuela Técnica Superior de Arquitectura de Madrid, Universidad Politécnica de Madrid, 28040 Madrid, Spain
2 Innovation and Technology for Development Centre, Universidad Politécnica de Madrid-itdUPM, 28040 Madrid, Spain
3 Escuela Técnica Superior de Ingeniería de Montes, Forestal y del Medio Natural, Universidad Politécnica de Madrid, 28040 Madrid, Spain; sara.romero@upm.es
* Correspondence: manuel.almestar@upm.es; Tel.: +34-646571955

Abstract: Climate adaptation in urban environments is often constrained by rigid institutional rules and fragmented governance, which limit inclusive and context-specific planning of public spaces such as schoolyards. This study addresses this challenge by examining how collaborative planning can transform schoolyards, from asphalt-dominated, monofunctional spaces into green, climate-resilient community assets. The research employed the Institutional Analysis and Development framework within a qualitative case study design. Two public schools in the San Cristóbal de los Ángeles neighbourhood of Madrid served as case studies, with data collected through document analysis, participant observation, and interviews with municipal officials, urban planners, educators, and community members. Results indicate that the collaborative planning process reshaped rules in use, expanded the network of actors, and transformed decision-making processes. Existing rules were flexibly reinterpreted to allow new uses of space. Children, teachers, and residents became co-producers of the public space, expanding the governance network, where new deliberative practices emerged that improved coordination across people and organisations. These institutional changes occurred without formal regulatory reform, but with the reinterpretation of the game's rules by each organisation. Thus, schoolyards can serve as laboratories for institutional innovation and participatory climate adaptation, demonstrating how urban experiments have the potential to catalyse not only physical transformations but also transformations in urban management.

Keywords: institutional analysis development; school playgrounds; climate adaptation; collaborative planning; urban experimentation; institutional innovation

1. Introduction

Climate change poses increasing challenges for cities, which must adapt to more extreme, recurrent, and unevenly distributed weather conditions. This challenge is particularly urgent in school environments, where vulnerable groups such as children are concentrated and where high temperatures and a lack of shade or green infrastructure affect thermal comfort and students' physical, emotional, and social well-being [1,2]. The transformation of these spaces should not be understood solely as a technical exercise of bioclimatic architecture, but also as an opportunity to rethink the relationships among public institutions, citizens, and territory in search of socially legitimised and sustainable solutions over time [3,4].

In this context, the role of proximity as a spatial logic and the diffusion of activity hubs within neighbourhoods are receiving growing attention in the scientific literature [4–6]. Transforming school environments into multifunctional infrastructures within daily urban systems contributes to decentralising public services and reinforces territorial equity. As Moreno's '15-min city' model highlights, proximity enhances access and promotes socio-spatial cohesion and ecological resilience [4]. This interpretative framework underpins the relevance of schoolyards as relational nodes capable of concentrating environmental, educational, and civic functions, especially in vulnerable urban areas.

City councils worldwide have implemented urban regeneration strategies and plans. Specifically, since 2020, the Madrid City Council has advanced a new urban agenda focused on the ecological regeneration of school environments. The aim is to transform school playgrounds and entrances into 'climate-social oases' [7–9]. However, the traditional design procedure for this type of intervention has historically been led by municipal technical teams and external consultants, with limited participation of the educational communities, let alone the children, their families, and communities, the users of these public facilities. Although effective from an operational point of view, this techno-institutional approach tends to render invisible situated knowledge, local needs, and community aspirations, reproducing dynamics of fragmentation and inequality in the use and appropriation of public space [10,11].

The European project LIFE PACT (People-Driven: Adapting Cities for Tomorrow) [12] proposes an alternative model for intervening in school environments and schoolyards. Based on urban experimentation, the model consists of active listening, co-creation, participatory prototyping, evolving communication, and evaluation. Implemented in Madrid between 2021 and 2025, the project was carried out in the San Cristóbal de los Ángeles neighbourhood (Villaverde district), one of the areas with the highest levels of social, economic, and climate vulnerability in the city [13]. A local consortium—including schools, municipal technicians, cultural and civil society organisations, and universities—was formed to test a new procedural model that introduces collaborative planning to co-design resilient school infrastructures [14].

Although collaborative planning has proliferated globally, it often remains confined to small-scale 'niches' due to significant barriers to expansion, including the challenges of interdisciplinary collaboration [15,16] and the difficulty of transitioning into established regimes without being co-opted [17]. Similarly, establishing effective multi-actor collaborations [18] or integrating institutional innovation into existing organisational structures [19,20] remains an ongoing challenge.

This paper addresses the following research question: How do collaborative planning processes and actor networks in the transformation of schoolyards reshape institutional arrangements? This question responds to an emerging need to analyse the physical outcomes of nature-based interventions and the institutional mechanisms that enable or hinder their implementation, diffusion, and continuity over time.

To address this question, the article draws on the experience of Madrid and analyses it through the lens of the institutional analysis and development (IAD) framework, developed by Elinor Ostrom [21–24]. The IAD framework enables the observation of how 'rules in use', community attributes, and the biophysical characteristics of the environment collectively shape an 'action situation'—that is, an institutional space where actors interact, make decisions, cooperate, or conflict [22]. By applying this analytical framework to the case study, this research identifies how the configuration of the actor system influences institutional arrangements, the incentives for collective action, the rules governing access and deliberation, and both the intangible and physical outcomes of the intervention.

This approach is part of the growing interest in applying theoretical frameworks from multistakeholder governance analysis to the study of urban innovation [24,25]. In contrast to linear or technocratic views of institutional change, the IAD framework allows us to capture informal transformations that are not always reflected in formal regulations or structures, but impact the legitimacy, sustainability, or adaptive capacity of public actions [22]. Moreover, the case study connects with an emerging stream of experimental urbanism studies where the shared production of knowledge, the validation of ideas in authentic contexts, and openness to uncertainty are seen as drivers of institutional transformation [14,26].

The article is structured as follows. Section 2 presents the conceptual framework, focusing on reframing schoolyards through institutional innovation and collaborative urban planning. Section 3 describes the materials and methods used in the Madrid case study, including documentary analysis, direct observation, and interviews with key stakeholders. It also details the IAD framework developed by Elinor Ostrom, which was applied to assess the institutional dynamics of the case study. Section 4 presents the results, offering a comparative analysis of the traditional planning procedure and the experimental model introduced. Section 5 discusses the findings, reflecting on the implications of these transformations for urban policy, participatory planning, and the design of institutional arrangements. Finally, Section 6 concludes by summarising the main insights and outlining recommendations.

The study reveals that the introduction of collaborative planning broadens the number and diversity of actors engaged in the intervention while also reshaping the rules that structure their interactions. Rather than following a closed process with predetermined decisions, collaborative planning fosters dynamics of deliberation, mutual learning, and co-production of expert and situated knowledge. These dynamics in turn enhance all participants' sense of ownership, relevance, and legitimacy.

2. Reframing Schoolyards: Collaborative Planning and Institutional Change in Urban Ecosystems

Adapting schoolyards and their surroundings to climate change offers a strategic window for rethinking and innovating urban management practices [4]. Traditionally conceived as functional spaces for children's recreation and sports practice, these spaces have historically been managed under centralised technical logic, with little institutional flexibility and little participation of the educational community [26–28]. However, the climate emergency, the growing concern for children's health, and the need to build more inclusive cities are revaluating these environments' role in urban regeneration and social resilience [26,27,29].

Innovation in the urban design of school playgrounds and environments does not lie exclusively in the technical–material dimension but also in the modes of governance and institutional management that allow these transformations to occur in a more democratic, contextualised, and sustainable manner [14,22,30]. The transition from top-down planning—centralised in the municipal administration, but fragmented in decision-making by its different departments—to increasingly collaborative urban planning, decentralised and co-led, implies a transformation in how decisions are made, how rules are negotiated, and who has the initiative in public interventions [31–34]. Collaborative urban planning requires integrating community voices into the deliberative and regulatory dimensions, thus bridging the gap between grass-roots action and institutional procedures [5,35].

2.1. From Functional Spaces to Climatic and Social Infrastructures

The design of school environments and playgrounds during much of the twentieth century was profoundly shaped by the principles of functionalist urbanism [36–38] emerging from the modernist movements of the early 1900s and particularly codified through the Athens Charter [39] promoted by Le Corbusier and the Congrès Internationaux d'Architecture Moderne (CIAM). This paradigm advocated for a zoning of urban functions—housing, work, recreation, and circulation—each separated and optimised according to technical rationales [39–41]. Schools, extensions of this logic, were designed following highly standardised models, where playgrounds were treated as empty, asphalted areas dedicated to physical exercise, surveillance, and discipline [26,36]. The physical environment reflected and reinforced a pedagogical model centred on order, repetition, and control, limiting opportunities for exploration, diversity of uses, or ecological interaction [6,40]. In this conception, children and the educational community were not considered active agents in shaping their environment. Instead, they were positioned as passive users within predetermined spatial arrangements, echoing the broader modernist aspiration to regulate urban life through technocratic design [36,41]. The legacy of the functionalist paradigm remains evident in many contemporary school environments: extensive impermeable surfaces, a lack of shade or ecological diversity, the prioritisation of movement over rest, and a limited understanding of children's social, emotional, and environmental needs [26,41].

In response, numerous European cities are making significant efforts to reverse this trend through ambitious urban policies that aim to reimagine school environments as climate-resilient and socially inclusive infrastructures. Examples include Barcelona's Protegim les Escoles programme [42], Pontevedra's Ciudad de la Infancia [43], Paris' initiatives Rues aux Écoles [44] and La Cour Oasis [45], and London's Climate Resilient Schools project [46]. Despite these efforts, a persistent imbalance remains between regulation, design, and social negotiation in urban interventions [21–24]. Institutional inertia and pre-established procedural norms limit meaningful citizen engagement in defining, managing, or transforming these public infrastructures. In many cases, the most significant barriers are not technical or economic, but institutional: fragmented competencies across municipal departments, rigid regulatory frameworks, lack of participatory channels, and insufficient intersectoral collaboration all hinder the systemic integration that school environments could enable [26,27].

As illustrated in Figure 1, these spaces intersect with multiple urban dimensions, from health and education to mobility, biodiversity, and democratic governance. Addressing these challenges requires a systemic approach to the problem to open the door to the multiple potential of these spaces, recognising school environments and playgrounds as more than isolated challenges, but as leverage points within a complex urban ecosystem [27,47]. The transformations of school playgrounds generate benefits that transcend the microclimatic sphere, such as thermal regulation and reduction in pollutants, and impact physical and mental health, community cohesion, and the development of social skills [48]. Conceiving school environments as climatic and social infrastructures demands activating systemic urban transformations that intertwine health, equity, sustainability, and education [27]. School playgrounds, as laboratory spaces, offer fertile ground for testing these dimensions in an integrated way [49]. These transformations improve the quality of public space and reformulate the urban language, orienting it towards a more liveable, inclusive, and resilient city [8,9].

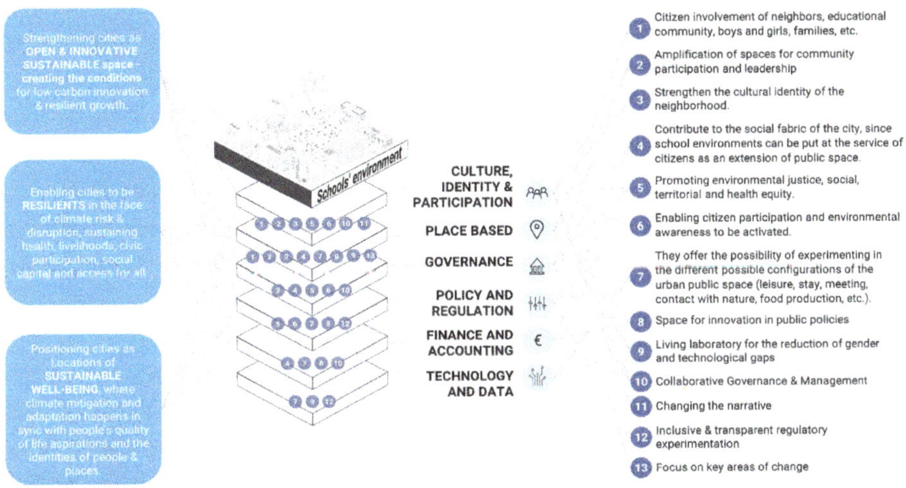

Figure 1. Multilayered analysis of school environments as spaces of urban transformation [26].

2.2. *Paradigm Shift: Experimenting with the Adaptation of School Playgrounds Through Collaborative Urban Planning and Adaptive Regulatory Frameworks*

The urban experimentation is the deliberate introduction of new rules of the game (procedures or forms of organisation) within an urban system to collectively test, observe, and reflect on socio-spatial innovations in an open way [14,50,51]. According to Ostrom [24], "the rules of the game" structure human action. Altering these rules—for example, by enabling children and community members to co-design their school environments—implies a profound reconfiguration of power dynamics, administrative routines, and urban governance practice [25,50]. This approach resonates with the 'experimental urbanism' literature, where the co-production of knowledge and situated learning are essential public innovation mechanisms [51]. Notable examples include studies on temporary urbanism, living labs, and tactical interventions, which challenge the linearity of formal planning and promote hybrid processes of institutional transformation [17,19,26,50].

The transformation of schoolyards must be understood within the broader planning system and regulatory frameworks that govern public space. Urban planning is structured around the intricate and sometimes conflicting interplay of regulation, design, and negotiation [52,53]. Intervening in school playgrounds, therefore, requires architectural or participatory innovation and engagement with land-use rules, planning instruments, maintenance responsibilities, and an interdepartmental coordination mechanism [17]. Collaborative urban planning offers a strategic opportunity to realign conventional procedures by creating deliberative arenas where institutional actors, such as municipal departments and schools, engage alongside civil society and local communities to co-define priorities and modes of intervention [18,19]. Rather than a mere participatory add-on, it constitutes a pathway to renegotiate institutional roles, redefine public responsibilities, and reform existing regulatory frameworks [54].

While collaborative planning offers multiple benefits, many authors have highlighted inherent ambiguities. For instance, Savini and Bertolini [55] caution that collaborative planning risks becoming a superficial innovation strategy disconnected from substantive institutional transformation without a deeper engagement with structural inequalities. Similarly, Swyngedouw [35] criticises that these practices are often promoted by governmental or technocratic elites, avoiding conflict and depoliticising urban decisions under the discourse of the 'possible' or the 'prototypable'. Swyngedouw adds that many participatory initiatives tend to depoliticise urban action, masking conflict under technocratic consensus [9]. Purcell stresses that without clear structures of equity and accountability, these processes can be co-opted by dominant actors [56]. Therefore, collaborative planning must transcend symbolic or non-binding engagement, anchoring participation in legally robust and inclusive mechanisms [57]. A critical and situated perspective is essential to ensure that collaborative planning not only innovates in technical terms but also addresses entrenched power structures, reshapes institutional priorities, and reconfigures modes of urban production.

3. Materials and Methods

This study adopts the IAD framework, developed by Elinor Ostrom and collaborators [21–24], to understand how collaborative planning processes and actor networks in schoolyard transformation reshape institutional arrangements. The IAD framework provides a robust conceptual tool to examine the institutional arrangements that shape collective decision-making processes, their internal dynamics, and their capacity for adaptation and transformation.

3.1. Conceptual Framework: Institutional Analysis and Development Framework

The IAD framework, developed by Elinor Ostrom in the 1990s, provides a conceptual tool for understanding how people make decisions, interact, and manage common resources in complex contexts. Although it initially emerged from analysing natural resource systems, its scope has been progressively extended to public policy, urban planning, and governance of social–ecological systems [22,25]. In its revised version, Ostrom [24] emphasises the analysis of institutional change and the capacity of governance arrangements to adapt to dynamic, multi-scale contexts. This broadening of the framework allows for looking at current governance structures and their trajectory, openness to learning, and potential for transformation (see Figure 2).

The core of the IAD framework is the 'action situation', conceived as the social space where different actors make decisions, negotiate, cooperate, or enter into conflict. This concept allows us to focus the analysis on the immediate structures that shape collective behaviours and the outcomes of an institutional process [21–25]. The IAD framework proposes we decompose these situations into their key elements to identify patterns, explain dynamics, and guide possible reforms in institutional arrangements.

Three major components condition the action situation: (1) the biophysical environment, which defines the material and ecological conditions that frame collective action; (2) community attributes, which include culture, social capital, shared norms, and history of cooperation; and (3) the rules in use, both formal and informal, that determine who participates, what they can do, with what information, and under what criteria decisions are made. These rules are organised into various analytical categories. Position rules delimit the roles that actors can play; access rules establish the criteria for entry or exit into those roles; and others regulate the actions allowed, the information available, the collective decision-making procedures, the applicable incentives or sanctions, and the outcomes that are considered legitimate. These rules shape the institutional framework within which actors interact [25]. While the IAD is a robust framework for analysing and mapping institutional relationships, it also has limitations. One is its tendency to simplify the multi-scale complexity characterising many decisions in contemporary cities. Urban practices are conditioned by influences that cut across different institutional and territorial levels, which requires complementing the analysis with a situated and relational view of the context [55]. A qualitative analysis based on interviews with project actors was incorporated to mitigate this bias and provide a systemic view at the institutional level and the broader context influencing management decisions and outcomes [58].

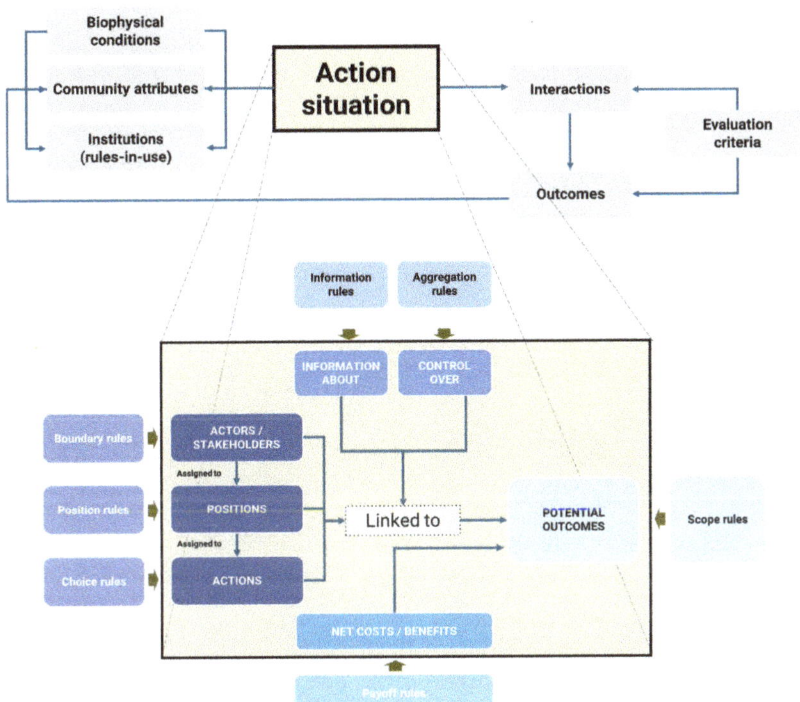

Figure 2. Institutional analysis and development framework and detailed 'action situation' analysis structure elements [22].

This analytical framework could generate contributions such as (i) identifying how the introduction of an urban experiment of collaborative planning transforms the rules in use into rigid institutional processes, allowing the recognition of new actors, knowledge and languages within urban design, and (ii) evidencing the emergence of collaborative governance mechanisms within and outside public administrations that overcome the traditional sectoral fragmentation in management. These contributions, rooted in the case study, allow us to understand institutional innovation not as a disruptive leap, but as a progressive transformation of the formal and informal rules that regulate urban public action.

With regard to the challenges pointed out by Ostrom, this article addresses (i) the need to adapt institutional arrangements to changing contexts, which is reflected in operational and regulatory flexibility; (ii) the importance of polycentric governance, observed in the articulation between diverse actors and different scales (municipal, district, school, neighbourhood); (iii) the potential replicability of the model; and (iv) the internal reflection of public institutions on their capacities to innovate from situated action.

Finally, the IAD framework was particularly suited for this study, as it critically examines how institutional arrangements are informally transformed through practice, without requiring formal regulatory change. In a complex, multi-actor intervention such as the participatory redesign of school playground, the IAD framework makes it possible the analysis of how power, rules, and responsibilities are renegotiated in situ. Its focus on 'rules in use' provides analytical traction to capture the institutional innovation that emerges through urban experimentation beyond prescriptive planning models.

3.2. Method

This study adopted a qualitative research approach based on a case study methodology, which is particularly suited for investigating complex social phenomena and exploring the relationships among context, agency, and institutional change [58,59]. As Eisenhardt [60] highlights, case study research is especially valuable in theory building, as it enables the researcher to generate in-depth and context-rich insights from real-world interventions. The case analysed corresponded to the collaborative planning for the renaturalisation of two public school environments (See Figure 3). This case was selected not only due to its relevance as a site of climate–social experimentation but also because of the direct involvement of the research team in the process, which enabled the consolidation of relationships of trust with both institutional and community actors.

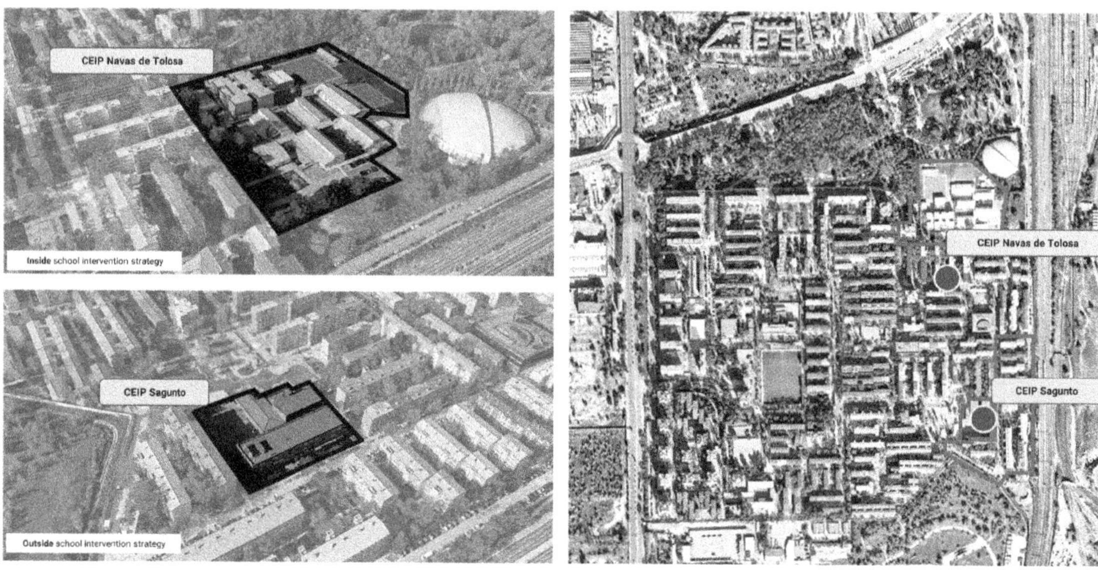

Figure 3. Plan and axonometry of the location of the case study (Navas de Tolosa School, Sagunto School, Villaverde).

Data collection combined primary and secondary sources. Five semi-structured interviews were conducted with key stakeholders, including municipal technicians, teachers, members of the consortium, and design facilitators [Appendix A]. These interviewees were selected through purposive sampling to ensure a diversity of perspectives relevant to the institutional dimension of the experiment, though they did not constitute a statistically representative sample [61,62]. The data were analysed using a deductive–inductive coding strategy informed by the categories of the IAD framework, while remaining open to emergent themes grounded in the fieldwork.

Complementary sources included documentary analysis of strategic, regulatory, and technical materials (such as implementation projects, diagnostic reports, and municipal policy plans), participant observation during design sessions, community events and interdepartmental meetings, as well as the review of audiovisual records, including a video-manifesto, photographs, and graphic documentation of the participatory process [See Supplementary Materials]. This methodological triangulation allowed for the identification of visible changes in the physical configuration of the school environments and less tangible shifts in institutional rules, power dynamics, and organisational learning patterns.

The methodological approach is framed as collaborative management research [63], in which researchers and system actors co-produce actionable knowledge [64,65]. This mode of inquiry made it possible to generate situated and practice-oriented insights, reduce interpretation biases by engaging with diverse stakeholders, and participate directly in key moments of the experimental process. Although the study was based on a single case (two schools in Madrid), the consistency of patterns observed across actors, levels, and outputs suggests the presence of trends that may be transferable to other urban contexts while acknowledging the inherent limitations of a singular case analysis.

3.3. Materials: Case Study

The case study analysed in this research focused on the renaturalisation of two public school environments located in the neighbourhood of San Cristóbal de los Ángeles, in the district of Villaverde, Madrid. With a population of 16,886 inhabitants [66,67], San Cristóbal is considered one of the areas with the highest levels of climate vulnerability in the city [13]. It also registers the lowest average net household income in the municipality, and it presents a high proportion of residents with only basic education (28.21%), which reflects a significant risk of the consolidation of social exclusion conditions [67]. According to the Municipal Register of 2021, 35.83% of the population is of foreign origin, predominantly from Latin America, sub-Saharan Africa, and Eastern Europe [66,67]. Despite these limitations, it has a rich neighbourhood network, with many self-deployed associations and services [66,67].

Within this context, the two public schools selected—Sagunto School and Navas de Tolosa School—served as test beds for exploring innovative approaches to the design of educational spaces that are not only greener and more climate-resilient, but also socially inclusive and better integrated into the neighbourhood's urban and community dynamics. Their selection was especially pertinent given their strategic role as daily points of encounter for vulnerable children and families, their limited environmental quality, and their potential to catalyse broader processes of ecological regeneration and community empowerment at the neighbourhood scale.

The project to renaturalise school playgrounds was supported by a multi-actor ecosystem composed of several municipal departments, educational institutions, civil society organisations, academic institutions, and design facilitators, whose coordinated involvement was essential throughout the different phases of the process. The Department of Energy and Climate Change of the Madrid City Council provided institutional leadership, aligning the intervention with climate adaptation policies and securing public investment. The participating schools—Sagunto School and Navas de Tolosa School—acted as territorial anchors, mobilising their educational communities and facilitating the integration of pedagogical, social, and spatial perspectives. Local civil society organisations and cultural associations contributed their proximity to the territory and knowledge of community dynamics, helping to mediate participation and support vulnerable groups. The Innovation and Technology for Development Centre team at the Technical University of Madrid (itdUPM) played a dual role as process facilitator and research partner, designing the participatory methodology, coordinating actor engagement, and documenting institutional learning. Finally, external design and engineering teams collaborated throughout the process to translate the co-created visions into technically viable execution projects. This distributed governance structure enabled a dynamic interplay of roles and responsibilities, fostering trust, mutual learning, and shared ownership across organisations.

A multiphase methodological architecture was structured around the project to bridge technical development with civic engagement. It combined three cross stages (CSs)—active listening, evolutionary communication, and evaluation and scaling—with three iterative stages (Ss): co-creation, prototyping, and technical drafting of the implementation project (Figure 4). These stages articulated participatory urban experimentation with formal planning instruments, enabling institutional reflection, collective creativity, and situated learning. This architecture allowed for integrating pedagogical, political, and design perspectives into a cohesive process of urban transformation, positioning the schoolyards not merely as objects of design, but as platforms for testing new governance models and climate–social infrastructures [Appendix B].

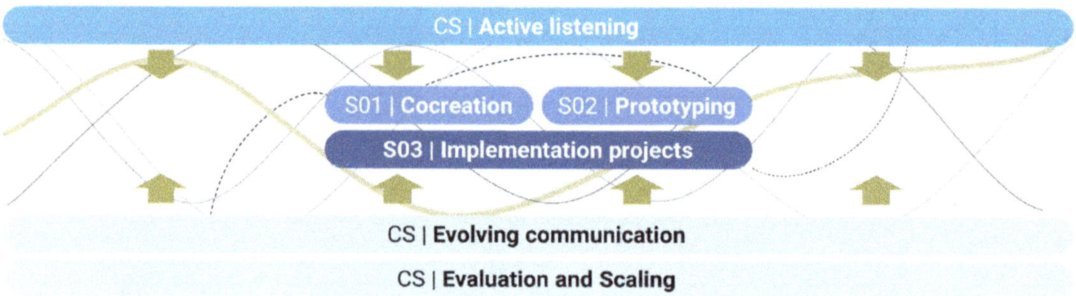

Figure 4. Stages of work implemented in the school playground project.

Each stage of the process is detailed below.

- **CS | Active listening:** From the very beginning of the project, a strategy of active listening [68] was implemented, including not only stakeholder mapping (Figure 5) but also the identification and analysing of existing projects, studies, and initiatives in the territory (Figure 6). This knowledge allowed a deeper understanding of the stakeholders in the area and their needs, tensions, and opportunities. This stage included participatory diagnostics with schools, interviews with local actors, urban tours with technical teams, and the analysis of municipal strategic documents and previous neighbourhood studies. Through this approach, spatial conflicts that were not visible were identified, such as those related to the lack of shade, inequality in the use of the playground according to gender, or the perception of insecurity in school access and environments. Active listening was key to situating the design in a relational framework, where the school space was not understood in isolation, but as part of a broader urban ecosystem. The mobilisation of the ecosystem of actors by the facilitator team also made it possible to connect the experiment with another seven projects from the public sector, academia, and civil society, establishing synergies and complementarities and generating a greater impact on the territory and better allocation of resources to each of them.
- **CS | Evolving communication:** The second cross-cutting stage consisted of designing a communication strategy as a pedagogical, political, and cultural tool that did not limit itself to informing, but accompanied the process, critically documented it, and allowed for shared learning [69]. Audiovisual materials were produced (such as the video manifesto La Asamblea de la Infancia and the documentary In Praise of the Schoolyard), collective publications, graphic records of the workshops, and educational resources adapted to different ages and audiences [Appendix C]. In addition, public events, school presentations, and technical forums were held, where progress was shared, impressions were gathered, and approaches were fine-tuned. This evolv-

ing communication legitimised the process, broadened its scope, and strengthened community ownership.

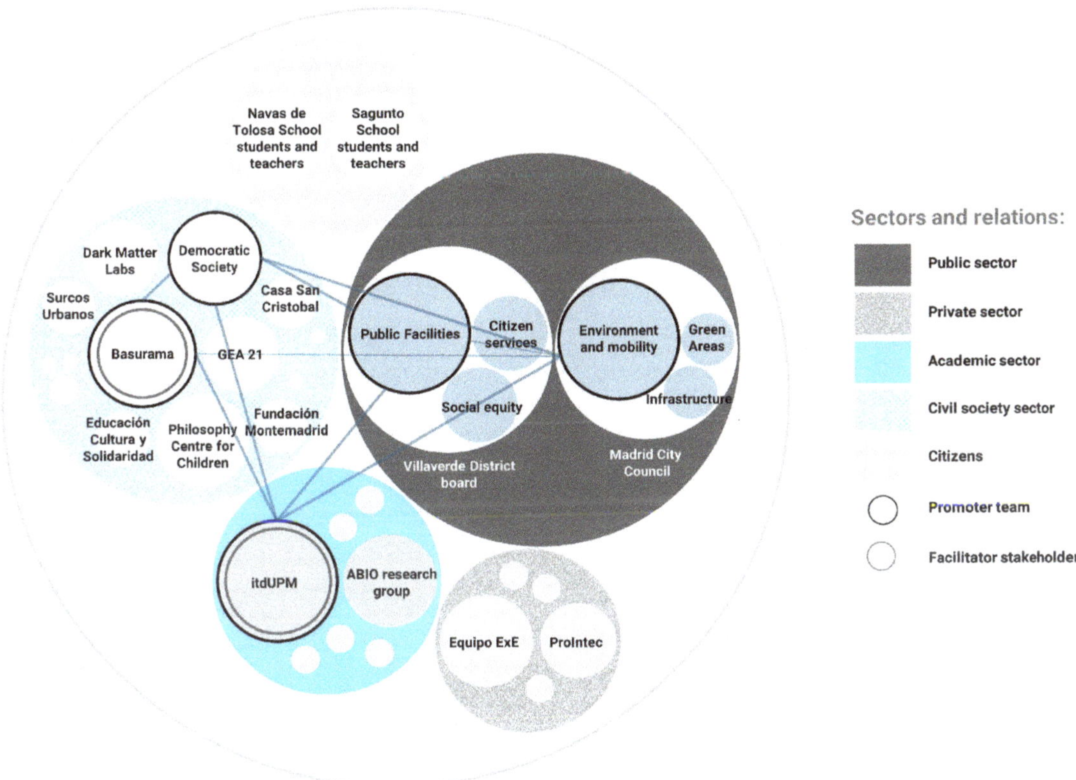

Figure 5. Mapping of the ecosystem of actors involved in the experiment.

- **CS | Evaluation and scaling up:** The third cross-cutting stage consisted of the evolutive evaluation of the learning generated, with the aim of systematising methodologies, identifying institutional barriers, and exploring possibilities for replicability. Unlike traditional evaluations, which are usually carried out only at the final stages, this approach proposed a dynamic and adaptive evaluation throughout the project's life cycle [70]. The evaluation was participatory and multi-scale: it included sessions with students, interviews with municipal technicians, meetings with school management teams, and reflective analysis by the research team. Moreover, the evaluation process extended beyond the project's formal conclusion by establishing a collective monitoring mechanism. The post-execution monitoring aimed to assess the resilience and social appropriation of the interventions over time, incorporating feedback from school communities, neighbourhood organisations, and municipal services. Regular dialogues were initiated with other educational centres interested in replicating the approach, facilitating peer learning and promoting the diffusion of the methodological framework to new contexts. Based on the findings generated throughout this stage, synthesis reports, policy recommendations, and proposals for integrating the approach into municipal strategies were produced. These included intervention guidelines for school environments and playgrounds with climate change adaptation criteria and new technical specifications for playground maintenance.

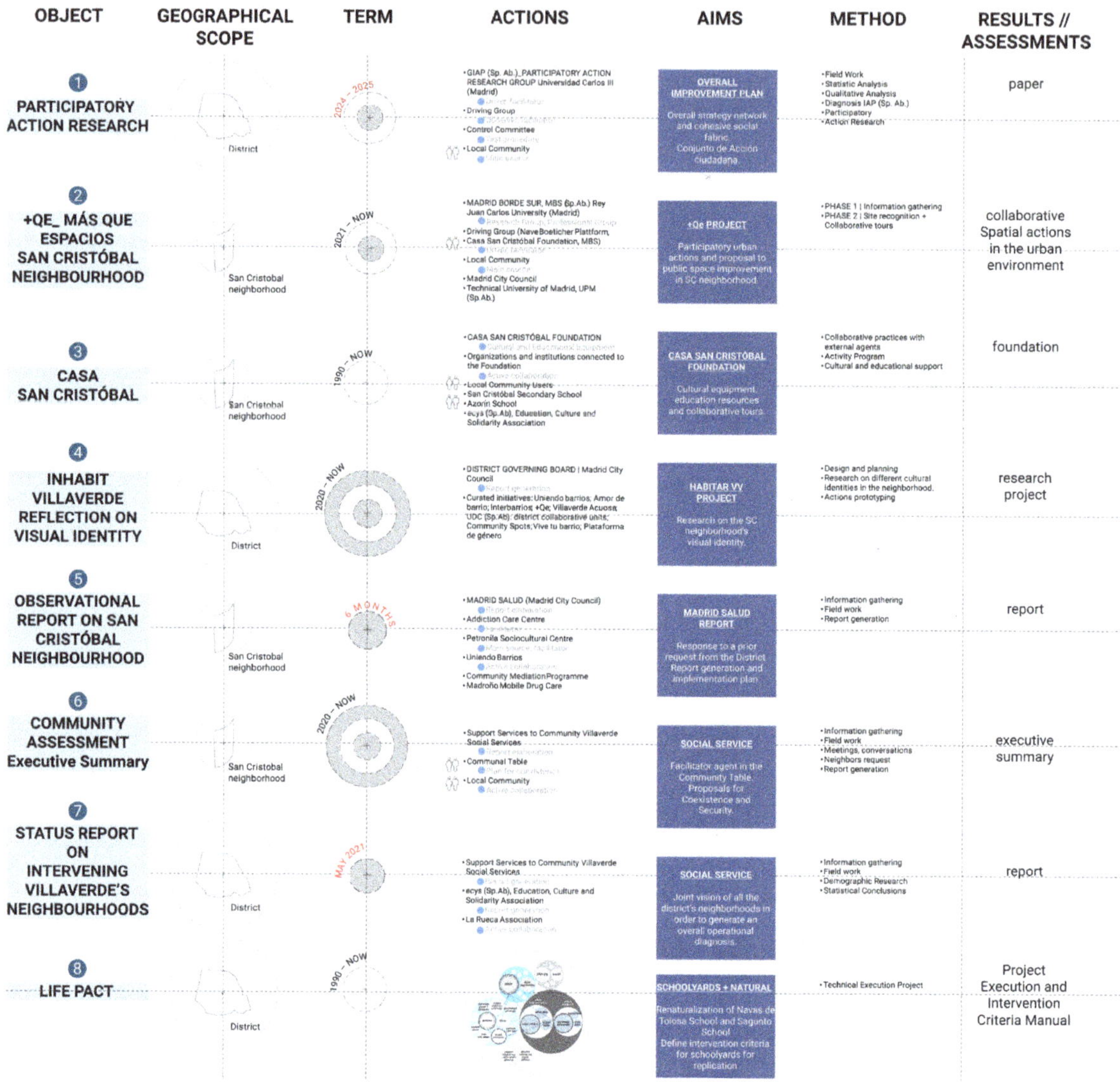

Figure 6. Mapping of existing projects, studies, and initiatives in the territory.

- **S01 | Co-creation:** Co-creation stood at the heart of the process, serving as both a methodological and political device that fostered the participation of diverse actors to jointly generate value through equality, transparency, dialogue, and trust among participants [71,72]. According to Trischler [73], co-creation transforms the traditional supplier–customer relationship, turning the citizen into an end user and a designer. Through more than 24 sessions with students from both schools, creative writing, ecological imagination, and participatory design strategies were deployed. Dynamics included games, emotional mapping, commented tours, and dramatisation of desired futures. Exercises such as the 'Patio Portrait' or 'Utopia is Possible' allowed for construction of a collective narrative of the school space and its shortcomings and potential (Figure 7). Co-creation was also extended to teachers, families, social organisations in the neighbourhood, and municipal technicians, creating a multistakeholder and multidisciplinary meeting space.

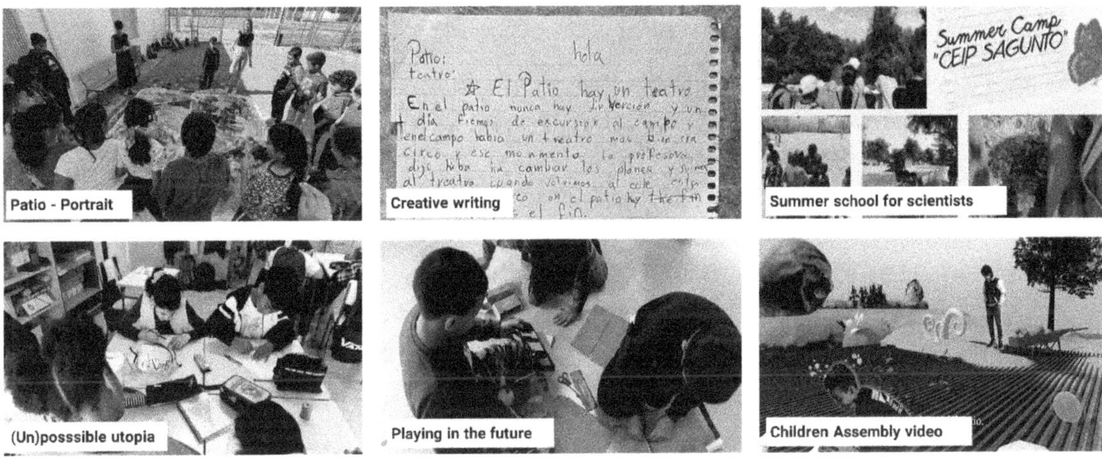

Figure 7. Co-creation and prototyping sessions of the experiment.

- **S02 | Prototyping:** Based on the results of the co-creation stage, a prototyping stage was carried out, in which spatial, material, and functional proposals were specified. Ideas were represented through ephemeral installations and uses were tested in the schoolyard and the immediate environment. This people-centred approach ensures that the prototypes provide contextually relevant and culturally appropriate solutions [74]. This phase allowed for the validation of solutions with criteria of climatic comfort, diversity of uses, accessibility, and identity. In addition, prototyping facilitated the conversation between technical and non-technical languages by translating children's imaginings into formats understandable to architectural teams and public procurement officers (Figure 7).
- **S03 | Drafting of technical projects:** The last stage consisted of the drafting of the technical execution project based on the synthesis of the learning generated. The final design incorporated proposals such as the replacement of impermeable floors with draining paving, the planting of adapted trees, the creation of shaded areas, the installation of fountains and inclusive furniture, and the improvement of accessibility. The school entrances were also redefined and the playgrounds were connected to the adjacent school squares, creating a climate infrastructure on a neighbourhood scale. The municipal teams validated these technical projects and integrated them into the City Council's investment plans, leveraging close to EUR 5 million of public investment for their implementation. [Appendix D; Appendix E]

Finally, based on interviews conducted with local stakeholders during the evaluation process and the systematisation report, the project generated significant impact in both environmental and social respects. The intervention covered more than 30,000 m^2, improving thermal comfort for students and reducing surface temperatures in schoolyards. Over 130 trees were planted, increasing local biodiversity and creating new shaded areas. The participatory process engaged more than 600 individuals on the social front, including students, teachers, families, municipal staff, and neighbourhood organisations. Finally, the lessons learned and the criteria developed for interventions in school environments were systematised and compiled in the Patios Escolares + Naturales guide published by Madrid City Council, providing technical guidelines to support the potential replication of the model in other urban contexts of the city.

4. Results

The results of applying the IAD framework to the case study are presented below. Figure 8 summarises the analysis of the action situation [23]. For this research, the traditional process is referred to as the 'current situation' and the collaborative planning as 'the experiment'. The current situation refers to the traditional process of designing school environments, characterised by a hierarchical, expert-led approach with limited participation from the educational community and a focus on technical compliance. The experiment denotes collaboratively redesigning and renaturalising two schoolyards through active listening, co-creation, participatory prototyping, and collective monitoring, integrating children's, teachers', and community members' knowledge and perspectives into public decision-making. The comparative analysis reconfigures the conditions that structure collective action and public decisions in the urban realm.

Firstly, there is a substantial broadening of the range of participants in the experiment (actors/stakeholders in Figure 8). While in traditional procedures, the relevant actors are mainly municipal technicians, policymakers, and contractors, the experiment actively involved students, teachers, neighbourhood associations, art collectives, and urban facilitators. More than 600 people and 33 organisations from different sectors participated. The mobilisation of the ecosystem of actors also enabled the experiment to establish connections with seven other projects from the public sector, academia, and civil society, fostering synergies and complementarities that amplified the territorial impact and optimised the allocation of resources across initiatives. This articulation was not merely nominal, but involved a substantive redistribution of agency and decision-making power over the space, reconfiguring the institutional ecology involved in planning and managing the school environment.

The configuration of roles also underwent significant changes. The traditional process is characterised by a hierarchical and rigid distribution of roles between the public sector (municipal technicians) and the private sector (architects, designers, contractors), excluding the space's direct users. In contrast, the experiment allowed the emergence of new positions: children were recognised as valid design agents; teachers adopted a mediation role between the pedagogical and the spatial dimensions; and the experiment's facilitators played key roles in translating technical and community languages. The institutional action situation was redefined by acknowledging students as users and co-authors of the intervention. This redistribution of the field of positions opened up new possibilities for deliberation and co-responsibility. Students' contributions shaped both spatial proposals and normative criteria, evidencing their ability to influence decision-making arenas. From identifying conflicts such as gender-based use inequalities to suggesting alternative uses and aesthetic preferences, their inputs introduced affective, symbolic, and experiential dimensions that broadened the planning rationale beyond technical–functional considerations.

The rules of choice also changed in the experiment. In traditional processes, technical regulations strictly delimit possible actions without room for contextual adaptation. The experiment introduced dynamics of iterative deliberation, where decisions were progressively adjusted according to collective learning, evidence, and the imaginings projected by the children and the community. Children's imaginings functioned as a basis for evaluating design options, modifying the game's rules. Although no formal changes in regulations were established, the municipal technicians could technically and legally justify the proposals in the different co-design spaces by relying on interpretations of existing norms. An example is the inclusion of water games in the school playground, the first to be installed in a public school playground in Madrid.

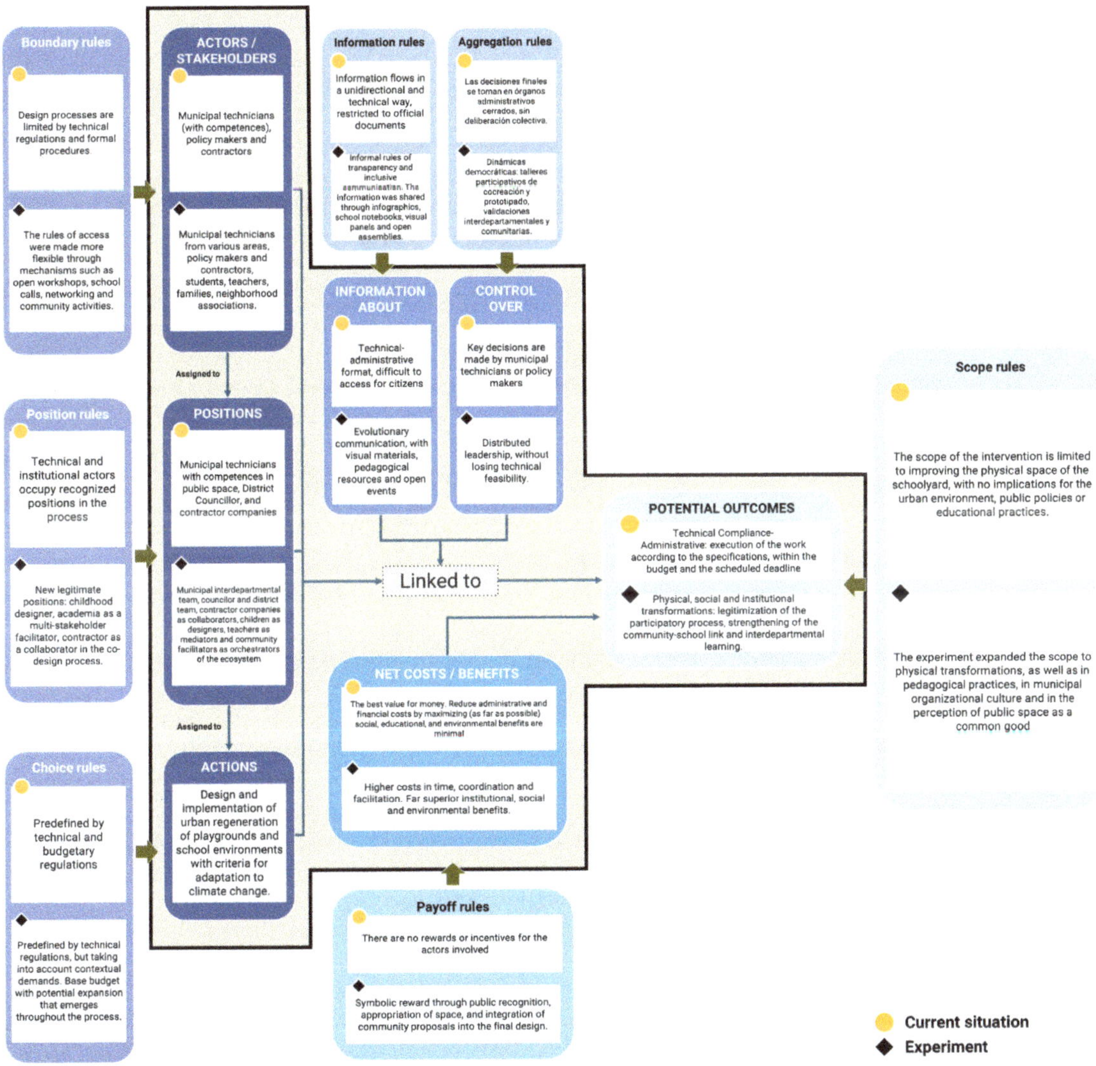

Figure 8. Analysis of the action situation using the IAD framework.

Another central issue was the level of information available during the process. Standard procedures restrict access to information, concentrating it in technical hands and reducing circulation to administrative documents of difficult non-technical interpretation. In the experiment, evolving communication strategies were developed that included visual materials, pedagogical tools, public events, and resources adapted to different ages and audiences. This pluralistic information circulation facilitated understanding and ownership of the process, informed participation, and greater institutional transparency.

Decision-making control was also altered. While this control rests almost exclusively with technical and political bodies in traditional processes, shared decision-making mechanisms were enabled in the experiment. Co-creation workshops, community validations, and participatory prototypes allowed for redistributing decision-making power, transforming urban design into an open, deliberative space. This openness did not eliminate existing institutional frameworks, but made them more flexible to incorporate other rationalities and sensibilities.

Regarding positional rules, the experiment shows how redefining the informal normative frameworks that determine who has the right to intervene in urban design is possible. The institutional recognition of children and the community as legitimate actors breaks a long tradition of technocratic planning. Likewise, the contractor moves from a unilateral transactional relationship (subcontracting) to an active agent in the collaborative design process, translating the ideas and imaginings of the participants into viable technical solutions. The facilitating team reconfigured the boundary rules by orchestrating the progressive entry and exit of stakeholders into the experiment, thus expanding the boundaries of the planning process. Aggregation rules, which in traditional procedures usually delegate decision-making to closed instances, were replaced by more horizontal aggregation mechanisms in the experiment. Consensus building took place through collective validations, interdepartmental agreements, and collaborative design sessions, where the final decision resulted from integrating multiple perspectives. In addition, reporting rules were transformed by adopting more accessible, inclusive, and visual formats, which democratised access to technical knowledge and reduced asymmetries between actors. Reward rules, traditionally non-existent in conventional procedures, were also modified. The participation of the network of actors in the experiment generated symbolic (recognition, appropriation of space) and organisational (institutional legitimacy, internal learning, public visibility) incentives, helping to sustain collective commitment throughout the process. These incentives were validated in the project's dissemination sessions, and video documentaries on the urban transformation were developed [Appendix D].

The traditional model is limited to the physical execution of the work in terms of the intervention's scope. Thanks to the collaborative planning and the direct involvement of municipal technicians and political agents, the experiment expanded the scope of the intervention from 8000 to almost 30,000 square meters, leveraging EUR 5 million of public resources for its implementation. In this sense, the experiment broadened the scope towards physical transformations and pedagogical practices in the municipal organisational culture and the perception of public space as a common good.

Although the IAD framework does not formally include the evaluation of results as one of its core analytical components, this study incorporates it as a complementary lens to capture emergent outcomes related to learning, legitimacy, and social impact. Important differences emerged between the traditional process and the experiment in this regard. In the conventional model, evaluation is primarily based on quantitative technical indicators, focusing on compliance with the project specifications, adherence to deadlines, and budgetary efficiency. In contrast, the experiment incorporated a participatory, qualitative, and multi-criteria evaluation approach. Beyond the physical execution, it assessed dimensions such as climatic comfort, children's ownership of the space, symbolic impact, quality of the collaborative process, and the potential for replicability. This shift broadened the scope of what was considered an outcome. While the traditional approach narrowly measured success in technical terms, the experiment recognised social, pedagogical, and institutional outcomes as equally relevant.

In terms of costs and benefits, although the experiment required greater investments of time, coordination, and institutional flexibility, it generated significant qualitative returns: it strengthened the social fabric, enhanced the legitimacy of participants within institutional frameworks, promoted organisational learning, and improved both the climatic and educational environment of the schoolyards. Particularly in high-vulnerability contexts such as San Cristóbal, this approach's medium- and long-term benefits outweighed the initial investment, demonstrating the value of integrating broader criteria into evaluating public interventions.

5. Discussion

Applying the IAD framework to this case study allowed reflection on how collaborative urban planning processes can be transformed through situated experimentation. This section discusses the study's findings considering four significant challenges: (i) the need to adapt institutional arrangements to changing contexts, (ii) the importance of developing collaborative governance schemes, (iii) the intrinsic tensions and contradictions in the rules in use, and (iv) the generation of institutional capacities to innovate operational practices.

5.1. Institutional Arrangements in Complex Contexts

One of the key findings of the case study is its demonstrated capacity to operationalise the adaptation of institutional arrangements to the specific conditions of complex neighbourhoods characterised by intersecting social, urban, and climatic vulnerabilities. In these contexts, the rigidity of current administrative and regulatory frameworks in responding to the territory's needs is evident. Urban planning operates with standardised solutions, hierarchical procedures, and closed specifications that limit the possibility of integrating situated knowledge, community proposals, or adaptations during the project. In contrast to the traditional model, the experiment introduced a working logic based on multistakeholder collaboration, enabling greater flexibility in operational frameworks without departing from existing institutional structures. An iterative process was designed to combine cyclical stages (co-creation, prototyping, technical drafting) with transversal stages (active listening, evolutionary communication, evaluation, and scaling). This methodological architecture progressively incorporated new rules, actors, and deliberation mechanisms within the institutional framework, fostering more adaptive and context-sensitive forms of public intervention.

The observed flexibility did not come from formal normative reforms, but from a practical reinterpretation of the game's rules, which enabled new forms of action without modifying the current legal framework. This reinterpretation was made possible thanks to interdepartmental collaboration, political will, and the mediation of the facilitating team, which translated the technical language into understandable and adaptable formats. In this sense, the experiment responds to Ostrom's [22] call to build governance frameworks capable of adapting to changing local conditions through organisational learning. A crucial factor in this flexibility was the inclusion of students' perspectives, which grounded design and planning procedures in everyday school life and tangible user experiences. This factor is particularly relevant for designing public policies that address complex problems such as climate change adaptation and social inclusion. This operational flexibility is one of the keys to designing public policies that are technically feasible, socially relevant, and territorially adjustable.

5.2. Collaborative Governance: Articulation Across Scales and Actors

In contrast to the dominant top-down approach in urban planning, the experiment promoted a distributed and multi-actor leadership, articulated across different scales (school, neighbourhood, district, and city) and sectors (technical, educational, community, and academic). This polycentric approach did not imply the elimination of formal frameworks of competence, but rather their reorganisation to allow for more horizontal collaboration among stakeholders. The process mobilised a network of more than 600 individuals and 33 organisations who participated throughout the different stages of the project.

This broad and sustained collaboration was not spontaneous: it was enabled by the explicit agreement on a set of action rules, as analysed in Figure 7 through the IAD framework. The definition of payoff rules, control rules, boundary rules, and aggregation rules created a clear operational framework for communication, coordination, and decision-making, which proved fundamental to sustaining the experiment over time and across organisational boundaries. In particular, the aggregation rules played a crucial role by establishing deliberative mechanisms, such as collective validations, participatory prototyping sessions, and interdepartmental agreements, that replaced closed decision-making processes with open and negotiated ones. This approach increased transparency and ensured that diverse perspectives were effectively integrated into the outcomes. Children's participation in these deliberative arenas was not only symbolic: their viewpoints reshaped spatial priorities and challenged adult-centric planning logics.

By not depending on a single actor or administrative unit, the experiment demonstrated a capacity for resilience, sustaining itself despite changes in political agendas, staff turnover, or shifting contextual conditions. This resilience is consistent with McGinnis [25] and Ostrom [24], who argue that distributed systems enhance organisations' ability to adapt, learn, and respond to complex and dynamic environments. Similarly, the fact that the experiment was embedded within a European project with a four-year horizon provided a crucial structural advantage: it created spaces for reflection, iterative learning, and distributed decision-making beyond the typical financial cycles of public institutions—something that would have been very difficult to achieve under standard administrative conditions.

5.3. Replicability of the Model and Public Capacities to Innovate from Situated Action

Unlike many pilot projects encapsulated in their singularity, the case study was conceived as an open, documented, and transferable model designed for scalability. This was made possible through the continuous systematisation of the process, the development of pedagogical tools, the elaboration of methodological and training guides, and the active involvement of municipal technicians with competencies in urban planning and regeneration. Furthermore, all materials produced during the experiment—including methodological frameworks, educational resources, and communication outputs—were disseminated under Creative Commons licences, reinforcing the project's commitment to openness and knowledge sharing and facilitating its adaptation to other urban contexts.

The replicability of the model resides not only in the physical redesign of schoolyards but also in the institutional architecture that enabled their transformation. The stages of listening, co-creation, prototyping, and evaluation—together with the set of action rules agreed upon by managers and users—constitute a processual framework that can be adapted to diverse urban contexts, provided that its underlying principles of situated participation, iterative experimentation, and intersectoral collaboration are maintained. This orientation coincides with the proposals of Sabel and Zeitlin [75], who underline that practical institutional innovation does not come from fixed recipes, but from adaptive frameworks that learn from action. In this regard, students' feedback loops were essential for refining implementation strategies and aligning interventions with actual needs.

In addition to its replicability potential, the project generated significant organisational learning within the public administration. It enabled the implementation of new forms of interdepartmental collaboration, the revision of public procurement procedures, the incorporation of climate and social criteria into technical specifications, and the redefinition of the role of municipal teams as mediators rather than mere executors. These lessons have already been reflected in broader initiatives, such as the municipal programme for the ecological regeneration of school environments. Furthermore, the strategies, methodologies, and measures developed through the experiment are progressively being scaled across the city. Madrid has established the renaturalisation of schoolyards and their surroundings with climate adaptation criteria as a strategic priority. This commitment has been formally embedded in Madrid's Climate City Contract under the European Union's Mission framework for 100 climate-neutral and smart cities by 2030 [9]. This institutional anchoring illustrates how collaborative planning, when systematically documented, evaluated, and shared, can transcend its initial scope and shape long-term urban policy at a metropolitan scale.

The institutional transformation was not planned ex ante, but emerged due to the process. This learning by doing constitutes one of the most important contributions of the experiment. As Karvonen et al. [76] have pointed out, cities need technological innovation or advanced design and institutional frameworks that allow testing, correcting, scaling, and learning. Although the case study results are encouraging, their consolidation as a model of urban intervention requires overcoming several challenges. Among them is the need to institutionalise the learning generated and adapting the regulatory and budgetary frameworks so that experimentation does not depend solely on calls for international funds, exceptional financing, or individual wills. Developing institutional capacities—particularly within local governments—is crucial to managing these processes' complexity, conflict, and uncertainty [14,26,27]. Advancing towards a model of innovative and adaptive urban management requires articulating three key dimensions: (i) the technical, ensuring the environmental quality of interventions; (ii) the political, enabling open and equitable decision-making spaces; and (iii) the deliberative, fostering critical engagement with the territory and institutional actions.

However, its replicability does not rely solely on the methodological instruments developed, but rather on the presence of key enabling conditions: (i) a strong and active community fabric capable of sustaining trust-based relationships; (ii) institutional legitimacy to engage with the ecosystem of actors, along with the administrative competencies required to implement neighbourhood-scale transformations; and (iii) dedicated financial resources to support the facilitation of the entire collaborative planning process.

5.4. Tensions and Contradictions in the Rules in Use

While the experiment enabled more inclusive and flexible planning processes, it also revealed key tensions between co-produced dynamics and existing formal rules. A prominent example concerns public procurement. Although co-creation workshops generated innovative design proposals rooted in children's and community members' inputs, these ideas had to be reconciled with legal and technical standards of public tendering. The challenge lay in translating collectively imagined solutions into specifications that could pass legal scrutiny while preserving their original intent.

Similarly, maintenance responsibilities became a friction point. The informal agreements and spontaneous uses generated through participatory prototyping often exceeded what existing service contracts or maintenance protocols could accommodate. This friction created a mismatch between the intervention's adaptive nature and the rigidity of municipal maintenance routines. In some cases, municipal departments hesitated to assume responsibility for non-standard elements, such as artistic installations or ephemeral interventions. These tensions did not result in outright conflicts, but required continuous negotiation. Facilitating teams played a key role in mediating between institutional imperatives and the experimental ethos, often serving as translators between formal frameworks and informal practices. From the IAD framework perspective, these dynamics reflect frictions between 'rules in form' and 'rules in use'—with the latter evolving through situated action and collective learning.

Recognising these tensions is crucial for institutionalising similar interventions. Rather than viewing them as barriers, they should be understood as productive contradictions that reveal the limits of current governance structures and the need for reform. In this sense, the experiment acted as a testing ground not only for spatial designs but also for institutional resilience itself.

From the IAD framework, the introduction of collaborative processes transforms action situations by redefining roles, rules, and decision-making criteria while broadening the range of actors involved and the types of knowledge mobilised. Its institutional value lies in challenging established routines and fostering new forms of governance that are more flexible, inclusive, and conducive to organisational learning. This approach acts as a field of collective learning, where both the community and administrations develop capacities to manage uncertainty and respond to complex contexts. Children's involvement exemplifies how non-traditional actors can contribute meaningfully to governance innovation, primarily when mechanisms exist to translate their input into planning criteria. However, as Karvonen et al. [76] warn, its transformative potential depends on its success in influencing institutional structures and not being limited to pilot dynamics without continuity. In this sense, collaborative planning should be understood as a mechanism of progressive institutional innovation [75] whose impact lies in its integration into public logic and its capacity to reformulate the state's role as facilitator, mediator, and learner.

While the experiment illustrates the potential of collaborative planning to generate institutional innovation, it is important to acknowledge the risks and tensions highlighted by the literature critically. Scholars such as Savini [55] and Swyngedouw [11,35] have warned that collaborative approaches may be co-opted by dominant actors, depoliticise planning decisions, or reproduce existing power asymmetries under the guise of consensus. In this case, efforts were made to counter these risks through active facilitation, open-ended deliberation formats, and the deliberate inclusion of non-traditional actors in the co-design process. Nevertheless, the possibility of asymmetries in influence or institutional inertia should not be underestimated. Ensuring collaborative planning processes remain genuinely democratic requires continuous reflexivity, transparency in governance mechanisms, and the capacity to confront conflict rather than neutralise it. Recognising these challenges is essential for evaluating the outcomes of this experiment and guiding the design of future participatory interventions.

6. Conclusions

This case study, analysed through the lens of the IAD framework, demonstrates that incorporating collaborative planning into the design of schoolyards can trigger profound transformations in the rules in use, the distribution of roles, and the structure of decision-making within public institutions. These transformations respond directly to the research question, showing that collaborative planning processes and multistakeholder networks reshape institutional arrangements by introducing more flexible, inclusive, and context-sensitive modes of governance. The study demonstrates how these transformations reconfigure the institutional 'action situation', shifting the centre of gravity of planning from a hierarchical and technical logic towards more adaptive and situated forms of urban governance. This reconfiguration manifested in the redistribution of decision-making power between technical, educational and community actors; in the relaxation of previously rigid rules defining who could participate and with what kind of knowledge; and in the emergence of deliberative dynamics in which decisions were the product of dialogue rather than imposition. In this new framework for action, traditional urban planning ceased to be a closed technical exercise. It became a shared process of institutional construction in which the objectives, means, and criteria for success were co-defined by a multiplicity of actors with diverse interests, incentives, and knowledge.

By integrating its different stages, the case study articulated a multistakeholder and multilevel ecosystem that improved the physical results of the intervention and contributed to organisational learning, normative innovation from practice, and the institutionalisation of more democratic methodologies. The new emerging rules broadened the base of actors involved, diversified the knowledge considered valid, and redefined the criteria by which the results of a public intervention are evaluated. These institutional changes reveal the transformative potential of collaborative practices for advancing inclusive climate adaptation strategies. This institutional innovation, anchored in situated action and the recognition of the specific conditions of territories, offers relevant clues for designing urban public policies capable of responding to complex challenges, such as adapting to climate change, reducing territorial inequalities, and strengthening social cohesion. Rather than replicating closed models, the case study highlights the capacity of local governments to experiment, adapt, and learn by doing.

From a theoretical perspective, this study substantially contributes to the IAD framework and collaborative urban planning theory. First, it extends the application of the IAD framework to the field of participatory urban design by showing that institutional 'action situations' can be restructured through practice-based collaborative processes. These findings provide nuance to classical institutional theory, demonstrating that institutional innovation can occur progressively without formal regulatory change. Second, the study enriches collaborative planning theory by offering an analytical lens to understand its effects: empirical findings show that active involvement of schools and local actors in planning enhances spatial outcomes and reshapes power structures and rules within public institutions. It challenges the traditional consultative view of participation and repositions collaborative planning as a driver of institutional transformation aligned with polycentric governance principles.

The research also presents certain limitations. First, it focused on a limited number of case studies (two schoolyards), which restricts the generalizability of the findings. However, the consistency of institutional transformations observed in both sites provides valid and solid grounds to extract conclusions applicable to other contexts facing similar challenges. Although transferable elements have been identified, their application in other environments will require adaptation and contextual analysis. The model's scalability will depend on each territory's ability to ensure three key enabling conditions: an active community fabric, institutional capacity to coordinate and execute interventions, and targeted funding to support deliberative and collaborative planning processes. In this regard, further studies should explore mechanisms for institutional scalability of such experimental processes, identifying which factors enable their stable incorporation into public policy, how rules in use evolve over time, and what institutional capacities are needed to sustain inclusive governance models. To this end, we propose a set of concrete and actionable policy recommendations across governance levels: locally, municipalities should embed collaborative planning practices into schoolyard transformation programmes, ensuring the early involvement of schools and communities and allocating specific resources for co-production; nationally, education and urban policies should support these approaches with dedicated funding, regulatory flexibility, and technical guidance; and at the supranational level, entities like the European Union should promote these practices within urban resilience agendas, encouraging replication through targeted funding and knowledge exchange platforms.

This study also offers an operational insight: cities are not merely objects of planning. Nevertheless, they can act as agents of institutional change, capable of redefining their own rules based on lived experience and collective imagination. In a scenario of climate, social, and urban uncertainty, this approach is essential to building governance frameworks that are more resilient, inclusive, and transformative.

Supplementary Materials: The following supporting information can be downloaded at https://drive.google.com/drive/folders/1DdPrAuH40Apw9hQRzAeKDw_aivvzVTPq?usp=drive_link (Reviewed 26 April 2025).

Author Contributions: Conceptualisation, M.A. and S.R.-M.; methodology, M.A. and S.R.-M.; validation, M.A.; formal analysis, M.A.; investigation, M.A.; resources, M.A.; data curation, M.A.; writing—original draft preparation, M.A.; writing—review and editing, M.A. and S.R.-M.; visualisation, M.A.; supervision, S.R.-M.; project administration, M.A.; funding acquisition, M.A. All authors have read and agreed to the published version of the manuscript.

Funding: This research is part of the European project 'People-Driven: Adapting Cities for Tomorrow'—LIFE PACT (LIFE20CCA/BE/001710), which is funded by the European Union's LIFE Programme.

Data Availability Statement: Data will be made available on request.

Acknowledgments: The authors would like to acknowledge the daily work of the Madrid City Council staff: Juan Azcárate, Marisol Mena, Luis Tejero, and Irene García, from the Subdirección General de Cambio Climático y Energía, Área de Gobierno de Medio Ambiente y Movilidad. Thanks to the Universidad Politécnica de Madrid team working on the experiment: Nieves Mestre and Irene Ezquerra. Thanks to the team working on the European project: Juan López-Aranguren (Democratic Society), Baptist Vlaeminck (Leuven City Council), and Alicia Carvajal (Dark Matter Labs).

Conflicts of Interest: The authors declare that they have no conflicts of interest. The funders were not involved in the study design, data collection, analysis, and interpretation, the manuscript writing, or the decision to publish the results.

Appendix A. List of Interviewees

Sector	Organization	Role
Public Sector	**Madrid City Council**	Project coordinator in the Deputy direction of Energy and Climate Change, Urban Planning Government Area
	Villaverde District board	District coordinator
Academic sector	**Universidad Politécnica de Madrid**	Technical coordinator of the European LIFE PACT project in Madrid
	Navas de Tolosa School	Director of the educational center
Private sector	**Basurama Collective**	Architects responsible for the co-design process and drafting of the construction execution project.

Appendix B. Methodological Architecture of the Urban Experiment

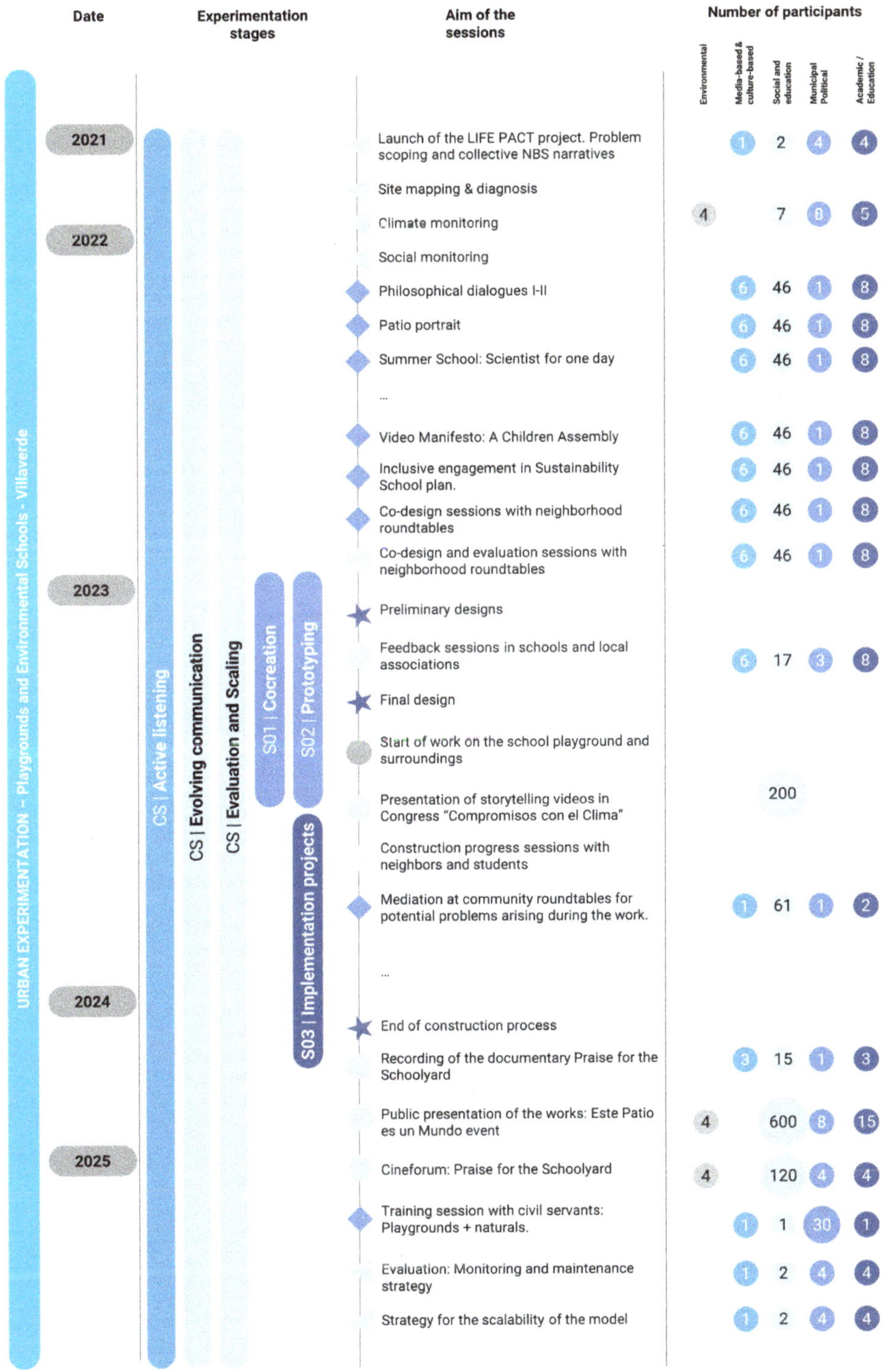

Appendix C. List of Dissemination Activities (Videos)

Tittle	Description	Link
This Playground is a World event	A reflection event and launch of climate micro-missions in schoolyards, based on the ClimaX Guide methodology	https://youtu.be/inGnOtlTIP8 (accessed on 26 May 2025)
Documentary In Praise of the Schoolyard	Interviews with neighbours, students, and school administrators following the intervention in the playground at the Navas de Tolosa Primary School	https://youtu.be/_j5ExJE_Nig?si=TO0dfY0CiL4f0-EQ (accessed on 26 May 2025)
LIFE PACT Stories—Madrid (Video 1)	Collaboration with public, academic, and community stakeholders in the co-creation of school spaces	https://www.youtube.com/watch?v=8KQKQemEUWk (accessed on 26 May 2025)
LIFE PACT Stories—Madrid (Video 2)	Generating collective narratives on climate adaptation through participatory design	https://www.youtube.com/watch?v=yHrsP4MdJQg (accessed on 26 May 2025)
LIFE PACT Stories—Madrid (Video 3)	Sensory portrait of the San Cristóbal neighbourhood, connecting biodiversity and urban ecosystems	https://www.youtube.com/watch?v=rW6JJQznMzw (accessed on 26 May 2025)
Children's Assembly	Participatory space for children to propose school transformations through art and philosophy	https://youtu.be/17KxmiyV1aY (accessed on 26 May 2025)
Compromisos por el Clima 2022	School Environments Micro-Mission: Importance of intervening in school environments to improve air quality and climate adaptation	https://www.youtube.com/live/WqpKKRu8Dfw?t=12692s (accessed on 26 May 2025)
Compromisos por el Clima 2023	School Environments Micro-Mission (Pilots): Presentation of progress and lessons learned from the nine school environment transformation pilots in Madrid	https://www.youtube.com/live/CJ1fMgrVlCs?si=KOxkDiPjq3JkhUSJ&t=2627 (accessed on 26 May 2025)
Compromisos por el Clima 2023	A Living Room for Sancris Impact of Co-Design at Navas de Tolosa Primary School as a Driver of Social and Urban Revitalisation in San Cristóbal	https://www.youtube.com/live/CJ1fMgrVlCs?t=11289s (accessed on 26 May 2025)

Appendix D. List of Design Recommendations Grouped by Typology and School

Recommendation Typology	Sagunto School	Navas de Tolosa School
Improvement of the Physical Space	Introduce topographical variations to alleviate the flatness of the plaza | Replace sections of the sand floor with more compact materials | Remove fences enclosing the playground.	Remove the interior fence dividing the schoolyard | Change the types of paving to create alternative uses for the spaces | Increase the variety of paving to differentiate areas of the playground | Remove the perimeter wall, making it more permeable and connecting it to the neighboring park. Demolish an unused building to create a meeting place serving the adjacent public facilities
Recreation and Leisure Areas	Include a distinctive element in the playground adapted to a wider age range. Maintain seating areas that facilitate social interactions. | Incorporate a centralized area for group activities (music, theater, etc.).	Expand the playground and separate it with natural elements. | Create more recreational areas in the yards for various recreational activities.
Sustainability and Nature	Increase the presence of vegetation and trees to generate shade and reduce temperature.	Increase shade with pergolas and adjustable awnings. | Maintain the school garden area. | Increase vegetation to create more pleasant spaces, promote biodiversity, and connect with nature.
Connection and Accessibility	Reduce the number of barriers and abrupt elevation changes to ensure smooth traffic flow.	Strengthen connections with neighboring facilities. | Create a connection with the neighboring park, improving accessibility from the school

Appendix E. Before and After Images of the Intervention at Navas de Tolosa School

References

1. Almestar, M.; Mestre, N.; Romero, S. *The Wild City: Collaborative Practices in Urban Renaturing*, 1st ed.; UPM Press: Madrid, Spain, 2024.
2. Baró, F.; Camacho, D.A.; Del Pulgar, C.P.; Triguero-Mas, M.; Anguelovski, I. School greening: Right or privilege? Examining urban nature within and around primary schools through an equity lens. *Landsc. Urban Plan.* **2021**, *208*, 104019. [CrossRef]
3. Scott, B.A.; Amel, E.L.; Manning, C.M. *Psychology for Sustainability*; Routledge: London, UK, 2021.
4. Moreno, C. *La Revolución de la Proximidad: De la «Ciudad-Mundo» a la «Ciudad de los Quince Minutos»*; Comercial Grupo Anaya: Madrid, Spain, 2023.
5. Healey, P. *Collaborative Planning: Shaping Places in Fragmented Societies*; Macmillan Press Ltd.: London, UK, 1997. [CrossRef]
6. Jacobs, J. *The Death and Life of Great American Cities*; Random House: New York, NY, USA, 1961; Volume 21, pp. 13–25.
7. Plan de Calidad del Aire y Cambio Climático. Plan A—Portal de transparencia del Ayuntamiento de Madrid (15 January 2025). Madrid City Council. Recuperado 26 de Abril de 2025. Available online: https://transparencia.madrid.es/portales/transparencia/es/Transparencia-por-sectores/Medio-ambiente/Aire/Plan-de-calidad-del-aire-y-cambio-climatico-Plan-A-2017-2020/?vgnextoid=fab664457127f510VgnVCM1000001d4a900aRCRD&vgnextchannel=33d9508929a56510VgnVCM1000008a4a900aRCRD (accessed on 26 May 2025).
8. Madrid City Council. Roadmap to Climate Neutrality by 2050 of the City of Madrid. 2022. Available online: https://www.madrid.es/UnidadesDescentralizadas/Sostenibilidad/EspeInf/EnergiayCC/06Divulgaci%C3%B3n/6cDocumentacion/6cNHRNeutral/Ficheros/RoadmapENG2022.pdf (accessed on 26 May 2025).
9. Madrid City Council. Madrid Climate City Contract. 2024. Available online: https://netzerocities.app/resource-4063 (accessed on 26 May 2025).
10. Miraftab, F. Insurgent planning: Situating radical planning in the Global South. *Plan. Theory* **2009**, *8*, 32–50. [CrossRef]
11. Swyngedouw, E. Governance innovation and the citizen: The Janus face of governance-beyond-the-state. *Urban Stud.* **2005**, *42*, 1991–2006. [CrossRef]
12. European Commission. LIFE 3.0—Life Pact People-Driven: Adaptive Cities for Tomorrow. (LIFE20 CCA/BE/001710). 2024. Available online: https://webgate.ec.europa.eu/life/publicWebsite/project/LIFE20-CCA-BE-001710/people-driven-adapting-cities-for-tomorrow (accessed on 26 May 2025).
13. Tapia, C.; Abajo, B.; Feliu, E.; Fernández, J.G.; Padró, A.; Castaño, J. *Análisis de Vulnerabilidad Ante el Cambio Climático en el Municipio de Madrid*; Ayuntamiento de Madrid: Madrid, Spain, 2015.
14. Evans, J.; Karvonen, A.; Raven, R. (Eds.) *The Experimental City*; Routledge: London, UK, 2016. [CrossRef]
15. Bursztyn, M.; Drummond, J. Sustainability science and the university: Pitfalls and bridges to interdisciplinarity. *Environ. Educ. Res.* **2014**, *20*, 313–332. [CrossRef]
16. Bammer, G.; O'Rourke, M.; O'Connell, D.; Neuhauser, L.; Midgley, G.; Klein, J.T.; Richardson, G.P. Expertise in research integration and implementation for tackling complex problems: When is it needed, where can it be found and how can it be strengthened? *Palgrave Commun.* **2020**, *6*, 5. [CrossRef]
17. Smith, A. Social innovation, democracy and makerspaces. *Rev. Española Terc. Sector* **2017**, *36*, 49–74. [CrossRef]
18. Stott, L. *Partnership and Transformation: The Promise of Multi-Stakeholder Collaboration in Context*; Taylor & Francis: Abingdon, UK, 2022.
19. Muñoz Sanz, V.; Romero Muñoz, S.; Sánchez Chaparro, T.; Bello Gómez, L.; Herdt, T. Making green work: Implementation strategies in a new generation of urban forests. *Urban Plan.* **2022**, *7*, 202–213. [CrossRef]
20. Romero-Muñoz, S.; Alméstar, M.; Sánchez-Chaparro, T.; Muñoz Sanz, V. The Impact of Institutional Innovation on a Public Tender: The Case of Madrid Metropolitan Forest. *Land* **2023**, *12*, 1179. [CrossRef]
21. Ostrom, E. *Governing the Commons: The Evolution of Institutions for Collective Action*; Cambridge University Press: Cambridge, UK, 1990.
22. Ostrom, E. *Understanding Institutional Diversity*; Princeton University Press: Princeton, NJ, USA, 2005.
23. Ostrom, E. Background on the institutional analysis and development framework. *Policy Stud. J.* **2011**, *39*, 7–27. [CrossRef]
24. Ostrom, E. Do institutions for collective action evolve? *J. Bioeconomics* **2014**, *16*, 3–30. [CrossRef]
25. McGinnis, M.D.; Ostrom, E. Social-ecological system framework: Initial changes and continuing challenges. *Ecol. Soc.* **2014**, *19*, 30. [CrossRef]
26. Alméstar, M.; Mestres, N.; Ezquerra, I. Cocreación a Través de la Experimentación Urbana: Los Entornos Escolares Como Oasis Climático-Social en Villaverde. [Comunicación Técnica]. Congreso Nacional de Medio Ambiente 2024, Madrid, España. 2024. Available online: https://www.fundacionconama.org/wp-content/uploads/conama/comunicaciones/7877/20240923_CONAMA_JMAU_word.pdf (accessed on 26 May 2025).
27. Alméstar, M.; Sastre-Merino, S.; Velón, P.; Martínez-Núñez, M.; Marchamalo, M.; Calderón-Guerrero, C. Schools as levers of change in urban transformation: Practical strategies to promote the sustainability of climate action educational programs. *Sustain. Cities Soc.* **2022**, *87*, 104239. [CrossRef]
28. Al Şensoy, S.; Sarı, R.M. Re-Design of Schoolyard for Effective Development of Child From a Universal Design Perspective//Etkili Çocuk Gelişimi İçin Evrensel Tasarım Perspektifinden Okul Bahçesi Tasarımı. *Megaron* **2019**, *14*, 443.

29. Sastre-Merino, S.; Zanarotti-Prestes-Rosa, C.; Izaguirres-Betancourth, M.-J. Analysis of initiatives to promote education for sustainability and climate action in schools in Madrid and proposals for improvement [Análisis de iniciativas para fomentar la educación escolar para la sostenibilidad y la acción climática en Madrid y propuestas de mejora]. *Rev. Española Pedagog.* **2025**, *83*, 105–123. [CrossRef]
30. Frantzeskaki, N.; Dumitru, A.; Anguelovski, I.; Avelino, F.; Bach, M.; Best, B.; Binder, C.; Barnes, J.; Carrus, G.; Egermann, M.; et al. Elucidating the changing roles of civil society in urban sustainability transitions. *Environ. Innov. Soc. Transit.* **2018**, *22*, 41–50. [CrossRef]
31. Ansell, C.; Gash, A. Collaborative governance in theory and practice. *J. Public Adm. Res. Theory* **2008**, *18*, 543–571. [CrossRef]
32. Sørensen, E.; Torfing, J. Making governance networks effective and democratic through metagovernance. *Public Adm.* **2009**, *87*, 234–258. [CrossRef]
33. Sørensen, E.; Torfing, J. Enhancing collaborative innovation in the public sector. *Adm. Soc.* **2011**, *43*, 842–868. [CrossRef]
34. Innes, J.E.; Booher, D.E. The Impact of Collaborative Planning on Governance Capacity. 2003. Available online: https://escholarship.org/uc/item/98k72547 (accessed on 26 May 2025).
35. Swyngedouw, E. The antinomies of the postpolitical city: In search of a democratic politics of environmental production. *Int. J. Urban Reg. Res.* **2009**, *33*, 601–620. [CrossRef]
36. Dudek, M. *Architecture of Schools: The New Learning Environments*; Routledge: London, UK, 2012.
37. Saint, A. Towards a social architecture: The role of school-building in post-war England. *Hist. Educ. Q.* **1987**, *28*, 650–653.
38. Woolner, P. *School Design Together*; Routledge: London, UK, 2015.
39. Le, C. *The Athens Charter*; Grossman Publishers: New York, NY, USA, 1945.
40. Mumford, L. *The City in History: Its Origins, Its Transformations, and Its Prospects*; Harcourt, Brace & World: San Diego, CA, USA, 1961.
41. Dudek, M. *Children's Spaces*; Architectural Press: New York, NY, USA, 2005.
42. Ajuntament de Barcelona. Protegim les Escoles | Urbanisme, Transició Ecològica, Serveis Urbans i Habitatge. (s. f.). 2023. Available online: https://ajuntament.barcelona.cat/ecologiaurbana/ca/que-fem-i-per-que/urbanisme-per-als-barris/protegim-escoles (accessed on 2 May 2024).
43. Concello de Pontevedra. Ciudad de la infancia: Concello de Pontevedra. Concello de Pontevedra. 2017. Available online: https://ok.pontevedra.gal/es/ciudad-de-la-infancia/ (accessed on 2 May 2024).
44. Ville de Paris. Les "Rues aux écoles " dans le 20e. Mairie Du 20e. 2024. Available online: https://mairie20.paris.fr/pages/les-rues-aux-ecoles-dans-le-20e-18452 (accessed on 2 May 2024).
45. Ville de Paris. Les Cours Oasis. 2024. Available online: https://www.paris.fr/pages/les-cours-oasis-7389 (accessed on 2 May 2024).
46. London City Hall. Climate Resilient Schools. 2024. Available online: https://www.london.gov.uk/programmes-strategies/environment-and-climate-change/climate-change/climate-adaptation/climate-resilient-schools (accessed on 2 May 2024).
47. Meadows, D.H. *Leverage Points: Places to Intervene in a System*; No. 99-6; The Sustainability Institute: Hartland, VT, USA, 1999.
48. Nesshöver, C.; Assmuth, T.; Irvine, K.N.; Rusch, G.M.; Waylen, K.A.; Delbaere, B.; Haase, D.; Jones-Walters, L.; Keune, H.; Kovacs, E.; et al. The science, policy and practice of nature-based solutions: An interdisciplinary perspective. *Sci. Total Environ.* **2017**, *579*, 1215–1227. [CrossRef] [PubMed]
49. Latour, B. Dadme un laboratorio y levantaré el mundo. In *La Ciencia Observada*; Knorr-Cetina, K., Mulkay, M., Eds.; Sage: New York, NY, USA, 1983; pp. 141–170.
50. Bulkeley, H.; Marvin, S.; Palgan, Y.V.; McCormick, K.; Breitfuss-Loidl, M. Urban living labs: Governing urban sustainability transitions. *Curr. Opin. Environ. Sustain.* **2019**, *22*, 13–17. [CrossRef]
51. Polk, M. (Ed.) *Co-Producing Knowledge for Sustainable Cities: Joining Forces for Change*; Routledge: London, UK, 2015.
52. Ultramari, C.; Ludwig, L.; Ciociola, D. Coexistence between design, regulation, and negotiation in urban planning: The case of Buenos Aires, Mexico City, and São Paulo. *Urbani Izziv* **2023**, *34*, 119–128. [CrossRef]
53. Alexander, E.R. Institutional transformation and planning: From institutionalization theory to institutional design. *Plan. Theory* **2005**, *4*, 209–223. [CrossRef]
54. Healey, P.; Healey, P. Systemic Institutional Design for Collaborative Planning. In *Collaborative Planning: Shaping Places in Fragmented Societies*; Springer: Berlin/Heidelberg, Germany, 1997; pp. 284–314.
55. Savini, F.; Bertolini, L. Urban experimentation as a politics of niches. *Environ. Plan. A Econ. Space* **2019**, *51*, 831–848. [CrossRef]
56. Purcell, M. Resisting neoliberalization: Communicative planning or counter-hegemonic movements? *Plan. Theory* **2009**, *8*, 140–165. [CrossRef]
57. Balboa, C.M. Urbanismo colaborativo. *Cuad. Investig. Urbanística* **2017**, *115*, 3–100.
58. Yin, R.K. *Case Study Research and Applications: Design and Methods*, 6th ed.; Sage Publications: New York, NY, USA, 2017.
59. Yin, R.K. The case study method as a tool for doing evaluation. *Curr. Sociol.* **1992**, *40*, 121–137. [CrossRef]
60. Eisenhardt, K.M. Building theories from case study research. *Acad. Manag. Rev.* **1989**, *14*, 532–550. [CrossRef]
61. Velasco, H.; Díaz de Rada, Á. *La Lógica de la Investigación Etnográfica*; Editorial Trotta: Madrid, Spain, 1997.

62. Díaz de Rada, Á. *El Taller del Etnógrafo: Materiales y Herramientas de Investigación Etnográfica = The workspace Of the Ethnographer: Materials and Tools for Ethnographic Research*; UNED: Madrid, Spain, 2011.
63. Lieberman, A. Collaborative research: Working with, not working. *Educ. Leadersh.* **1986**, *43*, 28–32.
64. Coughlan, P.; Coghlan, D. Action research for operations management. *Int. J. Oper. Prod. Manag.* **2002**, *22*, 220–240. [CrossRef]
65. Shani, A.B.; Mohrman, S.A.; Pasmore, W.A.; Stymne, B.; Adler, N. (Eds.) *Handbook of Collaborative Management Research*, 1st ed.; Sage Publications: New York, NY, USA, 2007.
66. Ayuntamiento de Madrid. *Padrón Municipal de Habitantes 2021—Distrito de Villaverde*; Ayuntamiento de Madrid: Madrid, Spain, 2021.
67. Distrito de Villaverde. *San Cristobal de los Ángeles: Diagnostico Comunitario*; Distrito de Villaverde: Madrid, Spain, 2022.
68. Engle, J.; Slade, S. Escuchar para promover el cambio social: Herramientas transformadoras que impulsan la innovación comunitaria. In *En Plataformas que Activan la Innovación. Un Nuevo Enfoque para la Cooperación Internacional al Desarrollo*; Dansk Design Center: København, Denmark, 2018; pp. 22–40. Available online: https://fundacionlacaixa.org/documents/234043/558289/work-4-progress-plataformas-activan-innovacion.pdf (accessed on 26 May 2025).
69. Moreno, I.; Perfetti, S.; Ramil, X. Communication in Social Innovation Platforms. *Rev. Diecisiete* **2020**, *2*, 105–122. [CrossRef]
70. Fisher, J. Consideraciones sobre el seguimiento, evaluación y aprendizaje en las plataformas de innovación social. In *En Plataformas que Activan la Innovación. Un Nuevo Enfoque para la Cooperación Internacional al Desarrollo*; Columbia University: New York, NY, USA, 2018; pp. 63–75. Available online: https://fundacionlacaixa.org/documents/234043/558289/work-4-progress-plataformas-activan-innovacion.pdf (accessed on 26 May 2025).
71. Stott, L. *Co-Production: Enhancing the Role of Citizens in Governance and Service Delivery*; Technical Dossier No. 4; Directorate-General for Employment, Social Affairs and Inclusion, European Commission: Brussels, Belgium, 2018. [CrossRef]
72. Voorberg, W.H.; Bekkers, V.J.J.M.; Tummers, L.G. A systematic review of co-creation and co-production: Embarking on the social innovation journey. *Public Manag. Rev.* **2015**, *17*, 1333–1357. [CrossRef]
73. Trischler, J.; Pervan, S.J.; Kelly, S.J.; Scott, D.R. The value of co-design: The effect of customer involvement in service design teams. *J. Serv. Res.* **2018**, *21*, 75–100. [CrossRef]
74. Obra Social "la Caixa"; itdUPM. *W4P—Guía III: Prototipado y Escalado*; Fundación la Caixa: Valencia, Spain, 2019; pp. 1–27.
75. Sabel, C.F.; Zeitlin, J. Experimentalist governance. In *The Oxford Handbook of Governance*; Levi-Faur, D., Ed.; Oxford University Press: Oxford, UK, 2012; pp. 169–183.
76. Karvonen, A.; Evans, J.; van Heur, B. The politics of urban experiments: Radical change or business as usual? In *After Sustainable Cities?* Hodson, M., Marvin, S., Eds.; Routledge: London, UK, 2014; pp. 105–114.

land

MDPI

Article

Innovative Participatory Practices in Three Sub-Regional Spatial Plans in the Valencian Autonomous Region (Spain)

Joaquín Farinós-Dasí [1,*], Ignacio Díez-Torrijos [2] and Pilar Lloret-Gual [3]

[1] Instituto Interuniversitario de Desarrollo Local (IIDL), University of Valencia, 46022 Valencia, Spain
[2] Urbanism Department, Polytechnic University of Valencia, 46022 Valencia, Spain; igdietor@urb.upv.es
[3] University of Valencia and Civil Servant in the Local Government in Liria, 46160 Valencia, Spain
* Correspondence: joaquin.farinos@uv.es

Abstract: Between 2017 and 2023, three sub-regional spatial plans were developed for specific areas of the Valencian Autonomous Region: the Alicante–Elche Metropolitan Area, Vega Baja del Segura County, and the Central Valencian Counties. Their main aim was to develop an envisaged territorial model as a binding framework for municipal urban master plans. During the elaboration of these plans, a set of activities was developed to understand the main consensus among stakeholders. The main axes of the planning process were addressed during territorial working groups conducted with relevant stakeholders, including those focused on green infrastructure, settlement systems, and infrastructure systems. Participants were selected from the public administration, non-governmental organizations, the economic sector, and the university. Drawing on the outcomes of the participatory activities, various factors are analyzed, including the ratio between invited stakeholders and actual participants in the territorial workshops, the contributions made by participants in each main axis of the plan, the inputs provided according to stakeholder type, the nature of these contributions, and the degree of alignment between the inputs and the objectives of the PAT. The present study reveals how contextual factors can influence the orientation of the participatory process. At times, contingency may emerge as an opportunity to energize a governance process. Similarly, the participatory technique is validated for its potential to enrich the process, while also highlighting the absence of voices not aligned with spatial planning in the participatory settings.

Keywords: land use planning; governance; participation; strategic spatial planning; urban and land policy; spatial sub-regional plans; Valencia autonomous region

1. Introduction

In 2018, two sectoral Territorial Action Plans (PATs)[1] were approved in the Valencian Autonomous Region: the "Plan for the Management and Revitalization of the Huerta of Valencia" and the "Green Infrastructure Plan for the Coastal Area of the Region". Between 2020 and 2023, a new series of PATs was developed, initially aimed at organizing the three biggest metropolitan areas (Valencia, Castellón, and Alicante–Elche) and subsequently other functional areas, including the Vega Baja del Segura and the Central Valencian Counties.

This study focuses on analyzing a set of public participation activities designed and developed with stakeholders involved in the drafting of the three comprehensive PATs under study: the PAT for the Alicante–Elche functional area, the PAT for Vega Baja del Segura County, and the PAT for the Central Valencian Counties. They were all developed in accordance with the legal framework in force in the Valencian Autonomous

https://doi.org/10.3390/land14061244

Region. The objective of the comprehensive plans was to harmoniously structure the three main components of the territorial system: green infrastructure, settlement system, and transportation infrastructure system. Their drafting and eventual approval represented a major challenge for the planning teams, of which the authors of this paper were part, given that, at that time, no comprehensive PAT had been approved in the Valencian Autonomous Region (only sectoral PATs had been approved).

An additional challenge was the broad territorial scope of the three spatial sub-regional plans, both in terms of surface area and population of the Valencian Autonomous Region. This was especially true for the PAT for the Central Valencian Counties, a cross-border plan which spans eight counties across the two provinces of Valencia and Alicante. Accordingly, for all three cases, a public participation plan was designed and implemented to meet the demands of each territorial context.

The opportunity to compare the same type of participatory activity, carried out by the same planning team over a four-year period, is relatively uncommon and allows for meaningful conclusions to be drawn about the effectiveness of participatory processes and their integration within spatial planning.

For each of the three PATs, a comprehensive set of participatory activities was carried out in parallel to reach diverse population profiles and relevant stakeholders. Many of these activities were designed to bring participation closer to where people live. The participatory activities analyzed in this study correspond to three key stages in the planning process: (a) the initial drafting phase; (b) a "check-in" moment with citizens and local actors to discuss preliminary territorial diagnoses developed by the planners' team; and (c) a final stage involving thematic or territorial workshops aimed at refining the diagnosis and providing input for the identification of feasible alternatives to be proposed by the plan. All these activities, which were not mandatory based on the current Valencian Law on spatial planning, took place prior to the formal public consultation phase (binding by law in this case) in order to reach agreement on the negotiated alternatives to minimize potential conflicts during the public review (consultation) process.

2. Theoretical and Contextual Frameworks: Public Participation in Spatial and Urban Planning in Spain

2.1. A Spatial Planning System Rooted in Urban Planning Traditions with a Strong Normative Approach

The effectiveness of spatial planning largely depends on public trust in institutions and in the work of public administrations to create the necessary conditions for the population's well-being, undertaking interventions that improve living conditions and support local economies. This trust is essential for the sustainability and continuity of spatial planning as a public policy [1]. Conversely, low levels of trust can lead to closed social contexts in which the legitimacy of institutions is questioned, posing serious challenges to territorial governance. This, in turn, threatens the possibility of ensuring governance that upholds citizens' rights and welfare within a social and democratic state governed by the rule of law [2]. Trust in institutions can be built by introducing and promoting active citizen involvement. Citizen participation when designing public policies legitimizes their implementation, facilitates execution, and allows for solving specific problems by harnessing collective intelligence.

In the Spanish context, public participation emerged later than in neighboring countries, and often in a rushed or superficial manner, misunderstood as an end in itself rather than as a means to achieving specific goals. Nonetheless, the integration of participatory processes has brought about changes in planning practices, fostering innovation and leading to the development of new practices and professional profiles among planners. This has not been a smooth path, nor have the results always been remarkable, due to deep-rooted factors (path dependence) that are difficult to overcome. Transforming this dynamic into a process of path creation through social innovation aims to ultimately lead to better spatial and urban planning.

Based on standard classifications of spatial planning styles across Europe, Spain belongs to the group of Mediterranean countries, with a strong land-use planning tradition [3]. In Spain, spatial and urban planning policies exhibit a clear regulatory bias and are highly institutionalized. The Spanish system is closer to the German tradition of spatial analysis (zoning-based land-use classification) than to the French model, which places greater emphasis on regional economic development and the integration of physical and socio-economic planning. In Spain, this regulatory and non-strategic nature dates back to the "Ley de Ensanche de Poblaciones" (1864) and was definitively consolidated with the first Land Act ("Ley del Suelo") (1956). This provided the foundation for subsequent regional spatial planning laws in the 1980s and, more significantly, for the new autonomous regions' urban planning laws in the 1990s. These new laws were necessary following the ruling of the Constitutional Court (STC 61/1997), which established that only the Autonomous Regions—and not the Spanish government—can legislate on spatial planning and urbanism[2].

This regulatory and institutionalized character still remains today, despite the clear influence of European Regional Policy—now Cohesion Policy—with its associated regulations, funding instruments, and tools: Community-Led Local Development (CLLD), Integrated Territorial Investments (ITI), Sustainable Urban Development Strategies (SUDS), later Integrated Territorial Development Strategies, Urban Initiatives, and Urban Agendas, in line with objectives 5.1 and 5.2 of the new 2021–2027 programming period. Unlike Spain's traditional approach, these initiatives are based on local territorial strategies that integrate public policies and heavily rely on participatory processes involving local stakeholders.

Two types of plans are used to address spatial planning in the Valencian Community. At the municipal level, general urban development plans (planes generales) aim to define the territorial model of a municipality. At a supra-municipal level, Territorial Action Plans (Planes de Acción Territorial, PATs) are used. These PATs can be either integrated or sectoral in nature, both with the same hierarchical status. Integrated PATs are designed to address the three main components of spatial planning, namely, green infrastructure, the settlement system, and the infrastructure system, whereas sectoral PATs focus on specific territories or particular spatial issues. The planning framework has been documented in reports aimed at establishing the state of spatial planning in the territory of the Valencian Community [4].

2.2. *The Fit of Participation in Spanish Planning Practice*

Despite all the above, and the enactment of successive participation laws in Spain (since 2006), as well as advances in supra-municipal scales for planning, public participation remains generally limited to formal mechanisms that involve taking part in the administrative processes that lead to the approval of the plan. This is the case with public consultation periods and the submission of objections [5], and more recently, with public participation during drafting of the Strategic Environmental Assessment (SEA) plan [6]. In this regard, Spain places excessive emphasis on administrative law to secure legal certainty and legitimacy, which ultimately limits permeability in the relationship between government and citizens[3]. This over-reliance on legally established procedures generates rigid participatory processes, only as a preliminary "information" level, according to the OECD Council Recommendation on Open Government [7].[4]

The transposition of "Directive 2001/42/EC of the European Parliament and of the Council of 27 June 2001, on the environmental assessment of certain plans and programs" into Spanish law was also conducted in a particular manner. It also followed an administrative procedural approach more typical of Environmental Impact Assessment (EIA) for projects rather than a comprehensive and cross-cutting sustainability vision that would permeate the entire planning process (see Section 4.2). As a result, this SEA also becomes institutionalized, with two main phases: the definition of the scoping of the Strategic Environmental Report when the plan's elaboration starts, and the Strategic Environmental Statement that "authorizes" the plan to proceed to its final approval.

Hence, the SEA is seen as a constraining element that complicates planning approval, leading to tendencies to simplify or relax it in order to facilitate interventions. Thus, broadly speaking, the need for participation from the perspective of the institutions and administrative bodies responsible for its implementation is interpreted more as an obligation to comply with a new legal requirement during the spatial plan's drafting and approval phases than as a facilitating element for better defining and selecting alternatives to agree and decide the final territorial model.

From the viewpoint of the regional and municipal entities that promote, elaborate, and approve the plan, they face a significant issue—a combination and integration of conventional binding regulatory planning with the more innovative strategic territorial planning and its methods. Some planning theories take for granted the possibility of reaching collective agreements to implement selected alternatives [8], based on [9]. These are more theoretical than realistic approaches, considering the frequent conflicts driven by competing (often contradictory) interests and power dynamics in the territory [10,11] cited in [12]. As a result, at least in the Spanish case, participation often ends up being not so much a mechanism for negotiation and agreement but, instead, a defensive pressure tactic to either ensure the demands of key actors with real influence or veto power are included in the plan or used as a means to voice the complaints of parties and less influential actors affected by the implementation of the plan.

Therefore, it is essential to introduce pedagogy into participatory processes and the concept of public welfare to help participants understand that there is a common space where all of them can benefit, and to contribute to reaching this, for example, by defining a territorial model using a spatial planning instrument agreed upon by participants.

It is easier to reach a broad consensus on global goals and general principles than on their practical implementation. The former allows for a more strategic and participatory approach. The latter, when transactional agreement is impossible, leads to responsible administrations and governments ultimately defining the range of possible choices and the rules of the game with due legal certainty. This, however, does not prevent—in the worst-case scenario—over-legislation or conflicting laws, which ultimately result in the judicialization of disputes, as is the case in Spain. Conflicts of interest are not resolved solely through scientific evidence but through agreements and, if these are unattainable, through political decisions and, ultimately, court rulings.

The final decision and control over the plan remain with the responsible planning authority. In the case of spatial plans, it is the Regional Government of each of the 17 Spanish Autonomous Regions that is the responsible authority; however, several binding reports from other administrations are needed during the process of obtaining final approval for the plan. When multiple authorities are involved, matters become more complicated due to the need for adequate cross-sectoral and multi-level coordination, which often fails due to the dominance of a siloed, sectoral approach in a Napoleonic-style administration. Add to this the fact that it is not mandatory to elaborate regional and sub-regional spatial plans in Spain (this is only mandatory for municipal urban master plans), when plans are contested, they are usually abandoned before their final approval or, if approved, are not implemented.

2.3. Participation Outcomes

The groups of stakeholders with the greatest veto power are ultimately the ones whose views are truly taken into account, as part of what could be called the "golden ring" of key actors; what is said matters less. This "golden ring" is composed of the responsible authorities or administrations, the economic actors responsible for public and civil works, and, more recently, environmentalist and conservation groups. This issue leads us to the debate on the usefulness of participation, including who participates, how[5] they participate, for what purpose (we assume to approve or help define a shared common spatial vision—for this, a good understanding of procedures and what spatial planning entails is needed for effective participation), and what capacity for real influence do they have (only "input legitimacy" or also "output generation").

Focusing on the "Who", we have already established a division between indirect co-decision-makers with significant influence who raise alarm among political leaders due to their ability to block the plans and potentially cause political damage. These include regional governments, municipal governments, construction companies, and environmental groups. This first ring would be followed by a second external ring comprising what we might call active participants, preferably organized, typically those whose interests are most affected by the plan, and who usually adopt a more reactive or preventive approach rather than one aimed at maximizing territorial potential. In a third outer ring are the remaining participants, either organized (preferable for validating the alternative leading to the territorial model proposed by the plan) or individuals.

These overlapping concentric groups, comprising actors from the administration, productive sectors, academia, or civil society, would ideally mix and propose joint options and (transactional) agreements in the participatory process. Some of these stakeholders are recurrent, which opens the debate on the representativeness and legitimacy of participatory processes. Frequent participation improves understanding of the processes and usefulness of both participation and planning[6]. The introduction of a civic lottery is a tool that could correct these weaknesses, as it offers impartiality in selecting participants to discuss the territorial issues at hand and provides a representative sample of the society involved.

This brings us to the question of the degrees and objectives of participation (how and for what), and its final level of influence on the decision. From the well-known "ladder of participation" proposed by [15], various approaches and adaptations to this idea have followed (see [16]). Here, we highlight two that share common points. On the one hand is [13], who identifies four phases based on the role participation plays in the decision-making process: public information, listening during the planning process, influencing the decision while seeking consensus, and accepting the decision through conflict resolution and negotiation. On the other hand [17], bases the objectives of participation and levels of influence on the decision, from which we synthesize the following:

1. Developing capacities for adequate public involvement: This implies, on the one hand, raising public awareness about planning and participation activities before actors engage, and on the other hand, and more crucially, offering education (training) to prepare people so that they can participate productively. As a preliminary phase, it would not directly influence decision-making; but it can do so indirectly as a result of citizen empowerment [18] by aligning actors at the planning stage, hence contributing to the development of a new territorial culture (in a 'learning by doing' process). This territorial intelligence, which [19] define as a collective intelligence based on the interaction of each individual with their environment and other community members, is necessary not only for decision-makers and institutions but also for civil society, individuals, and economic, social, and intellectual organizations [20]. This contributes to participation, proposition, and designing future proposals and new reference

frameworks that can be considered by political leaders. Collective intelligence also refers to common sensitivity and understanding of what is at stake in one's own territory and how to facilitate agreements in a collective way to provide efficient solutions for the local community. In the case of low participation culture, such as in the Spanish case, institutionalized public participation activities help to develop this collective understanding and feeling for the common living space [21,22]. In this way, institutionalized participation activities help participants to not only learn how to participate, but also why spatial planning is useful.

2. Generating input: This input can be used by planners to complement or verify official or baseline information with local knowledge (public experiences, attitudes, beliefs), thus contributing to better decisions. This leads us to the relationship between expert and non-expert knowledge, citizen science [23], post-normal science [24], and the link between epistemic communities and communities of practice [25]. If this input is ignored or misunderstood, its influence on decision-making is null. If it is ultimately used for decision-making (even as "input legitimacy"), it may have some influence.
3. Involving and committing citizens to decision-making, not only in formulating alternatives and designing criteria and methods for their selection, but also in the final decision on the chosen alternatives: What is agreed (consensus on objectives and action commitments to achieve them) will ultimately become normative and binding (as in the action plans of the new Local Urban Agendas [26]). If implemented, its influence is maximal; if not, the influence returns to zero. Non-implementation is frequent in the Spanish context. Approved instruments are not always developed in practice or, if they are, rarely fully as designed [27]. In this context, it is essential to move towards a change in the approaches of participatory processes, introducing an unorganized civil society using tools such as a civic lottery that can balance existing forces and prevent the formation of golden rings and decentralize elite power. On the other hand, it is crucial to clearly explain and define the exact margin of bindingness that the proposed measures will have in order to avoid participant disillusionment and unenforceable obligations for decision-makers.

3. Materials and Methods

3.1. Study Objective

From previous characterization of participation in the Spanish and Valencian context, several questions emerge about how participation occurs in practice: What is its regulatory and procedural framework? What methods or formulas are used? Which actors participate? How are responsibilities distributed? Above all, what are the results? It is important to identify the factors and key issues that affect the quality and outcomes of this participation as a means of evaluation.

The main objective of our research is to use case studies to derive conclusions and recommendations about participatory practices in the study area, considering three types of participation conditions: contextual, procedural, and outcome. The study examines how territorial roundtables are developed in the respective public participation plans in the three selected cases (methods, techniques, and procedures followed). The analysis is based on a series of evaluation indicators (number of participants by PATs and thematic axes, contributions by topics, contributions by types of stakeholders, usefulness of contributions regarding the PAT jurisdictions, alignment of contributions based on the nature and objectives of PATs, external contextual factors) that lead us to the results and, through their discussion, to the development of a series of conclusions and recommendations about how to implement participation and governance actions that improve spatial planning.

3.2. Participation Process Evaluation Indicators

To objectively evaluate the participation activities referred to as "Territorial Roundtables", we analyze a set of indicators. These also allow us to conduct a comparative analysis of the three plans.

The variables selected to analyze participation activities within the framework of sub-regional spatial plans in the Valencian Autonomous Region are as follows:

- Participation balance in thematic roundtables: The convening and attendance at the territorial roundtables are analyzed.
- Actor diversity: The roles of different participating actors are examined.
- Balance of contributions across each PAT axis: Contributions are studied in relation to the planning instrument's areas of competence.
- Type of contributions: Contributions are classified as informational, reflective, or propositional.
- Contributions by actor type: The relationship between actors and types of actions is established.
- Alignment of contributions with PAT objectives: An analysis of how proposals can drive the spatial plan is conducted.
- Aspects external to the PAT: Potential consequences of contextual conditions present in each PAT are evaluated as factors influencing the results obtained.

The variables developed by the authors of this study correspond to four main factors:

- Evaluation of the participatory process itself, with questions such as the following: How have the stakeholders responded to the call for participation?
- Factors associated with the methodology of the spatial planning process, which addresses three main areas of action: green infrastructure, settlement system, and infrastructure system. Questions explored include the following: Is there any particular axis in which stakeholders feel more comfortable participating?
- Factors reflecting the authors' concerns about how the contributions relate to the plan. For example, do the suggestions made by stakeholders actually have the potential to be implemented in the plan?
- External factors affecting participation: Does the context support or hinder the public participation process?

3.3. Information Sources

The information sources primarily comprise documents generated during the plan drafting process and the in-depth knowledge of the authors of this work, who are members of the drafting team of the three PATs.

3.4. Methodology

To analyze the function and results of the thematic and territorial roundtables, the methodology follows both a quantitative and qualitative approach. In the quantitative approach, measurable aspects obtained objectively during plan development are established, related to the number of attendees, attendance ratio, number of contributions, and so on (see Section 4). In the qualitative approach, aspects that relate to internal participation factors and external factors are highlighted, allowing us to have a more holistic vision of the process.

4. Analysis of Public Participation Experiences and Results

4.1. The Three PATs Under Analysis

Within the regional framework established by the Valencian Community Spatial Vision approved in January 2011, the Valencian government developed some subregional spatial plans between 2015 and 2023. Only some were approved (all sectoral, none integrated), including PATRICOVA, PATFOR, and, in 2018, PATIVEL and the Huerta PAT. This study analyzes three comprehensive PATs that were meant to define supra-municipal spatial planning: the Alicante–Elche Functional Area (PATAE), the Vega Baja del Segura (PATVB), and the Central Valencian Counties (PATCCCV) (see Figure 1), which, as of now, have not been approved.

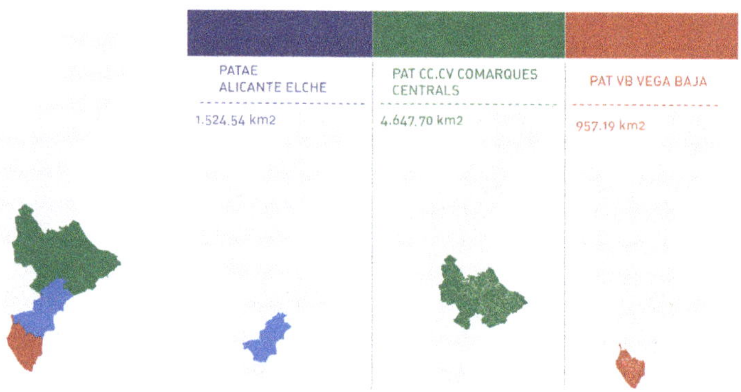

Figure 1. Territorial scopes of the three territorial action plans. Source: own elaboration.

PATCCCV is the largest, encompassing 175 municipalities across eight counties, with an approximate population of 850,000 people. PATVB covers 27 municipalities and contains nearly 360,000 inhabitants. PATAE consists of 21 municipalities with approximately 825,000 inhabitants. Both PATAE and PATVB mainly affect one county each.

All three plans were developed based on the definition of the three traditional types of land that configure the territorial system: green infrastructure models, settlement models, and communication infrastructure models, typical of spatial analysis. A fourth important area of analysis and diagnosis, the socioeconomic model, was added to the studied plans, bringing together economy and land (see Figure 2).

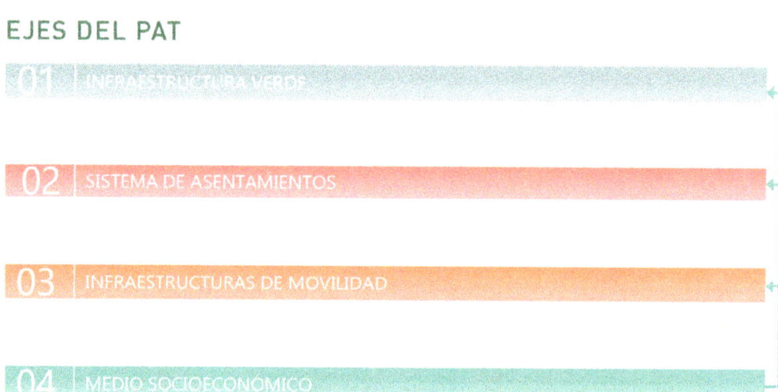

Figure 2. Competency axes and/or analysis–diagnosis axes of the three analyzed PATs: 01, green infrastructure; 02, settlement system; 03, infrastructures of mobility; 04, socioeconomic system. Source: own elaboration.

4.2. Thematic/Content Analysis of Legislation and Case Studies

A PAT follows a long and complex process in which participation is taken into consideration from the start of its drafting[7]. The three PATs were developed while the previous 2014 LOTUP, now replaced by Legislative Decree 1/2021, was in force, approving the consolidated text of the Law on Spatial Planning, Urbanism and Landscape, which contains numerous references to public participation.

From a regulatory perspective, according to the 2014 LOTUP, participation was mandatory and binding for certain phases of the plan development process, including the corresponding territorial and environmental assessment procedures; however, nothing was said about the preliminary phase during the drafting of the preliminary version of the plan prior to its initial approval. The phases were as follows (we have underlined steps where the regulations demand participation):

1. Consultation with public administrations.
2. Scoping document of the Strategic Environmental and Territorial Report.
3. Formulation, by the promoting body, of a preliminary version of the plan or program, which will include a Strategic Environmental and Territorial Report.
4. Submission of the preliminary version of the plan or program, along with the Strategic Environmental and Territorial Report, to initiate the public participation process, public information, and consultations.
5. Preparation of the plan or program proposal.
6. Strategic Environmental and Territorial Statement.
7. If necessary, adaptation of the plan or program to the Strategic Environmental and Territorial Statement.
8. According to the criteria established in this Law, in cases where modifications are introduced to the plan or program document, new information for citizens is required.
9. Approval of the plan or program and public exposition.
10. Implementation of the environmental and territorial monitoring plan following its approval. Compliance with environmental and territorial provisions is verified.

Thus, there are three mandatory phases of consultation and requirements for citizen participation in the planning procedure: 1, 4, and 8. At no time is participation contemplated prior to the drafting of the preliminary version of the plan. Despite the lack of express legal requirement for an ex ante public participation process to support the preparation of the preliminary version of the plan, there is a technical and political will for it during the drafting of the three PATs.

The participation activities analyzed in this study were conducted prior to the public information phase of the PAT:

- During the on-site information sessions and presentation of the plan at the start of its drafting, for the triangulation and validation with citizens and local actors of the initial territorial diagnoses prepared as desk work (see Section 4.3).
- During the workshops for landscape unit identification and assessment (resources and conflicts), leading to citizens' shared territorial requalification.
- During the thematic territorial roundtables for refining the desk diagnosis and receiving input for the definition and subsequent selection of alternatives for the planning report (see Figure 3). During the roundtables, the "focus group" technique was used, addressing the main axes structuring the plan (green infrastructure, settlement model, and communication infrastructure model) with the aim of obtaining bottom-up proposals that could be incorporated for development by the PAT.

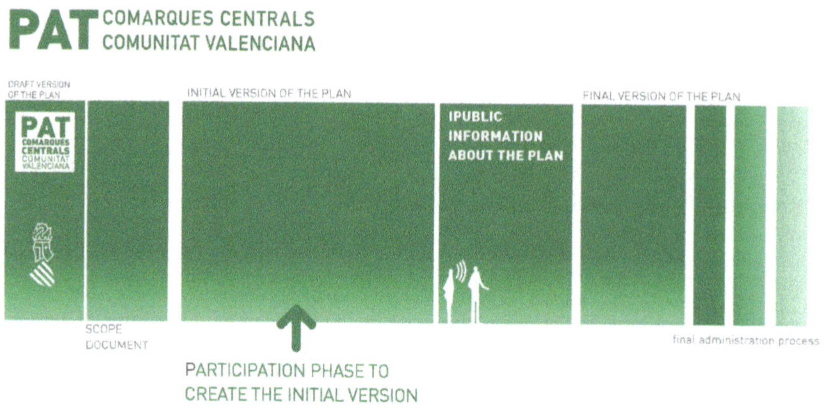

Figure 3. Initial phases of PAT drafting and integration of participation. Source: own elaboration.

The objective was to have preliminary versions of the PATs that incorporated the aspirations of the main stakeholders and the general public, thereby reducing potential conflicts and entry barriers. This would streamline the regulated procedures that require submitting the proposed plan to the public for initial approval. All this is performed before proceeding with the development of the mandatory Strategic Environmental Statement and Sectoral Reports. This is followed by obtaining the final (and formal) approval for the plan and its publication in the Official Gazette, making it a binding regulation for planning and individuals.

4.3. *Public Participation in the Drafting of the Initial Versions of PATs*

A participation plan was designed and developed for each of the three PATs from the beginning of the launch, and initially involved concerned municipalities in each PAT. However, participation was most intensive during the preparation of the initial version of the PAT. This initial version contains the complete PAT documentation, which is binding upon final approval. Approval of the initial version of the PAT is followed by a mandatory public consultation phase, during which interested parties and stakeholders submit comments to revise the plan before the final version is approved.

It is important to note that the number of activities conducted (i.e., the intensity of the participatory process) is directly correlated with the availability of financial resources to support such activities. Between 5 and 10% of the budget allocated for drafting the territorial plans was dedicated to the participation process. The drafting team intended to increase this share beyond 10%; however, given the limited available funding, a participation plan was designed to produce robust results that would enrich the decision-making process.

The participatory processes were concentrated within a two-month period, during which all activities were scheduled to maximize the effectiveness of communication efforts and the mobilization of key stakeholders and the general public. In the case of the PATVB, this period was extended due to the COVID-19 pandemic, necessitating an adjustment of the participatory timeline.

Prior to the most intensive phase of participation, a draft of the plan was prepared by the Generalitat Valenciana based on the territorial analyses conducted by the drafting team. For activities such as the territorial workshops described below, the draft plan was shared in the invitation, and the drafting team presented the work conducted up to that point.

The participation process was designed by the drafting team to fill a legal gap at the Spanish state level regarding how public participation should be addressed, specifically for urban and territorial plans. There are no national laws or regulations that specify the types of activities to be carried out or the approach to be followed. This legal gap is filled by guides or manuals, such as the Generalitat Valenciana's Landscape Studies Guide [28], which describes how to implement participation in landscape studies that require spatial planning documents.

All participation activities developed during the drafting phase of the initial version of each PAT were carried out in parallel with information sessions, as well as during landscape workshops and territorial roundtables, in order to obtain more robust and reliable results. These activities were as follows (see Table 1):

- Presentation and information sessions: These are meetings open to the general public, during which information about the initial document from which the PAT draft was developed is provided. The presentation session had a more institutional characteristic, while the information sessions had a more local characteristic and were held in different locations within the PAT territory.
- Landscape workshops: During these sessions, specific aspects of the population's relationship with their territory were addressed, including recognition and assessment of landscape units, identification of landscape resources or conflicts, and proposals for territorial requalification.
- Territorial or thematic roundtables: This is a discussion forum that addresses the aforementioned three competency axes of the plan, articulating the territorial system. It consists of experts and stakeholders from the territory.
- Meetings with municipalities: Once the planning alternatives had been noted, municipal entities participated in bilateral meetings with the PAT technical team to address the urban and territorial issues of each administrative area affected by the future PATs.
- Territorial survey: A questionnaire was distributed to collect the general perception of spatial planning in the PAT area in question, analyzing the three axes around which the territorial system was articulated. Additionally, for PATCCCV, since its promoting body was the Intercounty Council (bottom-up) and not the regional administration (as is the case for the others), a survey of economic agents in the said territory, who already had clear prior internal articulation, was conducted (see Figures 4 and 5).
- Feedback session: In the case of PATVB, a session was held to report the results of the participatory process, explaining how the conclusions from the process had been incorporated into the preliminary version of the proposals for PATVB.

Table 1. Participation activities conducted during the drafting of the initial version of the three PATs. PS, presentation sessions; IS, information sessions; LW, landscape workshops; MT, territorial roundtables; MM, meetings with municipalities; FS, feedback sessions. Source: own elaboration.

NUMBER OF PARTICIPATION ACTIVITIES						
	PS	IS	LW	TR	MM	FS
PATAE	1	4	4	4	21	0
PATVB	1	0	2	4	23	4
PATCCCV	7	4	4	4	49	0

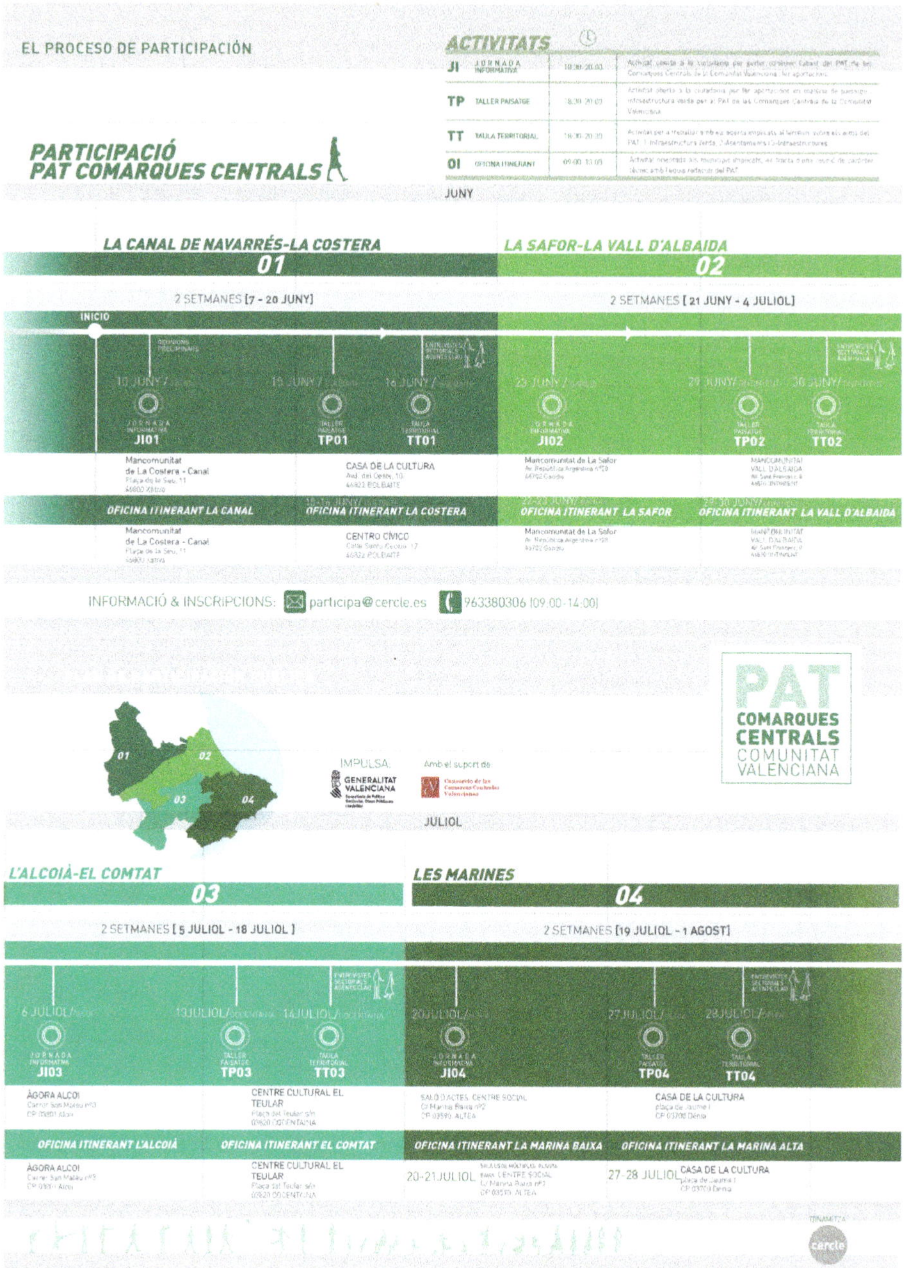

Figure 4. A set of activities developed for the four county groups during the drafting of the initial version of the Territorial Action Plan for the Central Valencian Counties (PATCCV). The activity schedule sequentially covers the entire territorial scope of the plan. Source: own elaboration.

Figure 5. Images of the deployment of activities developed for the eight counties involved in the Territorial Action Plan for the Central Valencian Counties (PATCCV). Source: own elaboration.

4.4. Participation Period

This section compares the territorial roundtable activities developed in the three PATs from 2018 to 2021, highlighting the relevance of testing the same participation tool developed in the same type of territorial instrument over a 4-year period. The dates of the territorial roundtables are shown in Table 2.

Table 2. Dates on which participation activities were developed. Source: own elaboration.

	PERIOD OF REALIZATION OF PARTICIPATION ACTIVITIES
PATAE	May to June 2018
PATVB	November to December 2020
PATCCCV	June to July 2021

While the PATAE participatory process proceeded without significant constraints, the PATVB process was affected by COVID-19 restrictions, and the PATCCCV process was influenced by the summer conditions in this Mediterranean area, where intense heat coincides with vacations for most of the population (it was not possible to change dates due to procedural issues and the political agenda).

4.5. Participation Topics

The territorial roundtables were held prior to the drafting of the preliminary version of the PATs with the aim of incorporating the visions and contributions made by experts and organized interest groups from the civil society in each of the foundational thematic axes around which the PATs revolved. The planning alternatives were defined as follows:

- Green infrastructure and the planning and regulation of rural land.
- The settlement system and its orientation.
- Analyzing the infrastructure system to improve aspects such as mobility.
- Socioeconomic system.

These topics were approached slightly differently in each PAT to adapt the methodology to each specific territory. Due to the coastal position of the study area, the socioeconomic territorial roundtable in PATAE was oriented towards tourism as the main element. In PATVB, however, tourism was addressed more in the settlement roundtable due to the importance of residential tourism in the county. In PATCCCV, the three binding axes of the PAT (green infrastructure, settlements, and communication infrastructure) were addressed simultaneously in all territorial roundtables. In this plan, the socioeconomic axis appeared transversally in the debate when specific contributions were made on topics such as tourism, agriculture, or industry within the corresponding binding axis.

4.6. Call for Sessions

It is important to acknowledge the challenges of convening stakeholders from extensive territorial areas for in-person sessions. This aspect was considered in order to expand the resources and ensure a balanced representation of stakeholder groups in all participation activities.

Invitations were sent via email, convening previously selected stakeholders through a decantation process that involved reviewing previous processes in the same territory, contacting key agents who could provide guidance, and reviewing lists of official entities.

In the first, more theoretical part of this paper, we mentioned the participation "golden ring" of stakeholders. In the analyzed case studies, these were clearly identified as the usual key actors in previous plans: representative administrations at regional, local, and national levels; real estate developers; and environmental groups. However, the proposed participation methodology in the 3 CS aimed to expand this core ring by including additional stakeholders in the area [29,30].

The stakeholder groups convened were as follows:

- ADMINISTRATIONS: All participants from different political administrative levels with competencies in the issues identified in each territory covered by the PAT.
    - Spanish: Representatives of governments, hydrographic confederations, and national companies such as Renfe, etc.
    - Regional: Representatives from relevant regional administration departments with competencies in the different thematic roundtables.
    - Provincial: Representatives from provincial councils and their associated departments.
    - County: County associations, as intermunicipal cooperation entities, were considered of high interest.
    - Municipal: While urbanism technicians from municipalities participated in the territorial roundtables in PATAE, in PATVB and PATCCCV, a clear distinction was made between local administrations and local stakeholders; therefore, while no local-level agents participated in the territorial roundtables in PATVB, local development agents (of great importance in predominantly rural areas) were encouraged to participate in PATCCCV.
- ASSOCIATIONS AND NON-GOVERNMENTAL ORGANIZATIONS: Efforts were made to involve key groups in the spatial planning and management. The following were the main groups:
    - Associations and collectives associated with land stewardship.
    - Business associations.
    - Agricultural entities.
    - Unions.
    - Others.
- UNIVERSITIES: Specialists from various universities in the Valencian Community were encouraged to participate.
    - Universitat Politècnica de València (UPV): Staff from the Valencia, Alcoi, and Gandia campuses were invited. These campuses are located within PATCCCV.
    - Universitat de València (UV).

- Miguel Hernández University of Elche (UMH): Staff from the Elche (PATAE area) and Orihuela (PATVB area) campuses were contacted.
- The University of Alicante (UA), which is located in the PATAE area and affiliated with PATCCCV.

All these stakeholder groups received the call inviting them to participate in thematic roundtable meetings. The meetings were conducted in three stages, held over approximately two hours:

- LAUNCH: At the beginning of the session, the consulting team provided a brief introduction to the progress of PAT in each thematic axis with the objective of providing more details and the right framework for the session. When participants received the invitation to attend the session via email or phone call, they were encouraged to review the available draft plan document. This draft, along with the introductory talk, set the basis for the debate held during the session.
- DEVELOPMENT: Participants present their contributions on the thematic working axis of the sectoral roundtable. Several territorial actors were invited to each session based on the following criteria:
    - Association with the study area.
    - Diversity in terms of typology: citizen entities, universities, administrations, etc.
    - Representation of government levels: state, regional, provincial, or municipal.
    - Association with each session's topic/theme.
    - Representativeness based on gender perspective.

As the facilitator, the consulting team guiding the session collected and categorized all contributions based on their potential to drive the PAT; that is, whether they were consistent with the spatial planning instrument's capacity to establish binding determinations.

- CONCLUSIONS: To close the session, an ideogram developed "just in time" (during the meeting) by the facilitating team was shared with the attendees. It contained the actors' contributions based on their potential for integration into measures within the PATs' competency scopes. Similarly, recurrences were identified and possible relationships between contributions were established (see Figure 6).

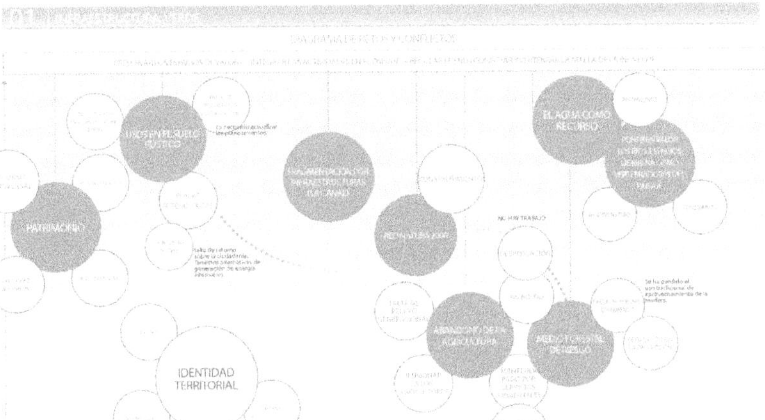

Figure 6. Example of a feedback ideogram developed within the framework of drafting the initial version of the PAT for PATCCCV at the Bolbaite territorial roundtable in June 2021. The contributions in the upper part of the panel are actions with greater potential to drive the PAT. Source: own elaboration.

5. Results and Discussion

5.1. Balance of Participation in Thematic Roundtables and the Addressed Topics

One of the main challenges of any participation process is ensuring that relevant information and input are gathered for the plan. In the case of thematic roundtables, rigorous and in-depth debate on PAT issues was conducted. This is why the sample of stakeholders participating in the roundtable should match the workshop dynamics. The number and attendance ratio (% of invited people) of attendees per session were also analyzed.

First, it should be noted that launching the roundtable call in extensive territories that lack a minimal governance structure requires significant effort. This task must be accompanied by substantial persuasion to overcome the reactivity that many people exhibit towards processes led by the regional administration. This is worse if there have been several previous attempts in the same locations that did not complete the approval process.

Table 3 shows that participation ranged from 6 to 18 people for the set of thematic roundtables, with an average attendance of 13–14 people. For the PATVB roundtables, which were conducted under COVID-19 restrictions, participation was also extended to people who could follow the session via "streaming".

Table 3. Attendance at each of the territorial roundtables. Source: own elaboration.

ATTENDEES AT EACH OF THE TERRITORIAL ROUNDTABLES.				
	GREEN INF.	SETTLEMENTS	INFRASTRACTURE	SOCIOECONOMIC
PATAE	18	9	16	6
PATVB	12	9	10	11
PATCCV	MT1 BOLBAITE	MT2 ONTINYENT	MT3 COCENTAINA	MT4 DENIA
	17	17	13	15

Regarding the topics addressed, while thematic roundtables were held based on the PAT action axes in PATAE and PATVB, due to the large area covered by the PAT in PATCCCV, the territorial roundtables were held in different counties. In each county, three action axes were chosen based on the areas that the PAT had the competency to address, i.e., green infrastructure, settlement model, and infrastructure model (see Table 3).

Analyzing the attendance ratios, which enable us to compare the proportion of invited agents versus those who actually attended, shows a significant decrease in PATCCCV, where about 70 involved agents were invited per session. Attendance ranged between 13% and 25% of the total invited persons (see Table 4).

Table 4. Attendance ratio per territorial roundtable. Source: own elaboration.

ATTENDANCE RATIO FOR EACH TERRITORIAL ROUNDTABLE.				
	GREEN INF.	SETTLEMENTS	INFRASTRACTURE	SOCIOECONOMIC
PATAE	0.69	0.35	0.52	0.22
PATVB	0.39	0.36	0.56	0.33
PATCCCV	MT1 BOLBAITE	MT2 ONTINYENT	MT3 COCENTAINA	MT4 DENIA
	0.25	0.25	0.18	0.13

5.2. Degree of Balance in the Contributions to Each of the Axes

Comprehensive PATs, such as those analyzed in this study, must respond to the thematic axes for which the plan has legal jurisdiction. To this end, participation was designed so that the actors could contribute to the thematic axes that the PATs could influence when defining the spatial model of the entire area.

The results in Table 5 show that in PATAE and PATVB, the issues represented by the axes were more balanced, highlighting infrastructure for PATAE and socioeconomic factors for PATVB, where agriculture and tourism are relevant factors. In PATCCCV, which has extensive territories with significant rural components, the proposals focused on the first block of green infrastructure, with suggestions to address issues such as agricultural landscapes, forest areas, and river corridors. It is striking how little attention was paid to socioeconomic aspects, despite the region being a territory with good internal articulation (with previous initiatives of interest such as the CONCERCOST Project, in the framework of the EU TERRA Programme in 1996) and having its own credit institutions (e.g., Caixa Ontinyent and Caixa Popular). However, it should be noted that those promoting the formulation of the PAT for this area were in the Steering Committee of the Consortium of Central Valencian Counties, where the main socioeconomic actors are represented.

Table 5. Percentage of contributions to each PAT based on thematic axis. Source: own elaboration.

% CONTRIBUTIONS RECEIVED BY MAIN LINE OF ACTION				
	GREEN INFRAST.	SETTLEMENTS	INFRASTUCTURE	SOCIOECONOMIC
PATAE	25.73	19.30	33.33	21.64
PATVB	27.2	20.08	19.67	33.05
PATCCCV	46.61	22.03	22.03	9.3

5.3. Contributions by Type of Actors

The same actor typologies participated in the three PATs, with minor differences. While members of the municipal urbanism technical team (who had already participated ex ante in bilateral meetings with the planning team when the PAT was launched by the Valencian regional government) were invited to attend the thematic roundtables in PATAE, participants in the working roundtables in PATVB were political officials. This is because the PAT roundtables in PATVB coincided with the development of a Strategic Climate Change Adaptation Plan promoted by the Valencian Government Presidency (Plan Renhace). The political and budgetary involvement of the Valencian Government in the execution of planned actions, together with the development of a multilevel governance model where all administrations could participate, was a catalyst for the involvement of actors who do not usually participate.

The stakeholders who participated in the territorial and sectoral roundtables belonged to the following groups:

1. Civil society: Understood to be the group of individuals who, in their own right, have the political status of Valencian citizenship or reside in the planning area. Within this category are also citizen entities or associations whose scope of action occurs within the PATs' thematic axes. Civil society can be organized, forming part of a collective with its own autonomy that does not originate from the institutions of the State, or operate individually as private citizens.
2. Academia: Universities and research centers with experts in the different subject areas of action.
3. Economic actors: Business people and developers whose scope of action corresponds to or affects the territory covered by the PAT.
4. Administration: Includes technicians from different political administrative levels, such as state, regional, provincial, or municipal. Only the second and fourth levels made contributions during the roundtables.
5. Professional associations: Concerned with or able to provide information on any of the contents or thematic axes in each PAT.

Similarities and differences are observed in the participation patterns (see Table 6). Organized civil society played a very active role in all three PATs, while unorganized civil society was less involved. Local administration technicians made the highest number of contributions in PATAE. A high level of contributions from the economic sector was also observed in PATVB and PATCCCV.

Table 6. Percentage of total contributions based on actor groups: 01, civil society; 02, academia; 03, economic actors; 04.1, regional administration; 04.02, municipal administration; 05, professional associations. Source: own elaboration.

PERCENTAGE OF CONTRIBUTION BASED ON TYPE OF STAKEHOLDER.						
	01	02	03	04.01	04.02	05
PATAE	26.19	7.14	4.76	23.81	35.71	2.38
PATVB	39.75	18.83	26.78	0.84	2.93	2.09
PATCCCV	43.22	5.93	27.96	0	12.71	10.16

The lack of contribution from administrations or entities at the county and provincial levels is noteworthy. Analysis of entities at the regional and national levels showed that members of the Valencian Government from departments other than the one promoting the PAT participated only in PATVB, specifically in a territorial area that had suffered flooding from the September 2019 DANA and where the Vega Baja Regeneration Plan (Plan Vega Renhace) had been launched.

5.4. *Type of Contributions*

The contributions made by different actors were classified into three categories:

1. • Informative: Providing data to enhance knowledge of the territory or specific problems, aiding in analysis and diagnosis.
2. • Reflective: Questioning certain aspects of the PAT's scope and initiating debate as a catalyst for reviewing diagnoses and proposals.
3. • Propositional: Proactive and concrete proposals that stakeholders wanted to be included in the PAT.

Comparison of the data in Table 7 shows that while informational contributions stood out in PATAE, in PATVB, actors made more propositional suggestions, potentially motivated by prior experience and preparation of the aforementioned Plan Vega Rehance and its corresponding action plan, which had been launched previously; in fact, many proposals were identical to those raised during Plan Vega Renhace. In PATCCCV, reflective and propositional contributions outweighed informative contributions. It should be noted that in this PAT, which had more condensed sessions, more propositional contributions were encouraged, probably due, again, to the territory's own articulation and prior experiences.

Table 7. Types of contributions. Source: own elaboration.

PERCENTAGE OF THE TYPES OF CONTRIBUTIONS.			
	INFORMATIVE	REFLECTIVE	PROACTIVE
PATAE	45.02	30.41	24.56
PATVB	17.57	25.94	56.49
PATCCCV	18.03	47.54	34.42

5.5. *Level of Alignment of Contributions with the PATs' Nature and Objectives*

Contributions were classified based on their potential to drive the PAT; that is, the extent to which they aligned with the PATs' objectives and whether they could be developed consistently based on the PATs' potential to establish binding determinations for urban planning and other derived planning that must be developed by municipalities and sectors. The following categories were considered:

A. Dissent: These aspects are contrary to the PAT's objectives, making their integration impossible.
B. Low potential to drive the PAT: This category includes contributions that, due to the scale of the PAT or legal jurisdictions, can influence the PAT only indirectly, and therefore have very limited scope.
C. Medium potential to drive the PAT: Aspects that could be integrated into the plan. Often, these are contributions referring to instruments or plans of a more limited scale.
D. High potential to drive the PAT: These are considerations that are fully aligned with the PAT's objectives and can be incorporated into urban planning.
E. Exceeds the PAT's competency scope: These include proposals that fall outside the corresponding PAT's reach.

Comparative analysis of these data showed how the debate framework was understood by the stakeholders (see Table 8). Sometimes contributions cannot drive the PAT; therefore, the team responsible for the participation plan during these stages of plan development, prior to initial approval, where participation is not mandatory, needs to sift through the information to identify proposals that can be integrated into and ultimately improve the PAT in order to better define the desired spatial model for the area for which the plan is being developed.

Table 8. Percentages of contributions and their alignment with PAT objectives: (A) dissent; (B) low potential to drive the PAT; (C) medium potential to drive the PAT; (D) high potential to drive the PAT; (E) exceeds the PAT's competency scope. Source: own elaboration.

PERCENTAGES OF CONTRIBUTIONS AND THEIR ALIGNMENT WITH PAT OBJECTIVES					
	A	B	C	D	E
PATAE	0	35	40	22.5	2.5
PATVB	6.32	13.68	28.42	27.37	24.21
PATCCCV	0	28.66	22.13	42.62	6.5
TODOS	3.19	15.96	30.32	30.85	19.68

More than 60% of the total contributions had the potential to drive the PAT, indicating the usefulness of the participation tasks. Up to 20% of the contributions exceeded the PATs' competency scope, while only 3% contradicted the PATs' objectives. This indicates the extent to which clearer information and better training on how to effectively participate in these sessions, and on why spatial planning is needed, should be provided in the future.

Examining the figures for each of the three PATs shows that PATAE had no contributions contrary to the objectives set at the start of the session; however, up to 35% of suggestions had little potential to be included in the PAT document as it is a spatial rather than an urban instrument (in this case, the participation of municipal urbanism technicians was prevalent).

In PATVB, some stakeholders made contributions that were contrary to the plan's objectives; additionally, 24% of the suggestions exceeded the PAT's competencies. Here, over 30% of issues raised could be implemented in this sub-regional spatial plan. One action proposed during the Strategic Plan Vega Renhace was to approve PATVB. This may explain why some stakeholders saw PATVB as an opportunity to reiterate similar measures suggested during the review of the Renhace Plan, despite them clearly being different planning instruments. Again, clearer information and better training on spatial planning are needed.

In PATCCCV, over 40% of suggestions had high potential to drive the PAT and could therefore be implemented. If we add medium-potential contributions, the figure increases to nearly 70%, making this the PAT for which stakeholder participation proved most fruitful, and which can be positively associated with the PAT's promoting body being the Central Counties Consortium itself (bottom-up sense), which had prior experience with this objective.

5.6. External Contextual Factors Influencing PAT Participation and Outcomes

Every participation process is influenced by the territorial and temporal context in which it occurs. This allows for a better comparison of citizen and stakeholder involvement in each PAT (see Table 9). Therefore, this section qualitatively evaluates some aspects associated with the participation context, distilled from reflections by the three authors who were members of the PATs' planning team. This complements the previous quantitative analyses.

Table 9. Exogenous factors influencing the participation process: (a) existence of prior participation; (b) participation intensity relative to territorial area (sense of belonging); (c) emergency situation (COVID-19); (d) existence of some process catalyst; (e) environmental risks; (f) good governance and political alignment. (++ > very positive impact on the participation process (PP); + positive impact on the PP; - negative impact on the PP; -- very negative impact on the PP). Source: own elaboration.

EXTERNAL ASPECTS OF THE PAT THAT INHIBIT AND REINFORCE THE PARTICIPATION PROCESS						
	(a)	(b)	(c)	(d)	(e)	(f)
PATAE		++		+		(--)
PATVB	++	++	(--)	+	++	(-)
PATCCCV				++	++	++

One factor that enriches the participation process and gives it more depth is having recently conducted similar work (not a large time lag). Plan Renhace had involved working with territorial stakeholders the year before as part of prioritizing actions to be launched in Vega Baja county after the 2019 floods; conversely, in some cases, certain actors used the same conceptual framework of contribution, diminishing the activity's effectiveness, as shown above.

Robust results are achieved, among other reasons, by deploying different complementary participation activities. The intensities of activities for the three PATs were comparable; however, PATCCCV is a very large area (covering eight counties in the Valencian Autonomous Region), and bringing together actors from all counties was a significant logistical challenge.

PATVB was undertaken during a time when significant COVID-19 restrictions were still in place, and public information sessions could not be held prior to the thematic roundtables. Additionally, the pandemic made participation in the roundtables more difficult as it was harder to separate the participants from the pandemic's impact such that they could focus on the tasks in the sessions.

An interesting aspect to consider is whether there are emerging processes in the territory that drive participation and reinforce the PAT. This was the case, for example, in the open debate on how photovoltaic energy should be implemented in the territory, given the current energy transition. While plans were being drafted in PATCCCV and PATVB, conflicts emerged between social groups and solar plant developers. This reinforced the recognition of the usefulness of such supra-municipal instruments to regulate this activity.

Another key issue driving participation in PATCCV was the CV60 extension project in La Safor County. CV60 is a highway that—while considered necessary by several authors—generated significant controversy by fragmenting an agricultural landscape with high cultural value. This topic attracted interested people to actively participate, and many contributions were collected through various activities on this topic.

Environmental risk perception by the population was also a relevant factor for participants. The 2019 DANA floods in Vega Baja, particularly in PATVB, and the forest fire risk in PATCCCV generated considerable interest among the participating actors, motivating them to contribute to the development of the PAT draft.

The governance framework has proven crucial for the proper functioning of the participatory process. While PATCCCV had a coordination body that became involved in the participation process itself and encouraged participation among territorial actors, the lack of political alignment in PATVB and PATAE, as well as differences between municipal governments and the regional administration, made the process more difficult and may even block it (as in PATAE).

6. Conclusions

As a main conclusion, participation in PATs—although regulated—can be beneficial and effective, both procedurally and substantively. The plans' content helps to better define the desired spatial model from the substantive point of view, while from the procedural point of view, it reduces barriers and streamlines processes. Obtaining approval for the plan is easier when conflicts and barriers to implementation are reduced; thus, conducting and improving participation during all stages of the plan elaboration process, both mandatory and voluntary (e.g., before the plan's initial approval), are useful in terms of obtaining input and promoting the plan.

Regarding the relationship between the resources employed to convene stakeholders and the final participation ratios, it should be emphasized that participation in places lacking a previous participatory culture requires additional effort. These efforts can sometimes exceed the resources allocated for the plan's development; for this reason, it is crucial to use ICT and optimize the method of participant selection to improve efficiency. If the covered area is extensive, then the challenge is bigger. In this regard, it is worth considering the need for greater involvement in communicating the plan itself and in generating sustained participation structures that can catalyze governance processes. Sometimes these structures already exist; as in the case of PATCCCV where, although this advantage was not observed in the participation ratios, it was key in dynamizing the process.

Although material resources may be constrained, participation can be increased by investing time in preparatory training sessions that emphasize the PAT's competencies and legal jurisdictions, provide information on how to participate, explain how stakeholders' proposals can be taken into account, and inform participants on why, when, and how they can ask for feasible transactions. It is all about introducing and establishing a territorial culture among citizens so that they clearly understand how they can contribute, as well as how these spatial planning instruments influence their daily lives.

Analyses of contributions by type of stakeholder and PAT thematic axes showed that civil society has significant weight in green infrastructure proposals. At the sub-regional level, green infrastructure is seen as a landscape and culturally important territory. Agricultural landscapes, heritage sites, and landscape resources are referred to in this section more frequently than other items related to territory connectivity or ecosystem service provision, which are more typical of green infrastructure but are more alien to the actors involved. Although economic actors had an uneven presence in the PATs, they are key to addressing the relationship between territory and economic activity. These actors' contributions often focus on infrastructure and economic activities; however, contributions related to settlements have been more distributed among groups of stakeholders when making proposals. Despite this evidence on the target of the agents involved and their "own" topics, it is necessary to investigate ways of encouraging the agents to step out of their comfort zones, as well as to confront scenarios regarding thematic axes that are not included in their initial aspirations.

Analysis of stakeholders' understanding of the true meaning and scope of participation revealed that less than a quarter of the contributions exceeded the scope of the PAT or went against the objectives set out in the draft document. On the other hand, more than 55% of the contributions fit within the PAT's scope and could support actions for which the PAT had the authority to propose management measures and alternatives. The debate among the stakeholders involved was fluid and resulted in a set of measures that could be implemented in a land use planning instrument of this nature. The ability of the proposals to drive the plan makes the participation process more efficient. Allowing room for maneuvering in the areas of discussion should be a primary objective in participation plans.

Participation can be greatly affected by contingencies and contextual conditions in complex processes, such as the development and approval of spatial planning instruments. An extraordinary event, such as a pandemic, can make a participatory process extremely difficult. Conversely, unforeseen circumstances may arise that can become catalysts for participation, if the involved stakeholders recognize that the planning tool can help mediate and advance the resolution of a conflict. However, these temporal hot topics (e.g., infrastructure construction and the proliferation of photovoltaic installations that have a significant impact on the territory) can capture the attention of the stakeholders during the debate and distract from the holistic vision. Overall, it is important to manage contingencies for the benefit of the planning process and avoid placing planning at the service of contingencies.

The relatively low level of conflict encountered when deciding the territorial model that the plan should propose in the participatory spaces developed in this non-mandatory phase contrasts with the strong opposition that plans face during the public disclosure phase (mandatory and binding). Certainly, more hegemonic actors avoid participating in these phases in order to influence the plan through other means, such as resorting to the courts and the judicialization of differences and conflicts. However, non-participation in these cases should be considered an indicator of a lack of willingness to engage in prior conciliation when the courts are required to issue a ruling (for instance, as already happens in labor laws, where attendance of pre-trial conciliation is binding), thus strengthening the already proven usefulness of such participation.

Finally, it is important to note that the plans were made during the participation process. In fact, before the roundtables, the only document shared with stakeholders was a draft version, which was very synthetic. The PAT is not modified by participation; it is derived from it and inspired by the pool of suggestions. Only one of the three PATs could follow the administration process, the PAT of Vega Baja, and none of them are approved at this moment.

Author Contributions: Conceptualization, J.F.-D. and P.L.-G.; methodology, J.F.-D., I.D.-T. and P.L.-G.; validation, J.F.-D., I.D.-T. and P.L.-G.; formal analysis, J.F.-D. and I.D.-T.; investigation, J.F.-D., I.D.-T. and P.L.-G.; resources, J.F.-D., I.D.-T. and P.L.-G.; data curation, P.L.-G. and I.D.-T.; writing—original draft preparation, J.F.-D., I.D.-T. and P.L.-G.; writing—review and editing, J.F.-D. and I.D.-T.; visualization, I.D.-T.; supervision, J.F.-D.; project administration, J.F.-D. and I.D.-T.; funding acquisition, J.F.-D. All authors have read and agreed to the published version of the manuscript.

Funding: This study is part of the Spanish research project "Propuesta de diseño institucional y comunitario para una ordenación del territorio integral en la transición hacia una economía sostenible"

(GOBEFTER-III), reference PID2021-128356NB-I00, financed by MCIN/AEI/10.13039/501100011033 and the European Regional Development Fund (ERDF).

Data Availability Statement: https://mediambient.gva.es/va/web/planificacion-territorial-e-infraestructura-verde/planificacio-territorial-en-tramitacio.

Conflicts of Interest: The authors declare no conflicts of interest.

Notes

1 This is the name for Subregional Spatial Plans in the Valencian Autonomous Region. In Spain, according to Article 148 of the Spanish Constitution (1978), the elaboration and approval of spatial plans is exclusively in the hands of Regional (Autonomous) Governments.

2 For an overview of the current situation regarding spatial planning instruments and legislation in all 17 Autonomous Regions of Spain, see [4].

3 These relationships are ensured and regulated by Law 39/2015, of October 1, on the Common Administrative Procedure of Public Administrations.

4 This is an initial level of participation characterized by a unilateral relationship in which the government produces and delivers information to citizens and stakeholders. It includes proactive measures to disseminate information, but there is no real feedback between the administration and the administered, as could occur during consultation or at active participation levels.

5 On methods, [13,14].

6 This does not usually happen, explaining why the demands and proposals raised in many participatory processes often have little to do with what the plan can or cannot do.

7 This is in accordance with the aforementioned Directive 2001/42/EC, transposed into Spanish law through Law 6/2001, of May 8 (currently Law 21/2013, of December 9, on environmental assessment), followed by various Spanish and Valencian Community public participation laws (Law 11/2008, of July 3, now repealed by the current LLEI 4/2023, of April 13, of Generalitat, on Citizen Participation and Promotion of Associative Life in the Valencian Autonomous Region).

References

1. Beunen, R.; Duineveld, M.; van Assche, K.A.M. Evolutionary Governance Theory and the Adaptive Capacity of the Dutch Planning System. In *En Spatial Planning in a Complex Unpredictable World of Change: Towards a Proactive co-Evolutionary Type of Planning Within the Eurodelta*; InPlanning: Driebergen-Rijsenburg, The Netherlands, 2016; pp. 98–116. [CrossRef]
2. Farinós, J. Gobernanza: Refinando significados desde el debate teórico pensando en la práctica. Un intento de aproximación fronética. *Desenvolv. Reg. Em Debate* **2015**, *5*, 4–24. [CrossRef]
3. European Commission (EC). *Directorate-General for Regional and Urban Policy: The EU Compendium of Spatial Planning Systems and Policies*; Publications Office: Luxembourg, 1997; Available online: https://op.europa.eu/en/publication-detail/-/publication/059fcedf-d453-4d0d-af36-6f7126698556 (accessed on 29 March 2024).
4. Dasí Farinós, J. *Marco Legal y Procedimental de la Ordenación del Territorio en España: Diagnóstico y Balance*; Thomson Reuters-Aranzadi: Navarra, Spain, 2020.
5. Lloret, P.; Farinós, J. *La Dimensión Participativa en el Diseño de Políticas Urbanas. El Caso Valenciano*; Gestión y Análisis de Políticas Públicas: Madrid, Spain, 2018; Volume 20. [CrossRef]
6. Almenar-Muñoz, M.; Antequera-Terroso, E. Acerca del procedimiento de evaluación ambiental estratégica y la incardinación de la salud en el proceso de elaboración de los planes. In *Marco Legal y Procedimental de la Ordenación del Territorio en ESPAÑA, Diagnóstico y Balance*; Farinós, J., Peiró, E., Farinós, J., Eds.; Thomson Reuters Aranzadi: Navarra, Spain, 2020; pp. 59–82.
7. OCDE. Recomendación del Consejo Sobre Gobierno Abierto. 2017. Available online: https://www.oecd.org/gov/open-government/recomendacion-sobre-gobierno-abierto.htm (accessed on 18 February 2022).
8. Healey, P. *Collaborative Planning*; Macmillan Press Ltd.: London, UK, 1997.
9. Habermas, J. *The Theory of Communicative Action*; Thomas, M., Ed.; Beacon Press: Boston, MA, USA, 1984.
10. Flyvbjerg, B. Phronetic planning research: Theoretical and methodological reflections. *Plan. Theory Pract.* **2004**, *5*, 283–306. [CrossRef]
11. Foucault, M. The ethic of care for the self as a practice of freedom. In *The Final Foucault*; Bernauer, J., Rasmussen, D., Eds.; MIT Press: Cambridge, UK, 1988; pp. 112–131.
12. Farinós, J.; Vera, O. Planificación territorial frenética y ética práctica. Acortando las distancias entre plan y poder (política). *Finisterra Rev. Port. De Geografía* **2016**, *51*, 45–69.
13. Creighton, J.L. *The Public Participation Handbook: Making Better Decisions Through Citizen Involvement*; Jossey-Bass: New York, NY, USA, 2005.

14. Latasa, I.; Lloret, P. *Técnicas de Participación Pública. Ordenación del Territorio y Medio Ambiente*; Farinós, J., Olcina, J., Eds.; Tirant Humanidades: Valencia, Spain, 2022; pp. 553–578.
15. Arnstein, S.R. A ladder of citizen participation. *J. Am. Inst. Plan.* **1969**, *35*, 216–224. [CrossRef]
16. Prieto-Martín, P. *Las Alas de Leo. La Participación Ciudadana del Siglo XX*. Asociación Kyopol—Ciudad Simbiótica. 2010. Available online: https://libros.metabiblioteca.org/bitstreams/a74bdfe5-62cf-490d-9e86-2fb2627b15c9/download (accessed on 28 March 2024).
17. Miskowiak, D. *Crafting an Effective Plan for Public Participation*; Center for Land Use Education, University of Wisconsin-Stevens Point: Stevens Point, WI, USA, 2004.
18. Turnhout, E.; Van Bommel, S.; Aarts, N. How participation creates citizens: Participatory governance as performative practice. *Ecol. Soc.* **2010**, *15*. Available online: https://edepot.wur.nl/157421 (accessed on 29 May 2025). [CrossRef]
19. Farinós, J. Bases conceptuales de la gestión territorial: Inteligencia territorial y ética práctica. In *Observatorios Territoriales Para el Desarrollo y la Sustentabilidad de los Territorios. Vol. 1: Marco Conceptual y Metodológico*; Cittadini, E., Vitale, J., Eds.; Instituto Nacional de Tecnología Agropecuaria: Mendoza-San Juan, Argentina, 2017; pp. 12–25.
20. Neffati, H.; Girardot, J.-J. L'intelligence Territoriale, un Concept Émergent. Les Cahiers d'Administration, Hors-Série de la Revue Administration (Supplément au n°243), Collection Territoires Pour Demain. 2014. Available online: https://shs.hal.science/halshs-01146237 (accessed on 28 March 2024).
21. Quaglia, A.; Guimaraes Pereira, Â. *Engaging with Food, People and Places*; Publications Office of the European Union: Luxembourg, 2021; Available online: https://publications.jrc.ec.europa.eu/repository/handle/JRC121910 (accessed on 29 May 2025).
22. Turnhout, E.; Metze, T.; Wyborn, C.; Klenk, N.; Louder, E. The politics of co-production: Participation, power, and transformation. *Curr. Opin. Environ. Sustain.* **2020**, *42*, 15–21. [CrossRef]
23. Nowotny, H.; Scott, P.; Gibbons, M. *Re-Thinking Science. Knowledge and the Public in an Age of Uncertainty*; Polity Press: Cambridge, UK, 2001.
24. Funtowicz, S.; Ravetz, J. Science for the post-normal age. *Futures* **1993**, *25*, 739–757. [CrossRef]
25. Gallego, J.R.; Pitxer, J.V.; Farinós, J. Marco general y narrativas predominantes que explican la forma en que se desarrollan los procesos: Una aproximación a la OT en España y las posibilidades de su conexión con la política del desarrollo económico regional. In *Evaluación de Procesos: Una Mirada Crítica y Propositiva de la Situación de la Política e Instrumentos de Ordenación del Territorio en España*; Farinós, J., Peiró, E., Rando, E., Eds.; Thomson Reuters Aranzadi: Navarra, Spain, 2021; pp. 45–82.
26. Hernández-Partal, S. Los Planes de Acción Local de la Agenda Urbana Española: Su papel en el urbanismo del siglo XXI. *Ciudad. Y Territorio. Estud. Territ.* **2023**, *55*, 829–858. [CrossRef]
27. Farinós, J. Presentación. Tratando de aproximar la teoría procedimental, con-textual y sustantiva de la planificación. In *Evaluación de Procesos: Una Mirada Crítica y Propositiva de la Situación de la Política e Instrumentos de Ordenación del Territorio en España*; Farinós, J., Peiró, E., Rando, E., Eds.; Thomson Reuters Aranzadi: Navarra, Spain, 2021; pp. 17–43.
28. Criado, A.M. *Guía Metodológica: Estudios de Paisaje*; Generalitat Valenciana: Valencia, Spain, 2012.
29. Bobbio, L. Designing effective public participation. *Policy Soc.* **2019**, *38*, 41–57. [CrossRef]
30. Chilvers, J.; Kearnes, M. *Remaking Participation: Science, Environment and Emergent Publics*; Routledge: London, UK, 2015.

land

MDPI

Article

Land Use Patterns and Small Investment Project Preferences in Participatory Budgeting: Insights from a City in Poland

Katarzyna Groszek [1], Marek Furmankiewicz [2,*], Magdalena Kalisiak-Mędelska [3] and Magdalena Błasik [4]

1 Independent Researcher, 09400 Płock, Poland
2 Institute of Spatial Management, Wroclaw University of Environmental and Life Sciences, 50357 Wroclaw, Poland
3 Institute of Management and Quality Sciences, Calisia University, 62800 Kalisz, Poland; m.kalisiak-medelska@uniwersytetkaliski.edu.pl
4 Independent Researcher, 55040 Domasław, Poland
* Correspondence: marek.furmankiewicz@upwr.edu.pl

Abstract: This article presents a spatial analysis of projects selected by city residents and implemented in five successive editions (2015–2019) of the participatory budgeting in Częstochowa, Poland. The study examines the relationship between the type of hard projects (small investments in public infrastructure and landscaping) and the pre-existing characteristics of the land use of each district. Kernel density estimation and Spearman correlation analysis were used. The highest spatial density occurred in projects related to the modernization of roads and sidewalks, recreation, and greenery, indicating a relatively high number of proposals within or near residential areas. Key correlations included the following: (1) greenery projects were more common in districts lacking green areas; (2) recreational infrastructure was more frequently chosen in areas with significant water features; (3) street furniture projects were mostly selected in districts with sparse development, scattered buildings, and postindustrial sites; (4) educational infrastructure was often chosen in low-density, but developing districts. The selected projects often reflect local deficits in specific land use or public infrastructure, but also stress the predestination of the recreational use of waterside areas.

Keywords: urban development; land use structure; social preferences; small-scale infrastructure; participatory budgeting; Poland

1. Introduction

The development of cities based on spontaneous processes or based solely on economic market rules often leads to their unsustainable development, excessive land-take and social inequalities in access to public infrastructure [1–4]. To counteract these types of problems, various participatory spatial planning initiatives and the partnership management of local public resources are often used, open to the needs of the local community [5–10]. These methods include, e.g., participatory budgeting (PB), in which residents propose and select projects for implementation using different types of voting procedures [11,12]. PB was initiated in Brazilian cities at the end of the 20th century [11,13]. In the early 2000s, these methods have spread virtually throughout the world [14–16]. PB has also been developing voluntarily in Polish cities and towns since the 2010s [17–20], and has been obligatory (based on the law passed by the Polish Parliament) in 66 cities with county rights (*poviat* in the Polish language) since 2018 [21].

The source literature on PB is dominated by analyses of social and political issues related to participatory democracy and social participation in PB [12,18,19,22,23]. The authors often describe the types of projects proposed and implemented in PB, but most commonly do not provide a detailed analysis that addresses the reasons for such a choice [24,25]. There are only a few attempts to determine the relationship between the social and demographic characteristics of the residents in the district and the PB results [26,27]. Analyses that attempt to explain the relationship between current land use or public infrastructure development in urban districts, and projects proposed by their residents are very rare.

The main purpose of our paper is to analyze the spatial (geographical) distribution of the PB hard projects (related to public infrastructure and landscaping) in the studied city (Częstochowa in Poland) and their relation with the existing land use structure, by applying basic statistical methods. The main research questions are as follows:

RQ 1: *What is the spatial distribution of projects submitted in the city's districts?*

RQ 2: *Does the structure of current land use in city districts correlate with the type of project chosen by residents?*

In this paper, we deliberately do not cover the issues of voting organization and general social activity in PB, although we refer to them in the discussion. The issues related to the organization of PB, and their social and political conditions have already frequently been addressed in the literature [16,22]. The novelty of our considerations is the attempt to find a relationship between land use and the type of projects chosen by residents.

The outcomes of our research may be useful to researchers in performing a comparative meta-analysis of project structures selected through PB procedures by residents in city districts and for planners and city officials as tips for more effective urban planning.

2. Possible Impact of Land Use and Infrastructure Development on PB Results

A general feature of the development of cities and regions in a market economy is uneven development, which can be analyzed on all geographical scales [28]. The development of city districts and neighboring suburbs often does not result from rational planning, but remains a characteristic of disorderly urban sprawl [1]. Urban development often results, above all, from the preferences of economic interest groups (e.g., developers and investors) or the spontaneous actions of the population migrating to the city [1,4,29,30]. Sometimes, the overexploitation of the assets of a public space by the economic sector is observed, neglecting other social needs [31]. In some areas, the needs of residents who wish to have access to both basic living infrastructure (e.g., water supply, sewage) and close access to the areas that passively provide them with various services (such as places for resting, recreation, sport activity, mobility, etc.) are not fully satisfied [32,33]. The need to use public spaces and infrastructure in a city also changes in the processes of historical, social, and functional transformation [1,34,35].

To meet the previously unmet and newly emerging needs of residents, various types of public consultations and dialogue, including participatory budgeting (PB), are organized. PB is intended to increase interest in city problems, make the residents more active, meet their needs for small-scale infrastructure, and help identify spatial planning needs [5,36]. PB often includes infrastructure investments (hard projects) and resident activity or public services (soft projects) [25]. In PB, local inhabitants propose missing infrastructure elements and land development methods, followed by voting on which proposals should be allocated with public funds. Research indicates that PB has a positive impact on social justice and well-being [27,37,38] and increases the effectiveness of public funds in meeting the needs of residents [39]. PB also shows the potential to counteract the shortcomings of the top–down approach in urban planning [40].

The creation of PB is of particular interest in less developed countries with an average democratic culture and relatively low trust in the authorities [41]. In rich democratic countries with high traditions of social participation, such as the European Nordic countries, PB is not so popular [42]—probably due to the high transparency of the local authorities' policies that take into account resident needs using other forms of social dialogue and consultations [16,43].

In the opinion of Szczepańska et al. [44], the scope of PB projects is most often for addressing the immediate needs of residents, and, therefore, they are predominantly implemented in their neighborhood. In Poland, projects concerning the development of urban infrastructure were among the most reported by residents in many cities [20]. Descriptive analyses suggest that the type of PB projects selected and their location are related to the characteristics of land use and the shortage of desirable local infrastructure [45].

In underdeveloped regions, PB helps meet the basic needs of the poorest social classes, including access to water pipes, sewage networks, or roads in poorer districts [15,33]. In Brazil, the adoption of PB increased public expenditure on sanitation, in line with demands in participatory forums, and was an important tool to reduce inefficiency in the allocation of public expenditure [46]. In Belo Horizonte, fragmented or inequitable infrastructure distributions, such as underdeveloped sewage or sidewalk networks, forced lower-income neighborhoods to focus on essential public works during PB cycles [47]. Similarly, in Porto Alegre, PB was also successful in providing basic infrastructure and housing for informal settlements [13,33].

In more developed countries, projects related to leisure, sports, recreation, and educational institutions are more popular because cities are better equipped with basic living infrastructure [38,48–50]. In Chengdu, China, recognition of current infrastructure investments led rural residents to use PB mechanisms to propose connectivity and basic amenities in newly integrated peri-urban zones [51]. Bernaciak and Kopczyński [52] found that, in less developed cities in eastern Poland, inhabitants preferred hard projects that raised the level of social and economic development, whereas in developed, highly urbanized regions, they preferred greenery for leisure. This indicates that PB results can be related to the level of infrastructure and economic development in a given district. Similarly, Martela [53] noted in the analysis of PB in cities in Poland that citizens prefer to vote for investment in public spaces (hard projects). Citizens were less interested in noninvestment (soft) projects, such as the organization of cultural or educational activities. He concluded that in Poland, PB functioned primarily as a tool for citizens and local authorities to make joint decisions on public space.

In contrast, in relatively affluent and centrally located areas characterized by high levels of development and social capital, PB initiatives tend to focus on projects that "beautify" the space, for example, on greenery, flower squares, art installations, murals [54]. However, such interventions are frequently small in scale and narrowly focused, which may inadvertently contribute to the fragmentation of urban space. This may exacerbate existing differences in the quality of public spaces and the level of development between different parts of the city. Ahn et al. [45] compared the densely built-up and densely populated central Margareten district, with the more peripheral and less built-up Simmering district in Vienna, Austria. The Margareten district showed a greater interest in environmental projects (including greenery) and a markedly greater interest in social-type projects (e.g., cultural events, sports facilities, seating). In contrast, the Simmering district showed greater interest in urban traffic projects (e.g., public transportation, bike paths, and traffic regulation).

Research on PB in cities in Poland most frequently found the following needs: (1) to expand the infrastructure related to the mobility of residents (roads, sidewalks, separated bicycle paths); (2) to develop green areas, where additional recreational infrastructure is often located (resting areas, playgrounds, etc.); and (3) to develop sports and recreational areas, mainly for children and youth, often next to educational institutions [21,25,44,49,53]. Numerous case studies suggest that the type of project selected depends on the characteristics of current developments in a given city district. The most basic principle concerns the frequent location of projects close to the place of residence, so in old residential areas [24]. According to Kociuba and Rabczewska [24], in Lublin (Poland), green projects were also implemented to reduce noise and the spread of exhaust fumes in areas that have a large share of the main communication arteries. In old post-industrial districts, investments were made in multigenerational meeting places. Projects related to recreational and leisure infrastructure, as well as greenery development, were also common in housing estates with numerous multi-family buildings. In the analysis of PB in Wrocław (Poland), a spatial concentration of projects referring to the development of backyards was observed in old, depreciated districts around the center, which posed a high risk of marginalization [55–57]. In turn, projects related to road and sidewalk modernization or construction were consistently selected in new peripheral districts characterized by a scattered expansion of housing estates, without the appropriate development of communication areas [56].

This review of the literature suggests that the types of project selected for implementation in PB may depend on the previous structure of land use in districts and the development of public infrastructure, thus expressing local social needs to reduce the deficiency in desirable types of public spaces. However, the literature lacks studies on how current land use affects the type of projects preferred by residents. In our case study, we use basic statistical methods to prove or challenge the hypothesis of possible correlations.

3. Materials and Methods

We have conducted research on the example of the city of Częstochowa (Poland). It is a city with significant industry (30% of the workforce in 2023) and a national center for religious tourism [58,59]. Częstochowa experienced social and economic problems related to limiting traditional heavy industry and the loss of regional (voivodship) capital city rights [60]. These changes led to the disappearance of some administrative functions [61] and to the depopulation from 259,135 inhabitants in 1995 to 203,615 inhabitants in 2024 [62]. The city government has been trying to introduce participatory methods of urban spatial policy for many years, including PB since 2015 [26,63,64]. Częstochowa has 20 districts in which the surveyed projects were submitted, of which the densely built and most populated are the districts located on the historical communication axis, approximately meridianally [60] (Figure 1).

In this paper, we use historical data of the projects proposed and selected for implementation in the five subsequent editions of PB in the districts of Częstochowa (2015–2019). Our goal is not to describe the city's current development, but rather to explore the relationship between land use and the specific projects residents choose to implement. In this context, the use of historical data is justified. The list of proposed and implemented projects in the districts was obtained from the City Hall of Częstochowa. The projects had to be completed in one year and had relatively low cost limits, which reduced their scope. Detailed organizational procedures for preparing and selecting projects have already been described in the source literature [26,63,65]. The projects were prepared and submitted by groups of city residents. After formalization, adoption by the city council, and public announcement, the city council organized a vote in which all residents could choose their preferred projects. On this basis, a ranking of projects was created. The projects with the most votes were qualified for implementation until the available budget was exhausted. The projects were implemented in the following year.

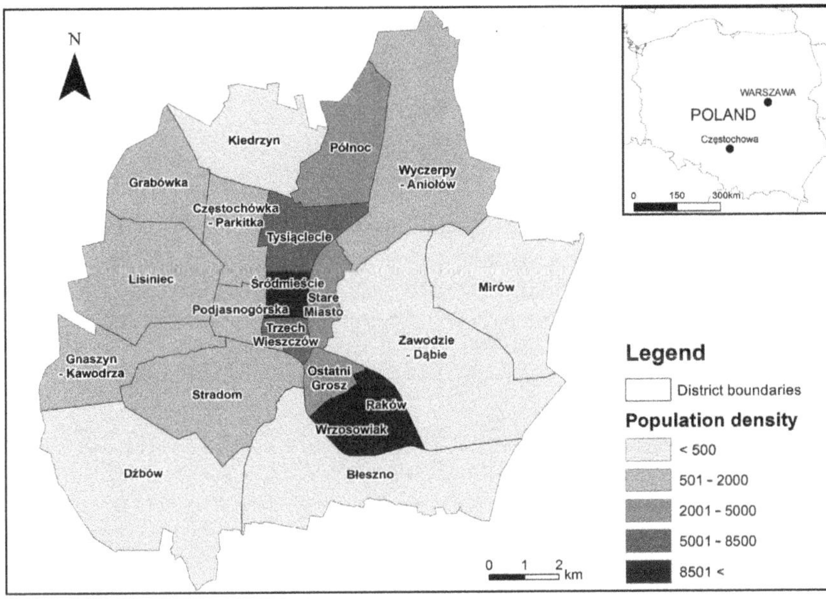

Figure 1. The location of Częstochowa in Poland and the population density of the districts analyzed. Source: Adopted from [26].

For analysis, we selected only the projects selected in district competitions. In the case of city-wide projects, where we did not have precise information about the location of individual activities, it would not be possible to analyze the correlation between the land use structure in a given district and the projects selected in that district. The division into city-wide and district-wide projects is typical in PB in Polish cities [66].

We have selected only hard district projects for the analysis (land development, modernization, or construction of infrastructure). We deliberately omitted soft projects associated with public services, events, training, and educational activities. We omitted soft projects because they do not involve land use or infrastructure development. The division into hard and soft projects is justified in the literature on research methodology [67] and has already been used in PB analyses [66].

In this paper, we use the same division into project types as in the article by Kołat et al. [26], in which the relationships between PB projects and the demographic structures of district populations was evaluated. That typology was created based on the project descriptions, assigning them to eight main research types: (1) road infrastructure; (2) pedestrian infrastructure; (3) bicycle infrastructure; (4) sports infrastructure; (5) recreation infrastructure and areas; (6) educational infrastructure; (7) development of urban green areas; (8) street furniture (so-called small-scale architecture, e.g., bins, benches, information boards, and lighting); and (9) an additional category "other", which included projects featuring different, unusual characteristics (they were not analyzed, e.g., city surveillance). Detailed criteria for assigning projects to a particular type are explained in Appendix A. We emphasize that this is a typology, not a classification. This stems from the fact that some projects may have included small scopes of investment work characteristic of other types. The assignment to a type was based on the characteristics of the dominant work in the project based on the project description. The typology adopted in this article is supported by the literature, particularly in Poland. For example, Madej [55] considers projects that involve the construction or modernization of roads, backyards, historic buildings, as well as pedestrian and bicycle projects in his analysis. Środa-Murawska et al. [34] distinguished the main types of projects such as safety, security and order, road infrastructure, culture and education, sport and recreation, environment and others.

In other countries, different divisions of project types were used. For example, in Chengdu, China, the projects were divided into four categories: (1) infrastructure and activities related to social needs (for example, culture, literacy, entertainment and fitness); (2) infrastructure for local economic development (for example, roads, drainage, irrigation, and water supply); (3) training in agricultural and business skills, and (4) village security, governance, and public services (including sanitation and solid waste collection). In a PB study in two districts of Vienna, Austria, Ahn et al. [45] distinguished between projects related to (1) environment (related to greenery and dog waste), (2) social (related to events, sports facilities, children's playgrounds, public toilets, social care services) and (3) traffic (public transport, traffic regulations, bike paths and parking lots). Chovanecek et al. [68] picked out 11 types of project, listing many more detailed examples of the scope of activities, but also combining projects related to services and infrastructure. Similarly, in Poland, Szczepańska et al. [44] analyzed hard and soft projects for Polish provincial capitals in categories such as: animal care, land management/public facilities, surveillance, lighting, transport, culture/entertainment; revitalization; playgrounds; sports/recreation; and urban greenery. Such divisions, which thematically combine hard and soft projects, would not be suitable for the purposes of our study, which focuses on the development of infrastructure elements and the development of public areas. Therefore, we used our own typology, which is more appropriate for the analysis and conditions of the city under study.

In the spatial analysis of the project locations (reflecting the local needs of many residents), we took into account all submitted projects (analysis related to RQ1). We applied the kernel density estimator (KDE) method to analyze the distribution and maximum spatial density of the proposed projects [69,70]. The KDE method reports where there are distinct clusters in the spatial data set of the location of projects implemented in the study area and has several advantages. KDE transforms discrete event locations (such as project locations) into smooth and continuous surfaces of density values [69,71]. This helps identify hot points and spatial patterns independent of administrative boundaries. Unlike choropleth density maps based on district borders, the KDE is not limited by arbitrary zones. It creates density surfaces that better reflect real spatial variations, which is particularly useful in assessing whether projects are distributed or clustered in an urban space [72]. The detailed principles and mathematical formulas of this method have already been described in detail in the literature [73–75] and we will not repeat them. This method is often applied in geographical studies that address the density of various types of infrastructure or the location of community projects [76–78].

The ArcGIS Desktop (ArcMap 10.7) program and its algorithms were used for the analysis [79]. The search radius (bandwidth) was computed specifically for the input data set using a spatial variant of Silverman's rule of thumb that is robust enough for spatial outliers (when some points are far away from the rest of the points) [70]. The search radius can affect statistical results by controlling the smoothness of the estimated density. Too long a search radius leads to too many generalizations; too small a search radius over-emphasizes local variation. Therefore, the search radius should vary depending on the density of points and the spatial scale of the investigation [72,74]. For this reason, individual maps in the result section should not be directly compared; they are separate analyses for each type of project.

For detailed correlation analysis, we downloaded data on land use from the European Urban Atlas (EUA). We used the data for year 2012 [80], that is, from the last available year before the first analyzed PB (later, the data from this source came only from 2018). We assumed that the structure of land use in the district (for instance, percentage share of communication areas, percentage share of densely built areas, percentage share of greenery) before the announcement of PB in 2015 may have had the main influence on the PB voting results.

The EUA provides reliable, intercomparable, high resolution land use and land cover data for 785 functional urban areas in 38 states of Europe, including Poland and Częstochowa. The minimum mapping width (MMW, narrowest detectable features) is 10 m wide, when the minimum mapping unit (MMU) in urban classes is 0.25 ha [80]. The use of EUA data, instead of national geodetic data, is justified so as to be able to compare the results with studies carried out in other European countries. Due to the scale of the study (the entire city), the level of EUA generalization is appropriate for the analyses carried out in our paper. High-resolution data used for geodetic purposes are not necessary. The use of EUA data is also justified in the literature and is used in studies carried out in many European countries [3,30,81]. A list of EUA land use types found in Częstochowa are listed in Appendix B. Only for the initial overview map, we used highly generalized data on land use types from the City Office, because showing all EUA types would be illegible with the scale used.

Shapiro–Wilk tests showed that some of the data did not have a normal distribution (Appendix C). For this reason, a Spearman correlation analysis was performed, which is also more robust to outliers [82]. Furthermore, Spearman's correlation gives correct results even for very small samples [83,84]. The Spearman correlation coefficient (r_S) was calculated for the percentage of a given land cover and the percentage of projects of a given type finally selected for implementation in the same district (N = 20 districts, analysis related to RQ2). Only correlations with statistical significance of at least $p = 0.05$ were taken into account for the analyses. The full table of correlation coefficients is shown in Appendix D.

The paper does not analyze the dependence on the type of land use inside the area in which the project was located, but rather the predominance of the type of projects selected in relation to the land use characteristics of the entire district of the city. Such an assumption is due to the mobility of residents, who can use a given type of infrastructure in many places close to where they live.

4. Results

Our research included 1224 hard (related to infrastructure or land development) projects submitted in districts. Of these, 349 projects were selected by voting procedures, as the most important to residents (Figure 2). The largest number of district projects in the period analyzed was submitted in densely built-up and densely populated central districts on the strip of the main historical north–south communication route (Tysiąclecie—136 projects; Śródmieście—89; Raków—87). Most of the projects are clearly located near residential areas. The largest number of projects was selected for implementation in the Tysiąclecie (31 projects), Północ (28), and Częstochówka-Parkitka (27) districts (complete data are provided in Appendix A).

The largest number of hard projects was submitted in the categories of road, recreational, pedestrian and bicycle infrastructure, but as a result of the vote, the road, educational and recreational infrastructures were identified as priorities (Table 1).

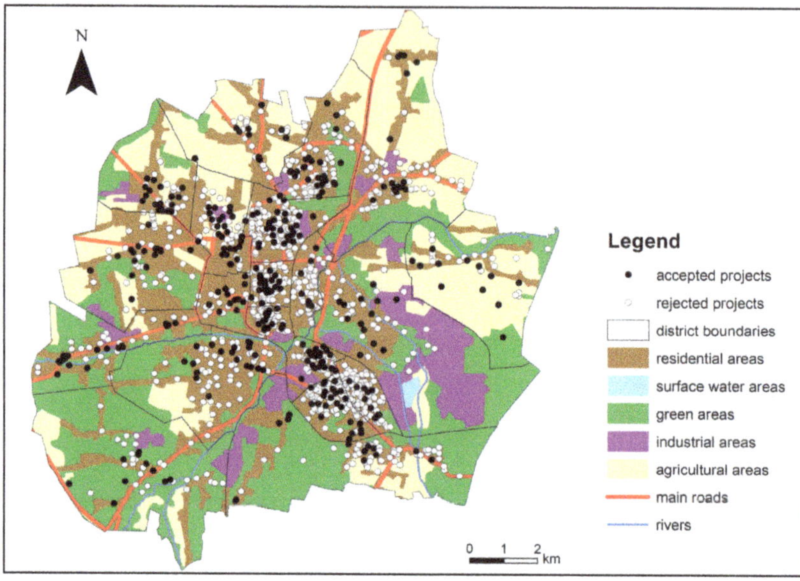

Figure 2. The highly generalized characteristics of land use in Częstochowa (based on municipal data from 2019) and location of analyzed projects (2015–2019). Rejected projects—projects that were not selected for implementation; accepted projects—projects that were selected for implementation. Source: prepared by K. Groszek.

Table 1. The analyzed PB projects submitted and selected for implementation.

No.	Type	Submitted 2015–2019	KDE_{max} of Submitted	Selected for Implementation	Percentage
1	Road infrastructure	397	20.54	90	22.7
2	Recreational infrastructure	177	13.05	45	25.4
3	Pedestrian infrastructure	128	13.84	30	23.4
4	Bicycle infrastructure	120	5.79	33	27.5
5	Educational infrastructure	113	5.38	54	47.8
6	Greenery	100	11.95	34	34.0
7	Street furniture	89	4.77	30	33.7
8	Sports infrastructure	36	2.95	11	30.6
9	Other (not classified)	64	7.52	22	34.4
	Total:	1224	-	349	28.5

Source: Authors' calculations based on the data from Częstochowa City Hall.

The highest number of projects proposed was in the category of road infrastructure. The maximum density index for this category was KDE_{max} = 20.54. Road infrastructure projects were proposed in all districts of the city, especially in central districts of dense construction (Figure 3a). However, this type of project was less frequent in areas with a high share of "other roads and associated land" ($r_s = -0.49$; $p = 0.028$), i.e., in areas where road density was already high.

The high density in the built-up areas (KDE_{max} = 13.84) was also recorded in the case of proposed projects with respect to the construction and modernization of pedestrian infrastructure. The map shows that most of these types of projects were implemented in densely built-up and populated central districts (Figure 3b). However, the number of projects selected for implementation was not related to the land use structures in the districts (no statistically significant correlations).

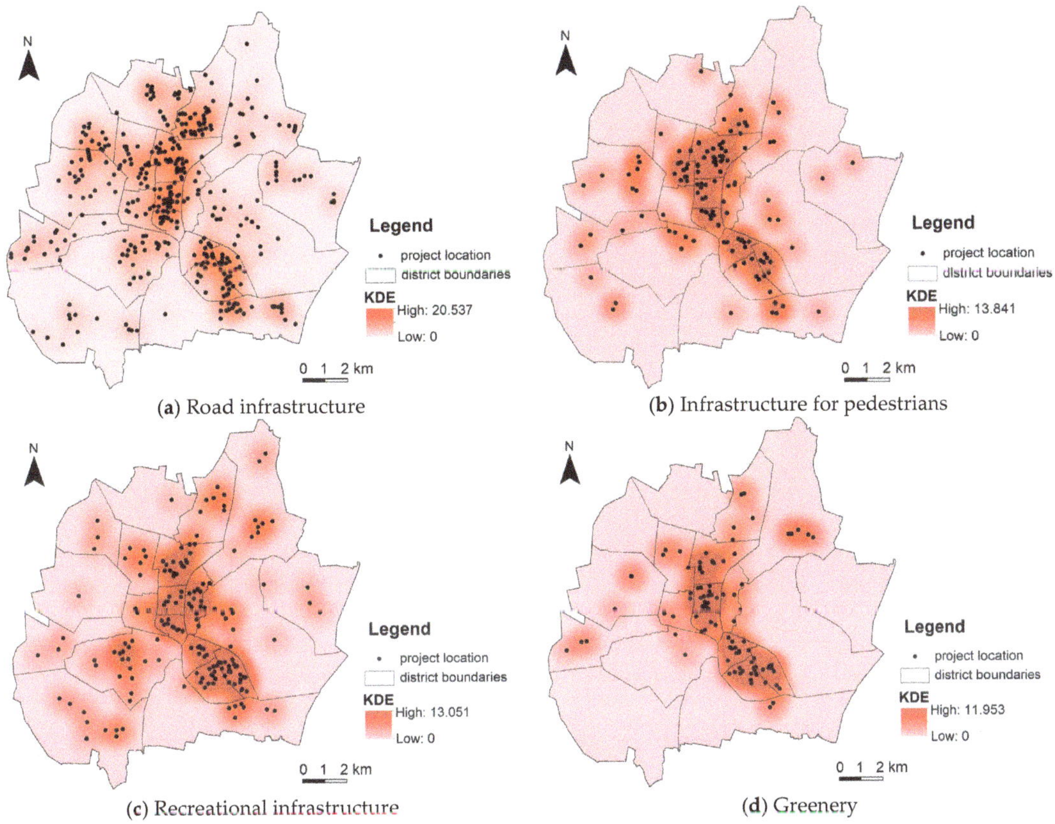

Figure 3. Location and kernel density estimation (KDE) of projects proposed in PB in Częstochowa. Source: prepared by K. Groszek.

The recreational infrastructure projects proposed by the residents also had a relatively high maximum density (KDE_{max} = 13.05). They were implemented in all districts, however, especially in the densely built-up central districts (Figure 3c). They were selected for implementation more frequently in districts with a high share of "water" areas ($r_s = 0.557$; $p = 0.00$), and much less frequently in the districts with a high share of "discontinuous dense urban fabric" areas ($r_s = -0.547$, $p = 0.01$).

Projects related to the development of green areas were selected in 15 districts, more frequently in districts with a high share of "other roads and associated land" ($r_s = 0.581$; $p = 0.00$), "railways and associated land" ($r_s = 0.624$, $p = 0.06$), while less often in the areas with a high share of "pastures" ($r_s = -0.465$, $p = 0.039$) and "forest" ($r_s = -0.553$, $p = 0.01$). Therefore, it is clear that the need for the development of greenery was expressed with great interest in the central areas with a low share of greenery, which is clearly visible in Figure 3d. Probably, unused land near transportation routes was especially developed.

All other type of projects had a visibly lower maximum density (Figure 4). The projects in the cycling category (KDE_{max} = 5.79) and sports infrastructure (KDE_{max} = 2.94) did not show any connection with the current development of the area (lack of statistically significant correlations). The cycling infrastructure was developed in all districts of Czestochowa (Figure 4a). In contrast, sports infrastructure (e.g., football pitches) requiring relatively large free areas was developed only in 10 districts (Figure 4b).

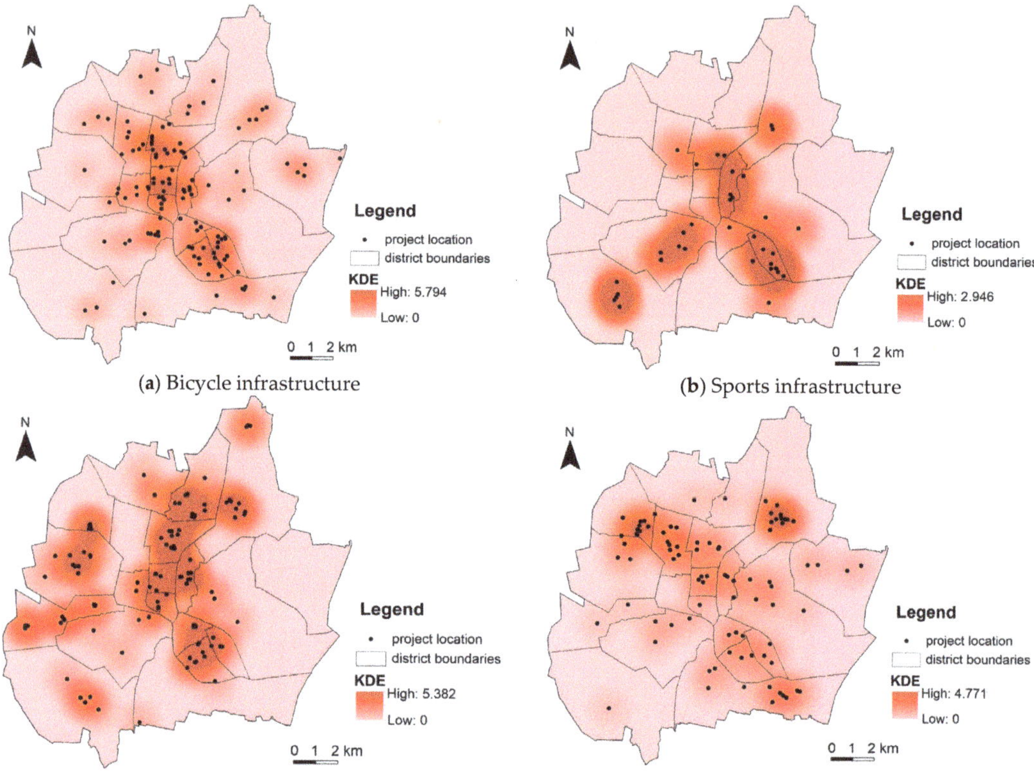

(**a**) Bicycle infrastructure (**b**) Sports infrastructure (**c**) Educational infrastructure (**d**) Street furniture (small-scale infrastructure)

Figure 4. Location and kernel density estimation (KDE) of projects proposed in PB in Częstochowa. Source: prepared by K. Groszek.

The projects focused on the development of areas related to educational facilities (usually related to schools and kindergartens) had a relatively low maximum density (KDE_{max} = 5.38), as they are among the basic services that are regularly located close to residential areas. Such projects were implemented in 18 districts (Figure 4c). These projects were selected more frequently in districts with a high share of "discontinuous very low-density urban fabric" ($r_s = 0.489$; $p = 0.029$) and rarely in peripheral districts with a high share of "land without current use" ($r_s = -0.567$; $p = 0.009$).

The street furniture was implemented in 19 districts (Figure 4d) and they were characterized by relatively low density ($KDE_{max} = 4.77$). The projects were chosen more frequently in districts with a high share of "mineral extraction and dump sites" ($r_s = 0.726$; $p = 0.00$), "isolated structures" ($r_s = 0.653$; $p = 0.00$), "discontinuous medium-density urban fabric" ($r_s = 0.609$; $p = 0.00$) and "discontinuous very low-density urban fabric" areas ($r_s = 0.577$; $p = 0.00$). "Mineral extraction and dump sites" are associated with the existence of former quarries and limestone pits, which are being developed for recreational and tourist purposes, including walking paths [85].

5. Discussion

We observed clear differences between the spatial density of proposed projects in city districts, which can result from the typical characteristics of densely built-up central districts and yet-to-be-developed, sprawling peripheral neighborhoods with larger areas of greenery. Traditionally, mobility-related issues constitute a major challenge in urban areas (roads and sidewalks). Residents also need public places for recreation. These projects were frequently reported in Częstochowa, reaching high density near residential areas, especially in central districts (response to RQ1).

However, a correlation analysis of the district land use patterns and the project structure selected for implementation reveals relationships that are not visible by analyzing the distribution of projects on the map. Our research confirms that some types of project selected by voters are correlated with the land use structure of the districts (response to RQ2). However, it can be related not only to the deficiency of some urban infrastructure in the districts, but also to the availability of areas for which it can be used in a specific way. In detail, we would like to highlight five main dependencies, which we also discuss in relation to the results of other authors' work. However, a comparison between the results of PB in Częstochowa and PB in cities in other countries may be difficult due to different historical, economic and cultural conditions for the development of cities [1].

Firstly, we have found that projects related to the development of green areas were selected more often in densely built-up districts with a deficiency in greenery. For the purpose of public greenery, wasteland near transportation routes was often developed. Green areas can be important as buffers against the adverse effects of transport (noise and exhaust fumes). It was also observed in other cities in Poland [24]. The excessive density of development in city centers resulting from economic factors (land prices) was also an incentive to develop recreation and meeting areas among greenery. Developers typically maximize development with buildings to make sales profits and are less interested in financing other infrastructure around such areas [1,19]. Municipal authorities in Poland, in order to obtain the highest possible tax revenues or to reduce costs, often allow the stimulation of investment, by including the commercial use of former green areas [4,86,87]. This is a typical problem for many fast developing cities in the world [1,88]. Public greenery and free recreational areas do not generate income, so they have been sometimes overlooked in urban development due to the strong influence of market forces [89]. Citizens try to solve these problems within the framework of PB, proposing and voting for the development of green areas (including city parks), as observed in many other cities [22,44,90]. Consequently, sometimes even a thematic "Green PB" is organized [40,55]. However, these actions are not always effective. Analyses of various urban development scenarios have shown that even in the scenario of ecological protection planning and control, urban growth can easily break through the ecological protection boundary [2]. In the years 2010–2018, the access to green spaces in large Polish cities decreased [91]. On the other hand, even small-scale, but carefully targeted green interventions can significantly improve quality of life in urban areas [92]. Interest in "green projects" indicates a growing demand among local communities for "nature-based solutions" [93,94]. Meanwhile, in many studies, relatively close access to wooded areas (parks and forests) is very important to residents [95]. A similar relationship was described in the comparative analysis of Vienna's districts, where in the densely populated and built-up central district, more greenery projects were reported than in the more peripheral district with less dense development [45]. However, these authors used the indicator of the amount of green space per inhabitant, not its share in a given district. In Lisbon, preference was given to projects that helped create a more sustainable, resilient and environmentally friendly city [40]. Such bottom-up processes can support and improve public management and ensure more responsive and sustainable public spaces that can serve as public goods.

Second, projects related to recreational infrastructure were selected more frequently in areas with a relatively high share of reservoirs and waterways. It confirms the significant role of areas that offer open water for leisure purposes. However, the limited possibility of developing other infrastructure in river floodplains may also be important. The restoration of riverside areas in Częstochowa has become of increasing importance in recent years [96]. Similar trends were observed in other Polish cities, where special attention was also paid to the "blue infrastructure" [97–99]. Riverside areas are often flood-prone areas, so it makes sense to create recreational and green spaces near them. Unfortunately, due to high land prices, cities allow dense development of such areas, which then increase the risk of flood damage [100].

Third, projects on so-called street furniture (small-scale infrastructure) were more common in peripheral districts with high-share mineral extraction and dump sites and sparsely built areas. These areas are often poorly developed, frequently peripheral, and characterized by the underdevelopment of such amenities. Green areas and low-quality communication routes that require the development of various types of street furniture are more popular there than in highly developed central districts [97]. In Częstochowa, this type of projects were especially associated with the revitalization of former limestone quarries, which were developed for recreational and tourist purposes [85,101].

Fourth, educational infrastructure projects were more popular in districts with a high share of discontinuous very low-density urban fabric, so relatively often in developing districts. The new apartments in the developing districts of Częstochowa are probably bought by relatively young families with children, which explains the need for infrastructure related to children and youth [26]. Furthermore, the age group of residents with children is often particularly active in PB and is also mobilized by schools to vote for PB [34].

Fifth, the selection of local road projects was negatively correlated with the share of transport (communication) areas. In cities, the need for public transport, including safe and convenient traffic areas, is one of the basic, most important, and obvious needs of residents [1,102,103]. Mobility projects are commonplace throughout the world in PB scenarios [15].

In the remaining cases (sidewalks, bicycle paths), we did not confirm correlations between land use and the projects selected for implementation in the PB procedures, although they seemed to be closely related in the densely built-up districts. Therefore, the need for their expansion actually occurred throughout the city area.

The main limitation of our results concerns the fact that correlation analysis does not conclusively indicate cause-and-effect relationships, so the mere fact that there is a statistical correlation does not mean that the land use structure in a given neighborhood was a major influence on the selection of the project. The needs of the residents may depend not only on the current state of development of the area, but also on the demographic and sociocultural characteristics of the population that resides in a given area [26,27]. The authors also emphasize that the structure of project types is influenced by the limitations of their material scope and the time period imposed by the city authorities in the PB rule; therefore, there is no full freedom to propose projects [104].

Municipal spatial planning regulations (urban development plans) in Poland can also play an important role in shaping the types of hard projects that residents propose and support in PB. PB procedures encourage the implementation of projects in line with the urban development plans. These planning frameworks determine land use designations, infrastructure priorities, and technical feasibility [100,105], directly influencing which citizen-proposed projects are admissible. When planning documents are clear and well prepared, they enable residents to propose realistic projects (such as greenery, roads or bike lanes) that align with municipal goals. In contrast, a lack of up-to-date or detailed urban development plan can limit feasible project areas and diminish civic participation, or lead to the original proposed projects not being admitted to the competition for formal and legal reasons. Kamrowska-Zaluska [106] highlighted that PB in Polish cities can be undermined by the goals of local government policy. Rogatka et al. [107] found that in small towns of the Kujawsko-Pomorskie Voivodeship in Poland, the spatial planning rules limited the effective use of participatory mechanisms due to the lack of adequate citizen knowledge on spatial planning. Szczepańska et al. [44] emphasized that PB often overlaps with spatial plans to improve public spaces, but their implementation depends on how these plans guide municipal investment priorities. When the project proposals align with municipal regulations, they are more likely to be selected.

The authors also point out that PB inputs in Poland actually have a negligible impact on solving the main deficits in the development of public infrastructure in cities, as they are usually allocated between 0.5% and 1% of cities' annual budgets. Rather, they serve to increase the activity and interest of residents in urban development issues and to stimulate their activity [106].

The results may also not fully reflect the needs of the majority of residents, but may be the result of irregular initiatives taken up by the most active social (interest) groups [15,19,44,55]. This means that the analysis of the projects selected for implementation may not be the best indicator to identify the actual and real structure of the needs of residents. In New York City, districts inhabited by relatively wealthy and educated residents were found to be particularly active, which does not make it easier to eliminate the city's biggest problems using PB [108]. For this reason, it is important to allocate separate funds for individual districts, e.g., in proportion to the number of inhabitants, to avoid PB funds being consumed only by the most socially active districts. In Indonesia, the authors highlighted the significant role of access to spatial information among urban residents. They believe that this can influence the scope of social participation and the activities proposed by residents [109]. Therefore, the education of residents can influence the structure of implemented projects.

Furthermore, the results of our case study may not be entirely universal, as they refer to a period with a specific level of infrastructure development and urban development. For example, using the example of Toruń, Środa-Murawska et al. [34] demonstrated that the percentage of infrastructure modernization projects was gradually decreasing in favor of new investments and soft projects. This behavior, according to the authors, was related to Maslow's pyramid of human needs model [110]. Initially, hard projects related to repairs or renovations were carried out. Only with time did the nominated projects concern the integration of local society and the personal development of the district's inhabitants. In Katowice, Poland, it was observed that in successive editions of PB, the percentage of projects related to renovations and repairs gradually decreased in favor of proposing new investments and soft activities related to the activity of residents [111]. In some Katowice districts, a decrease in the number of projects submitted was observed over time. In Wrocław, Poland, an unfavorable phenomenon was also the decrease in the interest of citizens in PB after several years of its existence [49]. This may mean that the most pressing needs of residents have been met or that public interest in participating in the creation and selection of projects for implementation diminishes over time.

The differences in the structure of project types in better and less developed areas reveal the adaptive dimension of PB. It reveals its ability to respond to local needs. In many cases, it reflects existing differences in the development and quality of urban space.

6. Conclusions

The research findings indicate that participatory budgeting is a practical tool that allows residents to reorganize space according to their needs. In this way, residents assume an active role as co-creators not only of the physical space but also of its social function [112]. Empirical research consistently demonstrates that the typology and orientation of participatory PB reflect the sociospatial disparities within cities.

The popularity of PB in many cities shows that the residents' needs in terms of access to some type of public space often were not met—either due to spontaneous urban sprawl or the ineffectiveness of spatial planning by city officials. Based on our analysis of participatory budgeting (PB) outcomes in Częstochowa, four key recommendations emerge to improve spatial planning policy in Polish cities.

- First, municipal plans should prioritize the preservation and strategic development of green spaces, especially in densely built districts with low greenery availability. Urban planning must protect the remaining natural areas and promote the conversion of underutilized land near transport routes into greenery.
- Second, cities should better integrate blue infrastructure, such as rivers, reservoirs, and floodplains, into land use planning. These areas have high recreational value and are often chosen by residents for development through PB, but are often overlooked in formal spatial frameworks. The prioritization of the development of green spaces and recreational areas in riverside areas also stems from the threat of flooding.
- Third, spatial plans must address infrastructure gaps in peripheral and postindustrial districts, where residents commonly request small-scale infrastructure such as information boards, benches, lighting, and sidewalks. These needs highlight the neglect of basic amenities in low-density or marginalized areas.
- Finally, planning policies in new housing zones should anticipate demographic changes and include provisions for educational and childcare infrastructure. New suburban neighborhoods often attract young families, whose participation in PB reflects a strong demand for facilities for children and youth.

The projects submitted by residents that are not selected could be part of public consultations and incorporated into larger city projects. This action could increase the effectiveness of city policies in meeting the needs of residents, as well as increase the efficiency of local authorities, which is an important factor in local development [113].

We agree that PB is not a perfect tool; it has both advantages and disadvantages [12,20,48]. The growing expectations of residents may be difficult to meet through their formula, for example, large infrastructure investment [22]. However, it can offer an auxiliary tool to examine needs and activate residents, thus increasing their interest in urban planning. We would also like to add that cities should ensure that their urban development plans are up to date and of high quality. Up-to-date and accessible planning documents improve the effectiveness of PB by allowing residents to propose feasible projects, while outdated or overly rigid plans risk disqualifying community-supported initiatives.

Our findings contribute meaningfully to the international discourse on PB by revealing how spatial patterns and land use structures influence residents' choices. We show that analyzing the location of PB projects can uncover infrastructure deficiencies overlooked by traditional planning, both in peripheral sprawling and central densely built-up districts. The findings also highlight the consistent preference for green and recreational spaces, particularly near water, strengthening calls for environmentally conscious urban planning. This research supports the argument that PB offers more than democratic participation; it provides insight into sustainable urban needs. While such a universal recommendation seems obvious, our PB analysis indicates that such needs in a case-study city have not yet been sufficiently met. This stems from the focus of local government and the economic sector being on economic growth.

Our article can contribute to the international literature as a relatively new analysis, comparing the characteristics of current land use with the need for additional investments reported by residents. Such analyses have been extremely rare to date. In the future, it would be valuable to create a statistical model of the correlations between the needs of residents expressed in PB, land development, and the sociodemographic characteristics of residents based on research conducted in several cities, as well as international comparisons.

Author Contributions: Conceptualization and methodology, M.F. and K.G.; software, K.G. and M.B.; investigation, K.G.; validation, K.G. and M.B.; formal analysis, K.G., M.B. and M.F.; resources, K.G.; data curation, K.G., M.B. and M.F.; writing—original draft preparation, M.F., K.G. and M.K.-M.; writing—review and editing, M.F., K.G. and M.K.-M.; visualization, K.G.; supervision, M.F. Estimated percentage engagement, K.G. 40%; M.F. 30%, M.K.-M. 20%, and M.B. 10%; data for the research were collected as part of the K.G.'s Master's thesis: "Spatial analysis of projects implemented under Participatory Budgeting in Częstochowa" prepared at the Institute of Spatial Management, Wroclaw University of Environmental and Life Sciences, Poland. All authors have read and agreed to the published version of the manuscript.

Funding: This research received no external funding. The APC was funded by MDPI within the framework of a 'Feature Paper' action—an individual invitation from the editors.

Data Availability Statement: Data are shown in the Appendices, available in the Częstochowa City Hall and in European Urban Atlas.

Acknowledgments: The authors should like to express their gratitude to: the Częstochowa City Hall for providing statistical data on the 2015–2019 PB.

Conflicts of Interest: The authors declare no conflicts of interest.

Abbreviations

The following abbreviations are used in this manuscript:

EUA	European Urban Atlas
KDE	Kernel density estimation
MMU	Minimum mapping unit
MMW	Minimum mapping width
PB	Participatory budgeting

Appendix A

Rules for assigning projects to specific research types, based on their titles and summary descriptions (we did not have access to a detailed description of the scope of work performed) are the same as in [26] and the number of projects selected for implementation.

(1) Road infrastructure: this type includes roads, pedestrian crossings, speed bumps, bus stops, bus shelters, parking lots, traffic lights, road lighting, under-road technology infrastructure (including water supplies and medical wastewater systems), drainage ditches that drain the road lane. This category also included tasks involving pedestrian and bicycle road modernization when they were an integral part of the upgrade or bodywork of an entire road modernization and could not be analyzed separately;
(2) Pedestrian infrastructure: separate pedestrian routes, modernized or constructed independently from the road lines (e.g., pavements, stairs, etc.);
(3) Bicycle infrastructure: separate bicycle paths, bicycle repair stations, bicycle stands, and the city bicycle rental points;
(4) Sports infrastructure: construction or modernization of sports facilities (such as public soccer fields, open basketball and volleyball courts) that are not related to education institutions (schools and kindergartens);
(5) Educational infrastructure: installation and supplementation of devices, development and revitalization of areas within educational institutions; such infrastructure may include school sports facilities;

(6) Recreational infrastructure: the development of new recreational areas and the construction of facilities for leisure and recreation, including children's playgrounds, outdoor gyms; modernization and supplementation of these facilities for residents;
(7) Street furniture: the installation of objects and pieces of equipment installed along streets, walking paths, and other public spaces for various purposes, e.g., benches, information boards, litter bins, and lighting;
(8) Greenery: development, cleaning, care, and enrichment of green spaces, including green spaces, parks, and fenced squares for walking dogs, pet waste stations, bird nest boxes, and others;
(9) Other: revitalization of backyards, the installation and renovation of public sanitary facilities, the renovation of public buildings, and monitoring of the city.

Table A1. Number of projects selected by citizens to be implemented and taken into consideration in the analysis.

Type Number District	1	2	3	4	5	6	7	8	9	Total	1 [%]	2 [%]	3 [%]	4 [%]	5 [%]	6 [%]	7 [%]	8 [%]	9 [%]
Błeszno	3	3	0	0	3	1	2	1	1	14	21	21	0	0	21	7	14	7	7
Częstochówka-Parkitka	6	7	2	1	4	0	3	2	2	27	22	26	7	4	15	0	11	7	7
Dźbów	3	0	1	1	1	2	2	0	0	10	30	0	10	10	10	20	20	0	0
Gnaszyn-Kawodrza	7	0	0	0	1	7	1	1	2	19	37	0	0	0	5	37	5	5	11
Grabówka	7	0	2	0	3	5	2	0	0	19	37	0	11	0	16	26	11	0	0
Kiedrzyn	7	1	0	0	0	2	0	0	1	11	64	9	0	0	0	18	0	0	9
Lisiniec	5	3	1	0	3	3	1	3	0	19	26	16	5	0	16	16	5	16	0
Mirów	4	1	1	0	0	0	4	0	2	12	33	8	8	0	0	0	33	0	17
Ostatni Grosz	3	1	1	0	2	2	5	6	2	22	14	5	5	0	9	9	23	27	9
Podjasnogórska	5	0	1	0	0	1	2	3	0	12	42	0	8	0	0	8	17	25	0
Północ	11	4	2	0	0	6	1	1	3	28	39	14	7	0	0	21	4	4	11
Raków	1	0	0	3	1	1	0	3	1	10	10	0	0	30	10	10	0	30	10
Stare Miasto	2	1	0	2	0	0	2	0	0	7	29	14	0	29	0	0	29	0	0
Stradom	2	0	4	2	4	3	4	0	0	19	11	0	21	11	21	16	21	0	0
Śródmieście	3	2	2	0	1	5	2	7	1	23	13	9	9	0	4	22	9	30	4
Trzech Wieszczów	2	1	2	0	0	2	3	1	2	13	15	8	15	0	0	15	23	8	15
Tysiąclecie	8	2	7	1	1	5	1	2	4	31	26	6	23	3	3	16	3	6	13
Wrzosowiak	5	3	4	0	1	4	4	2	0	23	22	13	17	0	4	17	17	9	0
Wyczerpy-Aniołów	2	0	1	1	4	3	3	2	1	17	12	0	6	6	24	18	18	12	6
Zawodzie-Dąbie	4	1	2	0	1	2	3	0	0	13	31	8	15	0	8	15	23	0	0

Source: Authors' calculations based on the data from Częstochowa City Hall.

Appendix B

The Urban Atlas provides comparable pan-European land cover and land use information for selected Functional Urban Areas (FUA) in Europe [81]. It is available for the 2006, 2012 and 2018 reference years including the change layers. We consider the following land use types found in the Częstochowa districts:

- 11100 Continuous urban fabric (S.L. > 80%);
- 11210 Discontinuous dense urban fabric (S.L.: 50–80%);
- 11220 Discontinuous medium-density urban fabric (S.L.: 30–50%);
- 11230 Discontinuous low-density urban fabric (S.L.: 10–30%);
- 11240 Discontinuous very low-density urban fabric (S.L. < 10%);
- 11300 Isolated structures;
- 12100 Industrial, commercial, public, military, and private units;
- 12220 Other roads and associated land;
- 12230 Railways and associated land;
- 13100 Mineral extraction and dump sites;
- 13300 Construction sites;
- 13400 Land without current use;

- 14100 Green urban areas;
- 14200 Sports and leisure facilities;
- 21000 Arable land (annual crops);
- 23000 Pastures;
- 31000 Forests;
- 32000 Herbaceous vegetation associations (natural grassland, moors etc.);
- 40000 Wetland;
- 50000 Water bodies.

In Urban Atlas, S.L. means 'soil sealing' (artificially surfaced areas). It represents the surface of the degree to which the land is covered by impervious materials, such as asphalt, concrete, or buildings, that prevent water from infiltrating the soil. Higher S.L. values indicate higher building and sidewalk density [81,114] (https://dd.eionet.europa.eu/vocabulary/landcover/uatl2012/view, last accessed on 24 June 2024).

Appendix C

Table A2. Tests of normality of distribution. When the *p*-value of the Shapiro–Wilk test is less than 0.05, it is recognized that the data tested are not normally distributed. In such a case, some statistical textbooks do not recommend using Pearson's correlation.

Variable	Statistics	df	*p* Value
Road infrastructure	0.920	20	0.101
Pedestrian infrastructure	0.885	20	0.022
Bicycle infrastructure	0.917	20	0.086
Sports infrastructure	0.575	20	0.000
Street furniture	0.884	20	0.021
Educational infrastructure	0.941	20	0.252
Recreational infrastructure	0.960	20	0.542
Greenery	0.804	20	0.001
Other (not classified)	0.866	20	0.010
Continuous urban fabric	0.788	20	0.001
Discontinuous dense urban fabric	0.986	20	0.987
Discontinuous medium-density urban fabric	0.815	20	0.001
Discontinuous low-density urban fabric	0.711	20	0.000
Discontinuous very low-density urban fabric	0.828	20	0.002
Isolated structures	0.828	20	0.002
Industrial, commercial, public, military, private units	0.825	20	0.002
Other roads and associated land	0.937	20	0.209
Railways and associated land	0.839	20	0.003
Mineral extraction and dump sites	0.642	20	0.000
Construction sites	0.680	20	0.000
Land without current use	0.824	20	0.002
Green urban areas	0.731	20	0.000
Sports and leisure facilities	0.834	20	0.003
Arable land (annual crops)	0.840	20	0.004
Pastures	0.860	20	0.008
Forests	0.696	20	0.000
Herbaceous vegetation	0.506	20	0.000
Wetlands	0.236	20	0.000
Water	0.782	20	0.000
Continuous urban fabric	0.788	20	0.001

Source: Authors' calculations based on the data from Częstochowa City Hall.

Appendix D

Table A3. A Spearman correlation table of the studied variables characterizing the districts of Częstochowa.

Type of Land Use (in Percentage, in District)	Coefficients [1]	Road Infr.	Pedestrian Infr.	Bicycle Infr.	Sports Infr.	Street Furnit.	Education Infr.	Recreational Infr.	Greenery	Other
Continuous urban fabric	r_s	−0.402	0.210	0.237	0.187	−0.357	−0.219	0.075	0.309	0.222
	p	0.079	0.374	0.315	0.430	0.122	0.353	0.752	0.185	0.346
Discontinuous dense urban fabric	r_s	−0.113	−0.191	0.143	0.083	0.077	0.207	−0.547 *	0.344	−0.075
	p	0.636	0.419	0.547	0.728	0.747	0.381	0.013	0.137	0.752
Discontinuous medium-density urban fabric	r_s	0.047	−0.360	−0.196	0.141	0.609 **	0.016	0.009	−0.061	−0.410
	p	0.843	0.119	0.407	0.553	0.004	0.946	0.969	0.799	0.072
Discontinuous low-density urban fabric	r_s	0.282	0.265	−0.096	−0.180	0.358	0.030	0.023	−0.426	0.029
	p	0.228	0.260	0.689	0.446	0.121	0.900	0.922	0.061	0.904
Discontinuous very low-density urban fabric	r_s	0.178	−0.107	−0.082	−0.028	0.577 **	0.489 *	−0.384	−0.304	−0.101
	p	0.453	0.653	0.732	0.906	0.008	0.029	0.094	0.193	0.671
Isolated structures	r_s	−0.005	−0.159	−0.179	0.121	0.653 **	0.066	0.084	−0.364	−0.126
	p	0.982	0.502	0.449	0.611	0.002	0.783	0.726	0.115	0.596
Industrial, commercial, public, military, and private units	r_s	−0.224	0.107	0.060	−0.116	−0.355	−0.263	0.160	0.403	0.046
	p	0.342	0.654	0.802	0.625	0.125	0.263	0.500	0.078	0.848
Other roads and associated land	r_s	−0.490 *	0.109	0.148	0.159	−0.247	−0.060	−0.183	0.624 **	0.131
	p	0.028	0.647	0.534	0.504	0.293	0.803	0.440	0.003	0.583
Railways and associated land	r_s	−0.296	−0.330	0.041	0.045	−0.165	−0.191	0.146	0.400	−0.059
	p	0.205	0.155	0.863	0.852	0.487	0.421	0.540	0.080	0.806
Mineral extraction and dump sites	r_s	0.034	0.006	−0.007	−0.054	0.726 **	0.125	0.009	−0.257	−0.432
	p	0.888	0.980	0.975	0.821	0.000	0.598	0.970	0.273	0.057
Construction sites	r_s	0.231	−0.122	−0.121	−0.079	0.141	0.249	−0.325	0.035	0.026
	p	0.328	0.608	0.611	0.740	0.553	0.290	0.163	0.882	0.915
Land without current use	r_s	−0.327	0.173	0.286	0.347	0.036	−0.567 **	0.297	0.116	−0.251
	p	0.159	0.466	0.221	0.134	0.880	0.009	0.204	0.626	0.287
Green urban areas	r_s	−0.035	0.388	0.208	−0.189	−0.302	−0.243	−0.024	0.352	−0.011
	p	0.885	0.091	0.379	0.425	0.195	0.301	0.920	0.128	0.964
Sports and leisure facilities	r_s	−0.040	0.128	0.174	0.263	0.141	−0.233	−0.106	0.066	−0.394
	p	0.867	0.591	0.463	0.262	0.554	0.322	0.656	0.782	0.086
Arable land (annual crops)	r_s	0.439	0.211	−0.235	−0.272	0.197	−0.041	−0.017	−0.429	0.143
	p	0.053	0.373	0.319	0.246	0.406	0.862	0.945	0.059	0.548
Pastures	r_s	0.256	−0.280	−0.102	0.117	0.428	0.233	−0.096	−0.465 *	−0.113
	p	0.276	0.232	0.668	0.624	0.059	0.322	0.686	0.039	0.635
Forests	r_s	0.283	0.039	0.133	−0.024	0.372	0.043	0.154	−0.553 *	−0.195
	p	0.226	0.871	0.577	0.920	0.106	0.856	0.517	0.012	0.411
Herbaceous vegetation	r_s	0.348	0.308	−0.157	0.073	−0.092	−0.145	0.120	−0.455 *	0.009
	p	0.133	0.187	0.510	0.761	0.701	0.542	0.615	0.044	0.971
Wetlands	r_s	0.100	−0.264	0.140	0.303	0.121	0.219	0.139	−0.264	−0.247
	p	0.676	0.260	0.555	0.193	0.611	0.353	0.558	0.260	0.294
Water	r_s	0.203	−0.203	−0.106	−0.009	−0.089	−0.256	0.567 **	−0.203	−0.275
	p	0.391	0.390	0.657	0.969	0.710	0.276	0.009	0.391	0.240

[1] Coefficients: r_s—Spearman correlation; p—statistical significance: * $0.01 \leq p < 0.05$; and ** $p < 0.01$. Authors' calculations based on the data from Częstochowa City Hall.

References

1. Pacione, M. *Urban Geography: The Global Perspective*; Routledge: New York, NY, USA, 2009.
2. Li, B.; Wang, Y.; Wang, T.; He, X.; Kazak, J.K. Scenario Analysis for Resilient Urban Green Infrastructure. *Land* **2022**, *11*, 1481. [CrossRef]
3. Salvati, L.; Tombolini, I.; Ippolito, A.; Carlucci, M. Land quality and the city: Monitoring urban growth and land take in 76 Southern European metropolitan areas. *Environ. Plan. B Urban Anal. City Sci.* **2018**, *45*, 691–712. [CrossRef]
4. Jadach-Sepioło, A.; Zathey, M. Alternative between Revitalisation of City Centres and the Rising Costs of Extensive Land Use from a Polish Perspective. *Land* **2021**, *10*, 488. [CrossRef]
5. Sanoff, H. *Community Participation Methods in Design and Planning;* John Wiley & Sons: New York, NY, USA, 2000; p. 306.
6. Tóbiás, K.; Boros, L. Participatory Planning and Gamification: Insights from Hungary. *Land* **2025**, *14*, 573. [CrossRef]
7. Furmankiewicz, M. Urban Local Action Groups in Poland: Monofunctional social activities within Community Led Local Development. In Proceedings of the 15th International Conference Liberec Economic Forum 2021, Liberec, Czech Republic, 13–14 September 2021; Antlová, K., Semerádová, T., Eds.; Technical University of Liberec: Liberec, Czech Republic, 2021; pp. 259–268.

8. Castro, D.G.; de Elizagarate Gutiérrez, V.; Kazak, J.K.; Szewrański, S.; Kaczmarek, I.; Wang, T. Nuevos desafíos para el perfeccionamiento de los procesos de participación ciudadana en la gestión urbana. Retos para la innovación social. *Cuad. Gest.* **2020**, *20*, 41–64. [CrossRef]
9. Deuskar, A.; Khozrevanidze, K.; Ozeren, Z.; Paradowska, I. Dialogue with the community in the planning process. How to use the participatory approach as a planning tool for the community's benefit? *Eur. Spat. Res. Policy* **2022**, *29*, 271–284. [CrossRef]
10. Furmankiewicz, M. Współrządzenie czy ukryta dominacja sektora publicznego? Koncepcja governance w praktyce Lokalnych Grup Działania LEADER. *Stud. Reg. i Lokal.* **2013**, *1*, 71–89. [CrossRef]
11. Avritzer, L. New Public Spheres in Brazil: Local Democracy and Deliberative Politics. *Int. J. Urban Reg. Res.* **2006**, *30*, 623–637. [CrossRef]
12. Sintomer, Y.; Herzberg, C.; Röcke, A. Participatory Budgeting in Europe: Potentials and Challenges. *Int. J. Urban Reg. Res.* **2008**, *32*, 164–178. [CrossRef]
13. Goldsmith, W.; Vainer, C. Participatory Budgeting and Power Politics in Porto Alegre. *Land Lines* **2001**, *1*, 7–9.
14. Sintomer, Y.; Herzberg, C.; Röcke, A.; Allegretti, G. Transnational Models of Citizen Participation: The Case of Participatory Budgeting. *J. Public Delib.* **2012**, *8*, 9. [CrossRef]
15. Cabannes, Y. Contribution of Participatory Budgeting to provision and management of basic services. In *Municipal Practices and Evidence from the Field*; International Institute for Environment and Development: London, UK, 2014.
16. De Vries, M.S.; Nemec, J.; Špaček, D. (Eds.) *International Trends in Participatory Budgeting: Between Trivial Pursuits and Best Practices*; Palgrave Macmillan: Cham, Switzerland, 2022. [CrossRef]
17. Martela, B.; Bubak, G.; Janik, L. *Barometr Budżetu Obywatelskiego. Edycja 2020 (Participatory Budget Barometer. 2020 Edition)*; Instytut Rozwoju Miast i Regionów: Warszawa–Kraków, Poland, 2021.
18. Sześciło, D. Participatory Budgeting in Poland: Quasi-Referendum Instead of Deliberation. *Croat. Comp. Public Adm.* **2015**, *15*, 373–388. Available online: https://hrcak.srce.hr/142728 (accessed on 20 June 2025).
19. Kębłowski, W.; Van Criekingen, M. How "alternative" alternative urban policies really are? Looking at participatory budgeting through the lenses of the right to the city. *Métropoles* **2014**, *15*, 1–23. [CrossRef]
20. Kozłowski, A.R.; Bernaciak, A. Participatory Budgeting in Poland. In *International Trends in Participatory Budgeting: Between Trivial Pursuits and Best Practices*; De Vries, M.S., Nemec, J., Špaček, D., Eds.; Palgrave Macmillan: Cham, Switzerland, 2022; pp. 163–185. [CrossRef]
21. Kociuba, D.; Bielecka, M. Wpływ zmiany ustawy o samorządzie gminnym na implementację budżetów obywatelskich w miastach wojewódzkich Polski (Influence of the Amendment to the Law on Municipal Self-Government on the Implementation of Participatory Budgeting in Voivodship Cities in Poland). *Stud. Reg. i Lokal.* **2021**, *1*, 84–111. [CrossRef]
22. Džinić, J.; Murray-Svidroňová, M.; Markowska-Bzducha, E. Participatory Budgeting: A Comparative Study of Croatia, Poland and Slovakia. *NISPAcee J. Public Adm. Policy* **2016**, *9*, 31–56. [CrossRef]
23. Schugurensky, D.; Mook, L. Participatory budgeting and local development: Impacts, challenges, and prospects. *Local Dev. Soc.* **2024**, *5*, 433–445. [CrossRef]
24. Kociuba, D.; Rabczewska, K. Rola budżetów partycypacyjnych w zagospodarowaniu przestrzeni publicznych polskich miast-studium przypadku Lublina (The role of participatory budgets on the development of urban public spaces-the Lublin case study). *Stud. Reg. i Lokal.* **2019**, *2*, 82–109. [CrossRef]
25. Mucha, A. Zróżnicowanie zwycięskich projektów budżetu obywatelskiego w latach 2013–2016 w polskich miastach (Diversity of the winning projects of participatory budgeting in Polish cities in the years 2013–2016). *Prace Geogr.* **2018**, *154*, 55–70. [CrossRef]
26. Kołat, K.; Furmankiewicz, M.; Kalisiak-Mędelska, M. What Are the Needs of City Dwellers in Terms of the Development of Public Spaces? A Case Study of Participatory Budgeting in Częstochowa, Poland. *Int. J. Environ. Res. Public Health* **2022**, *19*, 5171. [CrossRef]
27. Leśniewska-Napierała, K.; Napierała, T. Participatory budgeting: Creator or creation of a better place? Evidence from rural Poland. *Bull. Geography. Socio-Econ. Ser.* **2020**, *48*, 65–81. [CrossRef]
28. Christophers, B. Uneven Development. In *International Encyclopedia of Human Geography*; Kitchin, R., Thrift, N., Eds.; Elseviere: Amsterdam, The Netherlands, 2009; pp. 12–17. [CrossRef]
29. Kulczyk-Dynowska, A.; Stacherzak, A. The Impact of a City on Its Environment: The Prism of Demography and Selected Environmental and Technical Aspects Based on the Case of Major Lower Silesian Cities. *Sustainability* **2022**, *14*, 6393. [CrossRef]
30. Kazak, J.K.; Błasik, M.; Świąder, M. Land use change in suburban zone: European context of urban sprawl. *J. Water Land Dev.* **2022**, 92–98. [CrossRef]
31. Hołuj, D. Public Spaces and Cultural Heritage in Community Projects-The Example of Warsaw. *Eur. Spat. Res. Policy* **2017**, *24*, 127–144. [CrossRef]
32. Solecka, I.; Krajewski, P.; Krzyżanek, A.; Garczyńska, A. Citizens' Perceptions of Landscape Changes and Their Driving Forces: Evidence from Poland. *Int. J. Environ. Res. Public Health* **2022**, *19*, 1688. [CrossRef]

33. Pimentel Walker, A.P. Self-help or public housing? Lessons from co-managed slum upgrading via participatory budget. *Habitat Int.* **2016**, *55*, 58–66. [CrossRef]
34. Środa-Murawska, S.; Dąbrowski, L.S.; Smoliński, P. When dreams come true-urban land use and management trends desired by residents and participatory budgeting-a case study in Toruń. *Urban Dev. Issues* **2018**, *60*, 31–42. [CrossRef]
35. Sołkiewicz-Kos, N.; Tubielewicz-Michalczuk, M. Renovation and Adaptation of a Historic Modernist Facility to Contemporary Functional and Utility Needs. *Int. J. Conserv. Sci.* **2025**, *16*, 1111–1128. [CrossRef]
36. Nemec, J.; Špaček, D.; de Vries, M.S. Unraveled Practices of Participatory Budgeting in European Democracies. In *International Trends in Participatory Budgeting: Between Trivial Pursuits and Best Practices*; De Vries, M.S., Nemec, J., Špaček, D., Eds.; Palgrave Macmillan: Cham, Switzerland, 2022; pp. 287–313. [CrossRef]
37. Touchton, M.; Wampler, B. Improving Social Well-Being Through New Democratic Institutions. *Comp. Political Stud.* **2014**, *47*, 1442–1469. [CrossRef]
38. Schneider, S.H.; Busse, S. Participatory Budgeting in Germany—A Review of Empirical Findings. *Int. J. Public Adm.* **2018**, *42*, 259–273. [CrossRef]
39. Oh, Y.; Seong-ho, J.; Heontae, S. A Strategy for a Sustainable Local Government: Are Participatory Governments More Efficient, Effective, and Equitable in the Budget Process? *Sustainability* **2019**, *11*, 5312. [CrossRef]
40. Falanga, R.; Verheij, J.; Bina, O. Green(er) Cities and Their Citizens: Insights from the Participatory Budget of Lisbon. *Sustainability* **2021**, *13*, 8243. [CrossRef]
41. Fu, K.-J. Translating participatory budgeting into an administrative system: The case of Taipei City. *J. Asian Public Policy* **2021**, *16*, 312–331. [CrossRef]
42. Sobis, I. Participatory Budgeting in Sweden. In *International Trends in Participatory Budgeting: Between Trivial Pursuits and Best Practices*; De Vries, M.S., Nemec, J., Špaček, D., Eds.; Palgrave Macmillan: Cham, Switzerland, 2022; pp. 63–87. [CrossRef]
43. Dybał, M. Participatory budgeting: Is it a useful tool towards co-creation of public services? *Ekonomia—Wroclaw Econ. Rev.* **2022**, *28*, 43–58. [CrossRef]
44. Szczepańska, A.; Zagroba, M.M.; Pietrzyk, K. Participatory Budgeting as a Method for Improving Public Spaces in Major Polish Cities. *Social Indic. Res.* **2021**, *162*, 231–252. [CrossRef]
45. Ahn, B.; Friesenecker, M.; Kazepov, Y.; Brandl, J. How Context Matters: Challenges of Localizing Participatory Budgeting for Climate Change Adaptation in Vienna. *Urban Plan.* **2023**, *8*, 399–413. [CrossRef]
46. Gonçalves, S. The Effects of Participatory Budgeting on Municipal Expenditures and Infant Mortality in Brazil. *World Dev.* **2014**, *53*, 94–110. [CrossRef]
47. Lara, F.L. Beyond Curitiba: The rise of a participatory model for urban intervention in Brazil. *URBAN Des. Int.* **2010**, *15*, 119–128. [CrossRef]
48. Tomášková, E.; Buzková, R. Participatory Budgeting in Brno—Inspiration for Other Cites? *Eur. Res. Stud. J.* **2020**, *23*, 758–770. [CrossRef]
49. Bednarska-Olejniczak, D.; Olejniczak, J.; Svobodová, L. Towards a Smart and Sustainable City with the Involvement of Public Participation—The Case of Wroclaw. *Sustainability* **2020**, *11*, 332. [CrossRef]
50. Chruściński, J.; Palińska, I.; Kazak, J. Participatory Budgeting in the Management of Public Space (Budżet obywatelski w gospodarowaniu przestrzeniami publicznymi). *Archit. Kraj.* **2014**, *3*, 56–67.
51. Cabannes, Y.; Ming, Z. Participatory budgeting at scale and bridging the rural—urban divide in Chengdu. *Environ. Urban.* **2013**, *26*, 257–275. [CrossRef]
52. Bernaciak, A.; Kopczyński, F. Participatory budgeting—An indicator of social activity of residents and a tool of environmental protection in Poland: Spatial diversity in the east/west configuration. *Ekon. i Sr.–Econ. Environ.* **2019**, *2*, 8–22. [CrossRef]
53. Martela, B. Wpływ budżetu obywatelskiego na przestrzeń polskich miast (The impact of participatory budgeting on urban space in Poland). *Urban Dev. Issues* **2020**, *66*, 173–182. [CrossRef]
54. Lafuente, R.; Ganuza, E.; Paneque, P. Social Resistance to the Hydrological Transition in Southern Spain: Public Support for the Building of New Reservoirs. *Resources* **2020**, *9*, 22. [CrossRef]
55. Madej, M. Wrocław Civic Budget. Social Innovation Adapted in Polish Local Government. *Pol. Political Sci. Yearb.* **2019**, *43*, 583–601. [CrossRef]
56. Tatarowska, E.; Furmankiewicz, M. "Wrocławski Budżet Obywatelski 2014" jako przestrzenna forma identyfikacji potrzeb inwestycyjnych w mieście ("Wroclaw Civil Budget 2014" as a Spatial Form of Investment Needs Identification in a City). *Pr. Nauk. Uniw. Ekon. we Wrocławiu* **2018**, *504*, 120–133. [CrossRef]
57. Swiader, M.; Szewranski, S.; Kazak, J. The local development index as a tool for the evaluation of socio-spatial inequities. In *Hradec Economic Days Vol. 6(1). Double-Blind Peer-Reviewed Proceedings of the International Scientific Conference Hradec Economic Days 2016: "Economic Development and Management of Regions", February 2nd and 3rd, 2016, Hradec Králové, Czech Republic*; Jedlička, P., Ed.; University of Hradec Králové: Hradec Králové, Czech Republic, 2016; pp. 1055–1066.

58. Solkiewicz-Kos, N. Shaping of public space in downtown area on the example of the City of Częstochowa. *IOP Conf. Ser. Mater. Sci. Eng.* **2019**, *603*, 032065. [CrossRef]
59. Ruksza, B.; Kapsa, J.; Narolska, M.; Urbańczyk, M. *Strategia Rozwoju Miasta Częstochowa 2030+*; Referat Strategii i Rozwoju, Wydział Funduszy Europejskich i Rozwoju Urzędu Miasta Częstochowy: Częstochowa, Poland, 2016.
60. Kozak, M.; Noworól, A.; Noworól, K.; Hałat, P.; Leżańska, M.; Szecówka, M.; Staronie, A.; Grzybowski, P.; Ruksza, B.; Kapsa, J.; et al. *Program Rewitalizacji dla Miasta Częstochowy na lata 2017–2023. Aktualizacja*; Agencja Rozwoju Regionalnego w Częstochowie, Gmina Miasto Częstochowa: Częstochowa, Poland, 2020.
61. Przybyła, K.; Kachniarz, M.; Hełdak, M.; Ramsey, D.; Przybyła, B. The influence of administrative status on the trajectory of socio-economic changes: A case study of Polish cities. *Morav. Geogr. Rep.* **2022**, *30*, 179–191. [CrossRef]
62. Grotkowska, M.; Kalisz, P.; Owczarek, A.; Przybyła, M.; Sobieraj, J.; Stachańczyk, A. *Sytuacja Demograficzna Województwa Śląskiego w 2024 r. (Demographic Situation of Śląskie Voivodship in 2024)*; Urząd Statystyczny w Katowicach (Statistical Office in Katowice): Katowice, Poland, 2025.
63. Ślusarczyk, B.; Herbuś, A. Citizens' budget and local community: An insight from Czestochowa. *J. Pengur.* **2020**, *58*, 159–167. [CrossRef]
64. Ławińska, O. Public Consultation in Management of Technical Infrastructure Investments in Czestochowa in 2010–2015. In Proceedings of the 1st International Conference Contemporary Issues in Theory and Practice of Management CITPM 2016, Częstochowa, Poland, 21–22 April 2016; Okręglicka, M., Gorzeń-Mitka, I., Lemańska-Majdzik, A., Sipa, M., Skibiński, A., Eds.; Wydawnictwo Wydziału Zarządzania Politechniki Częstochowskiej: Częstochowa, Poland, 2016; pp. 268–275.
65. Krawczyk, A. Doskonalenie modelu budżetu obywatelskiego na przykładzie miasta Częstochowy (Enhancing the model of participatory budget on the example of the city of Częstochowa). *Zarządzanie* **2016**, *22*, 242–256. [CrossRef]
66. Kurdyś-Kujawska, A.; Zawadzka, D.; Kwiatkowski, G.; Rosiński, R. Participatory Budgeting in Polish Cities: Funds' Allocation Mechanism. In *Finance and Sustainability. Proceedings from the Finance and Sustainability Conference, Wroclaw 2017*; Bem, A., Daszyńska-Żygadło, K., Hajdíková, T., Juhász, P., Eds.; Springer International Publishing AG: Cham, Switzerland, 2018; pp. 107–120. [CrossRef]
67. Crawford, L.; Pollack, J. Hard and soft projects: A framework for analysis. *Int. J. Proj. Manag.* **2004**, *22*, 645–653. [CrossRef]
68. Chovanecek, J.; Panek, J.; Frlickova, B. What are the driving factors for implementing participatory budgeting? A case study from Czechia. *Eur. Plan. Stud.* **2023**, *32*, 1165–1184. [CrossRef]
69. Kalinic, M.; Krisp, J.M. Kernel Density Estimation (KDE) vs. Hot-Spot Analysis-Detecting Criminal Hot Spots in the City of San Francisco. In *Geospatial Technologies for All: Short Papers, Posters and Poster, Proceedings of the 21th AGILE Conference on Geographic Information Science, Lund University, Lund, Sweden, 12–15 June 2018*; Mansourian, A., Pilesjö, P., Harrie, L., van Lammeren, R., Eds.; Lund University: Lund, Sweden, 2018; pp. 1–5.
70. Silverman, B.W. *Density Estimation for Statistics and Data Analysis*; Chapman and Hall: New York, NY, USA, 1986.
71. Yu, W.; Ai, T. The visualization and analysis of urban facility POIS using network kernel density estimation constrained by multi-factors. *Bol. Ciências Geodésicas* **2014**, *20*, 902–926. [CrossRef]
72. Yu, W.; Ai, T.; Shao, S. The analysis and delimitation of Central Business District using network kernel density estimation. *J. Transp. Geogr.* **2015**, *45*, 32–47. [CrossRef]
73. Bełej, M. Analysis of spatial distribution of touristic accommodation in Poland with the kernel density estimation of POIs. *Acta Sci. Pol. Adm. Locorum* **2021**, *20*, 159–171. [CrossRef]
74. Carlos, H.A.; Shi, X.; Sargent, J.; Tanski, S.; Berke, E.M. Density estimation and adaptive bandwidths: A primer for public health practitioners. *Int. J. Health Geogr.* **2010**, *9*, 39. [CrossRef]
75. Lin, Y.-P.; Chu, H.-J.; Wu, C.-F.; Chang, T.-K.; Chen, C.-Y. Hotspot Analysis of Spatial Environmental Pollutants Using Kernel Density Estimation and Geostatistical Techniques. *Int. J. Environ. Res. Public Health* **2011**, *8*, 75–88. [CrossRef] [PubMed]
76. Jędruch, M.; Furmankiewicz, M.; Kaczmarek, I. Spatial Analysis of Asymmetry in the Development of Tourism Infrastructure in the Borderlands: The Case of the Bystrzyckie and Orlickie Mountains. *ISPRS Int. J. Geo-Inf.* **2020**, *9*, 470. [CrossRef]
77. Nieto Masot, A.; Cárdenas Alonso, G. The Rural Development Policy in Extremadura (SW Spain): Spatial Location Analysis of LEADER Projects. *ISPRS Int. J. Geo-Inf.* **2018**, *7*, 76. [CrossRef]
78. Więckowski, M.; Łaska, W.; Timothy, D.J.; Michniak, D.; Cyargeenka, A. From closed border to open tourist space: Landscape changes along the Polish-Slovak border. *GeoJournal* **2024**, *89*, 33. [CrossRef]
79. ESRI. How Kernel Density Works (ArcMap 10.7). Available online: https://desktop.arcgis.com/en/arcmap/10.7/tools/spatial-analyst-toolbox/how-kernel-density-works.htm (accessed on 3 March 2022).
80. *Urban Atlas Land Cover/Land Use 2012 (Vector), Europe, 6-Yearly. (Dataset)*; European Environment Agency: Copenhagen, Denmark, DAT-190-en. [CrossRef]
81. Montero, E.; Van Wolvelaer, J.; Garzón, A. The European Urban Atlas. In *Land Use and Land Cover Mapping in Europe*; Manakos, I., Braun, M., Eds.; Springer: Dordrecht, The Netherlands, 2014; pp. 115–124. [CrossRef]

82. de Winter, J.C.; Gosling, S.D.; Potter, J. Comparing the Pearson and Spearman correlation coefficients across distributions and sample sizes: A tutorial using simulations and empirical data. *Psychol. Methods* **2016**, *21*, 273–290. [CrossRef]
83. Corder, G.W.; Foreman, D.I. *Nonparametric Statistics for Non-Statisticians: A Step-by-Step Approach*; John Wiley & Sons, Inc.: Hoboken, NJ, USA, 2009. [CrossRef]
84. Bonett, D.G.; Wright, T.A. Sample Size Requirements for Estimating Pearson, Kendall and Spearman Correlations. *Psychometrika* **2000**, *65*, 23–28. [CrossRef]
85. Skreczko, S.; Wolny, M. Wykorzystanie nieczynnych kamieniołomów na wybranych przykładach obszaru Jury Krakowsko-Częstochowskiej (Use of abandoned quarries on selected examples from Cracow-Czestochowa area). *Pr. Kom. Kraj. Kult.* **2014**, *26*, 67–78.
86. Szmytkie, R. Suburbanisation processes within and outside the city: The development of intra-urban suburbs in Wrocław, Poland. *Morav. Geogr. Rep.* **2021**, *29*, 149–165. [CrossRef]
87. Kryczka, P.; Szmytkie, R. In the face of planning system reform: An overview of land use planning in large post-socialist cities in Poland after 2003. *Cities* **2025**, *166*, 106205. [CrossRef]
88. Hewitt, R.; Hernandez-Jimenez, V. Devolved Regions, Fragmented Landscapes: The Struggle for Sustainability in Madrid. *Sustainability* **2010**, *2*, 1252–1281. [CrossRef]
89. Nobles, E.C.; Moore, K. Barriers to equitable greening in urban environmental policies. *J. Environ. Policy Plan.* **2024**, *26*, 388–401. [CrossRef]
90. Maksymiuk, G.; Kimic, K. 'Green Projects' in Participatory Budgets-inclusive initiatives for creating city's top quality public spaces. Warsaw case study. In *Inclusive Exclusive Cities. Book of Proceedings from SINERGI Project International Scientific Conference*; Marina, O., Armando, A., Eds.; City of Skopje: Skopje, North Macedonia, 2016; pp. 120–135.
91. Wysmułek, J.; Hełdak, M.; Kucher, A. The Analysis of Green Areas' Accessibility in Comparison with Statistical Data in Poland. *Int. J. Environ. Res. Public Health* **2020**, *17*, 4492. [CrossRef]
92. Warchalska-Troll, A. Nature conservation as a non-obvious context for the compact city idea. *Urban Dev. Issues* **2023**, *74*, 125–131. [CrossRef]
93. Smaniotto Costa, C.; García-Esparza, J.A.; Kimic, K. Participatory Budgeting and Placemaking: Concepts, Methods, and Practices. *Urban Plan.* **2024**, *9*, 7162. [CrossRef]
94. Warchalska-Troll, A.; Pistelok, P. (Eds.) *Zieleń w Centrach Polskich Miast. Stan, Funkcje i Wyzwania*; Instytut Rozwoju Miast i Regionów: Warszawa-Kraków, Poland, 2023. [CrossRef]
95. Zagórska, K.; Krajewski, P.; Sylla, M.; Lebiedzińska, M.; Kołodyńska, I.; Furmankiewicz, M.; Czajkowski, M. Willingness to pay for landscape benefits: Examining variation by landscape type in Lower Silesia, Poland. *Ecol. Indic.* **2025**, *171*, 113129. [CrossRef]
96. Solkiewicz-Kos, N. Problems of Balancing the Urban and Natural Environment on Example of the River Areas of the City of Czestochowa. *IOP Conf. Ser. Mater. Sci. Eng.* **2019**, *471*, 102022. [CrossRef]
97. Bernaciak, A.; Rzeńca, A.; Sobol, A. "New" public urban space: Citizens' initiatives in participatory budgeting in Katowice, Łódź and Poznań. *Misc. Geogr. Reg. Stud. Dev.* **2018**, *22*, 197–202. [CrossRef]
98. Budzik, G.; Sylla, M.; Kowalczyk, T. Understanding Urban Cooling of Blue–Green Infrastructure: A Review of Spatial Data and Sustainable Planning Optimization Methods for Mitigating Urban Heat Islands. *Sustainability* **2025**, *17*, 142. [CrossRef]
99. Grzyb, T. Recreational use of the urban riverscape: What brings people to the river? *Morav. Geogr. Rep.* **2024**, *32*, 14–25. [CrossRef]
100. Koszewska, J.; Kuzak, Ł. The problems of spatial planning and natural determinants of urban development-the Case of Powiśle in Warsaw. *J. Water Land Dev.* **2021**, *50*, 1–9. [CrossRef]
101. Majgier, L.; Badera, J.; Rahmonov, O. Kamieniołomy w województwie śląskim jako obiekty turystyczno-rekreacyjne na terenach uprzemysłowionych (Quarries in the Silesia Voivodship as touristic-recreational objects on industrial areas). *Probl. Ekol. Kraj. Probl. Landsc. Ecol.* **2010**, *27*, 267–275.
102. Piepiora, Z.; Bagińska, J.; Witkowski, K.; Mielniczuk, K.; Piepiora, P. The impact of organizing a big sport event on the structure and level of city's expenditures. *Arch. Budo Sci. Martial Arts Extrem. Sports* **2023**, *19*, 51–64.
103. Leśniewski, M.; Świąder, M.; Kaczmarek, I.; Castro, D.G.; Kamińska, J.A.; Pilawka, T.; Kazak, J.K. Towards Transit-Oriented Development for Sustainable Urban Mobility: Insights from a Central European City. *Geomat. Environ. Eng.* **2021**, *15*, 39–53. [CrossRef]
104. Milosavljević, M.; Spasenić, Ž.; Benković, S.; Dmitrović, V. Participatory Budgeting in Serbia: Lessons Learnt from Pilot Projects. *Lex Localis-J. Local Self-Gov.* **2020**, *18*, 999–1021. [CrossRef] [PubMed]
105. Błasik, M.; Wang, T.; Kazak, J.K. The Effectiveness of Master Plans: Case Studies of Biologically Active Areas in Suburban Zones. *Geomat. Environ. Eng.* **2022**, *16*, 27–40. [CrossRef]
106. Kamrowska-Zaluska, D. Participatory Budgeting in Poland—Missing Link in Urban Regeneration Process. *Procedia Eng.* **2016**, *161*, 1996–2000. [CrossRef]
107. Rogatka, K.; Kowalski, M.; Starczewski, T. Less important space? Spatial planning in small towns in Poland. *Land Use Policy* **2023**, *130*, 106674. [CrossRef]

108. Shybalkina, I. Toward a positive theory of public participation in government: Variations in New York City's participatory budgeting. *Public Adm.* **2021**, *100*, 841–858. [CrossRef]
109. Indrajit, A.; Van Loenen, B.; Van Oosterom, P. Assessing Spatial Information Themes in the Spatial Information Infrastructure for Participatory Urban Planning Monitoring: Indonesian Cities. *ISPRS Int. J. Geo-Inf.* **2019**, *8*, 305. [CrossRef]
110. Kenrick, D.T.; Griskevicius, V.; Neuberg, S.L.; Schaller, M. Renovating the Pyramid of Needs: Contemporary Extensions Built Upon Ancient Foundations. *Perspect. Psychol. Sci.* **2010**, *5*, 292–314. [CrossRef]
111. Janiszek, M.; Majorek, A. Waloryzacja przestrzeni miejskiej na podstawie projektów zrealizowanych w ramach budżetu obywatelskiego (Valorisation of Urban Space Based on Projects Implemented within the Civic Budget). *Biul. KPZK PAN* **2017**, *268*, 167–185. [CrossRef]
112. Foster, S.; Iaione, C. The City as a Commons. *Yale Law Policy Rev.* **2016**, *34*, 281–349. [CrossRef]
113. Babczuk, A.; Kachniarz, M.; Piepiora, Z. Work efficiency of local governments. In *Hradec Economic Days, Vol. 7(1), Double-Blind Peer-Reviewed Proceedings of the International Scientific Conference Hradec Economic Days 2017*; Jedlička, P., Marešová, P., Soukal, I., Eds.; University of Hradec Králové: Hradec Králové, Czech Republic, 2017; pp. 20–28.
114. Amitrano, D.; Di Martino, G.; Iodice, A.; Riccio, D.; Ruello, G. Urban Area Mapping Using Multitemporal SAR Images in Combination with Self-Organizing Map Clustering and Object-Based Image Analysis. *Remote Sens.* **2022**, *15*, 122. [CrossRef]

land

MDPI

Article

Participation and University Teaching in La Paz: An Urban Diagnosis Through a 'Map of Gender Insecurity'

Sara González Álvarez [1,*] and Isidoro Fasolino [2]

1 Department of Urban Planning and Architectural Representation, University of Valladolid, 47014 Valladolid, Spain
2 Department of Civil Engineering/DICIV, University of Salerno, 84084 Fisciano, Italy
* Correspondence: sara.gonzalez.alvarez@uva.es

Abstract: This article presents the results of a participatory urban diagnosis conducted in District 2 of La Paz, Bolivia, as part of an educational cooperation project aimed at exploring the spatial and symbolic dimensions of urban insecurity. Drawing on feminist and intersectional frameworks, this research combined participatory action methods, digital surveys, and collective mapping to identify patterns of fear and exclusion in public space. The analysis revealed significant disparities in how insecurity is perceived and experienced by different social groups—especially women, Indigenous peoples, and LGTBQ+ individuals—highlighting the importance of spatial configuration, social presence, and care infrastructure in shaping everyday urban life. The project also served as a pedagogical innovation, integrating architecture students into a process of civic engagement and co-production of knowledge. The resulting 'Map of Gender Insecurity' contributed to local planning efforts through the "Seguras, No Valientes" initiative. While the limited representation of some groups restricts statistical generalization, the approach offers a replicable model for linking research, education, and public action in pursuit of more inclusive and safer cities.

Keywords: participatory urbanism; gender-sensitive planning; intersectionality; spatial insecurity; co-production of knowledge; inter-university cooperation; Bolivia

1. Introduction

The 'Map of Gender Insecurity' project in District 2 of La Paz (Bolivia) was developed during the 2021/22 academic year as an inter-university teaching innovation and collaborative research initiative. It emerged within the framework of the Master's Degree in International Development Cooperation at the University of Valladolid (Spain), in close cooperation with three Bolivian universities: Universidad Católica Boliviana 'San Pablo,' Universidad Privada de Bolivia, and Universidad del Valle La Paz. The project stemmed from a shared conviction: that urban planning must place citizen participation—especially the voices of historically marginalized groups—at the core of any agenda for democratic, inclusive, and sustainable urban development.

Bolivia is notable in Latin America for having adopted an early and progressive legal framework that promotes citizen participation and decentralization. The 1994 Popular Participation Law (Ley de Participación Popular, No. 1551) enshrined the right to community involvement in urban planning, budgeting, and development oversight [1]. However, despite this legal foundation, participatory practices in urban development remain fragmented and are often hindered by structural barriers such as political clientelism, weak deliberative mechanisms, and enduring social inequalities [2–4]. As a result, participatory urbanism—with its emphasis on co-design, spatial justice, and the right to the city—has not yet been fully developed in practice [5,6]. These limitations are also reflected in architectural education, where academic curricula continue to focus primarily on formal design, leaving little space for critical pedagogy or the integration of citizen perspectives [7–9].

This disconnect between academic training and urban realities underscores the need to explore new teaching approaches rooted in community-based research and engagement. Citizen participation is now widely recognized as a cornerstone of contemporary urban planning, essential not only for enriching urban design but also for legitimizing planning processes and promoting social justice [10–13].

Crucially, any effort to democratize the production of urban space must be approached through a gender lens. A gender perspective brings visibility to diverse civic voices and challenges patriarchal structures that continue to shape urban development. It also validates everyday experience and emotional knowledge as critical sources of insight [14,15]. Within this framework, Feminist Participatory Action Research (FPAR) offers a particularly valuable approach, as it centers the lived experiences and epistemologies of women and marginalized groups while emphasizing relationality, reciprocity, and care in the research process [16].

In the Latin American context, the intersection of gender and citizen participation acquires a particularly political character. Deep-rooted structural inequalities linked to gender, class, and ethnicity continue to hinder women's participation in formal decision-making spaces, despite their long-standing leadership in defending territory and shaping daily urban life [17].

In response, the project presented in this article designed and implemented a participatory urban diagnosis focused on mapping perceptions of insecurity in District 2 of La Paz, with particular attention to the experiences of historically marginalized communities. Based on tools such as collective mapping, digital surveys, and field observation, the resulting Map of Gender Insecurity aimed to reveal both material deficiencies (e.g., infrastructure, lighting, and visibility) and symbolic or affective perceptions (e.g., fear, avoidance, and spatial stigma).

This study is guided by two core research questions: 1. How do different social groups—such as women, men, LGBTQ+ individuals, and indigenous people—experience and spatialize perceptions of insecurity in the urban space of District 2 in La Paz? 2. What spatial, symbolic, and environmental factors contribute to the construction of urban fear, and how can participatory diagnostics help uncover and address these dynamics?

Although the primary focus of this article is the participatory diagnosis itself, the initiative was embedded within a broader educational framework that significantly shaped its methodology. The project was implemented through a series of theory–practice workshops and field-based activities grounded in service-learning and project-based learning approaches. These pedagogical models emphasize experiential learning as a means of fostering critical thinking, civic engagement, and social responsibility [18–21]. In recent years, service-learning has become a widely adopted methodology for linking academic training with local social needs, particularly in contexts of urban inequality and informality [22,23].

While this educational dimension is not the object of analysis, it served as an enabling structure for the participatory process—contributing to its depth, legitimacy, and civic value. In this sense, the initiative brings together multiple dimensions—gender-sensitive urban analysis, academic formation, inter-institutional collaboration, and collective knowledge production—through which the findings aim to contribute to ongoing scholarly debates.

2. Project Context: Institutions, Territory, and Participants

The 'Map of Gender Insecurity' project was developed during the 2021–2022 academic year as part of the Master's Degree in International Development Cooperation at the University of Valladolid (Spain). It was conceived as an educational innovation that combined learning participatory methodologies in urban contexts with the practical experience of a collaborative territorial diagnosis in La Paz, Bolivia. The initiative was inspired by the "Mapa del Miedo-Valladolid" project, launched in 2017 by the Valladolid City Council in collaboration with the University of Valladolid and various social organizations [24]. That pioneering experience—focused on spatially visualizing perceived insecurity among women—served as both a conceptual and methodological reference for adapting participatory mapping tools to the Latin American context, with attention to local socio-cultural and urban specificities.

The project in La Paz was also integrated into the "Seguras, No Valientes" municipal initiative, funded by Swiss cooperation through the PADEM program ("Programa de Apoyo a la Democracia Municipal") and Solidar Suiza within the challenge "Mi municipio lo puede hacer major". This initiative seeks to promote more inclusive, accessible, and safe urban environments for women and girls by strengthening municipal policies with a gender perspective[1]. As such, the academic experience not only served pedagogical and training purposes but also contributed concrete outputs to the participatory design of gender-sensitive urban diagnostics—aligning with broader public policy efforts in the city.

From the outset, the project was conceived as a joint venture between European and Bolivian academic institutions, working in close coordination with public bodies, neighborhood associations, community-based organizations, and professionals. The goal was to create a socially engaged and context-sensitive mechanism for experiential learning, civic empowerment, and territorial co-diagnosis. This collaborative framework was not only logistical but methodological: a core element of the participatory design process.

The project's intervention site—District 2 of the Central Macro-District of La Paz—is a densely populated area (356 inhabitants per hectare) with high urban dynamism but some spatial inequalities (Figure 1). Covering 2.7 km^2, it includes neighborhoods with diverse morphologies, income levels, and access to services. While it concentrates important political, cultural, and financial infrastructure, it also exhibits significant deficits in safety, mobility, and the equitable distribution of public space. According to the 2018 Macro-District Report, the district had 31.8 m^2 of public space per capita and only 7.3 m^2 of green areas per resident. Over 333,000 daily trips were recorded, 80% of which used public transport, 15% by walking, and the rest by private vehicle or bicycle [26]. These urban conditions, combined with active community dynamics and a strong presence of informal settlements, made District 2 a particularly relevant site for applying participatory and feminist spatial methodologies.

The project brought together a multi-sectoral, inter-institutional alliance that included four universities (University of Valladolid, Universidad Católica Boliviana 'San Pablo', Universidad Privada de Bolivia, and Universidad del Valle La Paz), public officials from the Sub-Mayor's Office of the Central Macro-District, and a range of local organizations and professionals. This horizontal and distributed network was essential to building a dialogical and context-grounded process. While the University of Valladolid led the methodological design, Bolivian universities played a central role in adapting pedagogical strategies to the local reality. The implementation involved students, professors, researchers, gender experts, urban planners, and community facilitators working as part of interdisciplinary teams.

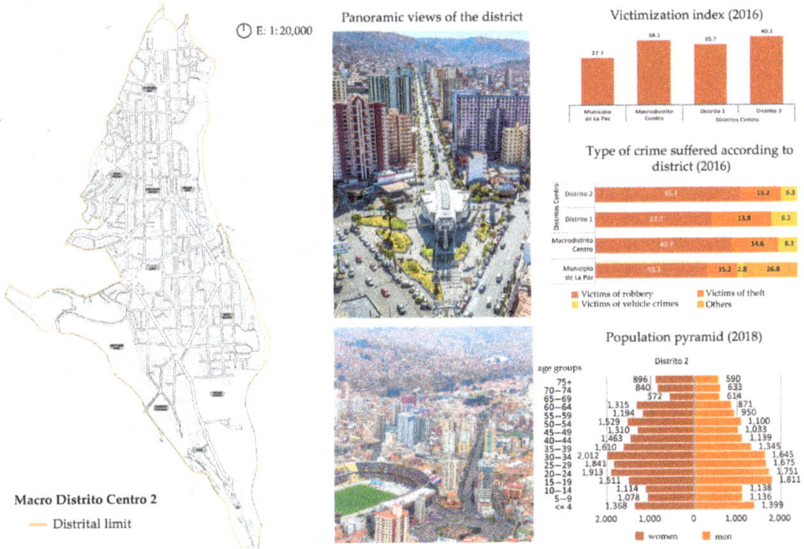

Figure 1. Spatial location, panoramic views, population pyramid (2018), and crime statistics (2016) of the project intervention area: District 2 of the city of La Paz (Bolivia). The authors from ref. [26].

Grassroots organizations with experience in feminist and participatory urbanism—such as "Pachakamani", "Paisaje Caminante", and "Movimiento Propacha"—were key allies in structuring and executing the fieldwork. Their collaboration enriched the critical dimension of the project and ensured meaningful community engagement. Local experts in architecture, planning, anthropology, and heritage contributed further rigor and cultural relevance to the intervention.

A technical partner, the "LabTecnosocial Collective", supported the project with the development of digital tools and the management of georeferenced participatory data. This technological layer allowed the translation of community knowledge into cartographic outputs usable by both citizens and local governments.

Far from being a one-directional transfer of knowledge, the project evolved as a space of co-construction, where multiple actors contributed from their respective trajectories, epistemologies, and practices. Together, they worked toward a shared goal: to strengthen civic capacities and rethink architectural education from an inclusive, participatory, and socially committed perspective.

3. Materials and Methods

3.1. Methodological Approach

This study adopts a participatory action research (PAR) approach, understood as a collaborative and transformative methodological strategy in which participants—in this case, students, instructors, professionals, and local actors—co-construct knowledge with the goal of critically engaging with reality [27–29]. This methodological choice reflects both the pedagogical nature of the project and its political and territorial dimensions, as it integrates training in architecture and urbanism with concrete practices of citizen participation.

Simultaneously, the project is grounded in the principles of project-based learning (PBL) and situated learning [30], which anchor theoretical university content in a real-life urban diagnosis experience in Municipal District 2 of La Paz, Bolivia. This methodological combination made it possible to integrate the educational dimension, inter-institutional collaboration, and participatory territorial action into a collaborative, reflective, and transformative process [31,32].

3.2. Phases and Analytical Instruments

The implementation of the participatory urban diagnosis was structured into three interrelated methodological phases: (1) theoretical–practical training; (2) participatory fieldwork; and (3) systematization and feedback. Although proposed in a logical and chronological sequence, in practice, the process was dynamic, marked by constant feedback between territorial experience, collective learning, and methodological adjustments made to adapt to contextual conditions. Flexibility and adaptability were therefore key operational principles, in line with the participatory and experimental nature of the project.

3.2.1. Phase I: Theoretical–Practical Training Workshops

The first stage of the project focused on the methodological and conceptual preparation of the participating team, particularly the students. An interdisciplinary training program was designed to strengthen analytical, technical, and ethical capacities necessary to intervene in complex urban contexts from a critical, contextualized, and gender-sensitive perspective. The group consisted of 30 students: 10 from U.C.B, 10 from the U.V, and 10 from U.P.B.

The training involved four intensive, in-person, and participatory workshop sessions held at Univalle's facilities, which provided the best overall accessibility. These sessions combined theoretical lectures, experiential exercises, analysis of previous experiences, and dialogue with institutional actors (Figure 2).

Figure 2. Training workshop sessions on citizen participation for the 'Map of Gender Insecurity' project at Univalle facilities in La Paz, Bolivia. The authors.

The first session centered on introspection and collective memory, based on the assumption that any territorial intervention must begin with a reflective understanding of one's own positionality. Through sensory exercises, personal memory evocation, and group sharing, participants explored how their trajectories, imaginaries, and emotions shape their view of urban space. This introspective approach was key to fostering openness and empathy in the fieldwork.

The second session addressed gender-sensitive urbanism, a central focus of the project. Theoretical inputs on spatial inequalities, symbolic violence, and inclusive urban design were combined with group exercises to identify potential insecurity points based on women's everyday experiences. This session directly bridged to the 'Map of Gender Insecurity' methodology, later applied in District 2.

The third session was devoted to the principles of citizen participation and training in social research tools. From a critical perspective, students explored the ethical and methodological challenges of working with communities, analyzed real cases, and developed practical skills in tools such as surveys, interviews, and stakeholder sociograms. This session solidified the understanding of urban diagnosis as a collective process, repositioning the role of the researcher as a facilitator.

The fourth session had a more institutional and political character. A dialogue was held with the Citizen Security Department of the Municipal Government, attended also by neighborhood representatives from District 2. This exchange was key to understanding the official framework of urban security management and comparing it with community perceptions and demands. It opened a communication channel that remained active throughout the rest of the project.

Overall, these training workshops were not only instrumental in preparing participants but also fundamental for building a shared ethics, trust, and reference frameworks—essential conditions for carrying out a truly participatory and transformative territorial process.

3.2.2. Phase II: Fieldwork and Participatory Process

The second phase involved the territorial implementation of the participatory urban diagnosis in District 2 of La Paz. This stage combined qualitative research techniques with participatory tools and was carried out through exploratory walks, interviews, and collaborative mapping exercises (Figure 3).

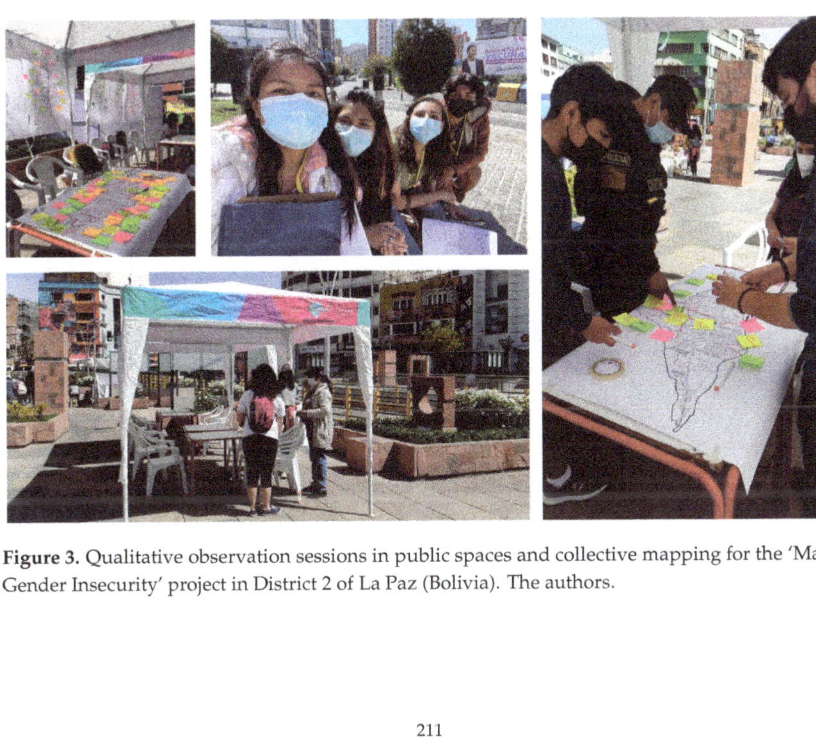

Figure 3. Qualitative observation sessions in public spaces and collective mapping for the 'Map of Gender Insecurity' project in District 2 of La Paz (Bolivia). The authors.

The fieldwork began with a direct observation and technical recording session aimed at producing an initial environmental and functional diagnosis. Six mixed teams composed of students from the three universities systematically surveyed different sectors of District 2 using a previously designed observation matrix. This tool considered variables such as type and use of public spaces, accessibility conditions, lighting, urban infrastructure, environmental quality, pedestrian flows, street furniture, and signage. Emphasis was placed on morphological configurations and spatial occupation patterns, as well as the identification of potentially problematic or underused areas. This observation was complemented by informal interviews and short walks informed by spontaneous interactions with users of public spaces. In parallel, a pilot test was conducted for the digital survey tool.

The two collective mapping sessions formed the methodological core of the fieldwork. Public tents were set up, displaying schematic maps of the district (using the 'Big Map' technique) to collaboratively identify problems, resources, and key urban relationships. Residents used drawings, keywords, symbols, and color markers to locate problematic zones, reference points, daily routes, and valued or avoided spaces. This exercise not only enabled the expression of complex perceptions in an accessible and visual format but also encouraged collective dialogue, shared territorial narratives, and critical reading of the urban environment, especially in terms of safety, accessibility, mobility, and care.

Another key tool was the digital survey, collectively designed by the academic team and administered semi-structurally by the students themselves (https://ee.kobotoolbox.org/x/fIuGiaMd (accessed on 19 August 2025)). The survey collected data on space usage, mobility, access to services, and perceptions of neighborhood centralities. It included both closed and open-ended questions, allowing for the combination of quantifiable data with rich qualitative insights, particularly regarding emotions, perceptions, and narratives linked to the territory. One of the tool's strengths was the inclusion of gender-sensitive questions, which revealed distinct patterns in how women, men, and gender-diverse individuals experience public space. Designed for mobile devices using accessible platforms and georeferenced maps, the survey ensured portability, automated systematization, and streamlined analysis. Acknowledging the technological limitations in some areas, a semi-assisted format was adopted, whereby trained students guided participants through the process in person. This hybrid format—digital and face-to-face—ensured accessibility and clarity, while strengthening trust and interpersonal connection during data collection.

The selection of respondents followed a non-probabilistic, purposive sampling strategy, focused on obtaining a diverse range of perspectives across gender, age, and socio-spatial backgrounds. Participants were recruited during collective mapping activities, through neighborhood organizations, and in strategic public spaces across District 2. This approach prioritized inclusivity and accessibility, particularly seeking to amplify the voices of women, gender-diverse individuals, and elderly residents—groups often underrepresented in formal participatory processes. While this method enriched the data with contextual and situated knowledge, it also introduced certain limitations in terms of representativeness, which are acknowledged in the discussion section. Nonetheless, the goal was not statistical generalization but a deeper understanding of the lived experiences of insecurity and urban space through participatory engagement.

3.2.3. Phase III: Systematization, Analysis, and Feedback

The final phase focused on organizing, analyzing, and communicating the data collected through the participatory process, and returning the results to both community participants and institutional stakeholders (Figure 4). The systematization combined qualitative, visual, and statistical approaches, organizing the data into thematic categories that reflected key spatial and social issues: insecurity, spatial fragmentation, lack of services, and the absence of community meeting spaces.

Figure 4. Data recording and systematization sessions for the creation of synthesis maps for the 'Map of Gender Insecurity' project in District 2 of La Paz, Bolivia. The authors.

A set of synthesis maps was created, integrating the outcomes of the collaborative mapping workshops, digital surveys, and field records. These maps aimed to represent not only physical conditions but also symbolic, emotional, and narrative dimensions of urban insecurity. The maps were complemented by visual and narrative documents that illustrated the main findings. Importantly, these were not conceived as definitive outputs but rather as open tools for discussion, validation, and shared planning.

The final 'Map of Gender Insecurity' is primarily based on the results of the participatory mapping sessions, during which residents collectively identified key danger zones, underused spaces, and symbolic territories. However, georeferenced data from the digital survey were also used to refine the spatial interpretation of recurrent concerns and validate overlapping perceptions. Although the survey data allowed spatial tagging of responses, the cartographic synthesis emphasized the value of collective, place-based narratives over purely quantitative distribution.

This hybrid approach—bridging participatory cartography and survey-based spatial data—prioritized local knowledge, intersubjective experience, and collaborative meaning-making. It contributed to constructing a multi-layered territorial reading, which informed both the academic analysis and institutional dialogue.

Ultimately, the participatory diagnosis of District 2 laid the groundwork for a subsequent Action Plan to be developed through the "Seguras, No Valientes" initiative, funded by Swiss cooperation via PADEM under the challenge "Mi Municipio lo Puede Hacer Mejor".

In terms of quantitative analysis, the digital survey data were first processed using descriptive statistical methods to identify overall patterns of insecurity perception, violence types, and spatial conditions. Responses were subsequently disaggregated by key identity variables—gender, sexual orientation, and Indigenous identification—following an intersectional approach. This disaggregated analysis allowed for the comparison of trends across different social groups, revealing differentiated risk factors and spatial experiences. Results were expressed in terms of absolute frequencies and relative percentages and were later cross-referenced with qualitative insights from the participatory mapping and workshop sessions. While some subgroups had limited representation, their inclusion contributed to a broader, politically informed interpretation of urban fear and spatial injustice.

4. Results

4.1. Participatory Urban Diagnosis: Factors of Insecurity and Territorial Perception

The diagnosis process carried out in District 2 of the Central Macro-District of La Paz enabled the collection, comparison, and analysis of a wide variety of quantitative and qualitative data. This was achieved through the implementation of an open digital survey, participatory workshops, collective mapping exercises, technical walks, and community feedback sessions. The use of this range of data collection and analysis tools allowed not only for the identification of conflictive zones but also for an understanding of the material, social, and symbolic conditions that shape such perceptions.

A total of 364 valid surveys were processed, 197 of which corresponded to residents or frequent visitors of the district. This empirical base served to analyze how perceptions of insecurity are constructed in urban space. Regarding the sociodemographic characteristics of the respondents (Table 1), most were young adults (average age: 34.9 years), with a predominance of women (59%). In terms of occupation, most were either employed (44%) or students (40%), reinforcing the idea that study and work environments within the district are significant in shaping perceptions of safety.

Table 1. Sociodemographic characteristics of survey respondents (n = 197). The authors.

	Category	Frequency	%
Gender	Female	117	59.4
	Male	80	40.6
	Other/Prefer not to say	0	0.0
Age	Under 18 years	10	5.1
	18–25 years	78	39.6
	26–40 years	38	19.3
	41–60 years	61	31.0
	Over 60 years	10	5.1
Occupation	Student	76	38.6
	Employed	87	44.2
	Unemployed	10	5.1
	Unpaid care work/Homemaker	9	4.6
	Retired	8	4.1
	Other/Prefer not to say	7	3.6
Indigenous Identity	Identifies as Indigenous—Aymara	10	5.1
	Identifies as Indigenous—Quechua	1	0.5
	Identifies as Indigenous—Other	1	0.5
	Does not identify as Indigenous	185	93.9
Sexual orientation	Heterosexual	178	90.4
	LGTBI+ community	11	5.6
	Other	4	2.0
	Prefer not to say	4	2.0

From a quantitative standpoint, one of the most consistent findings was that 59% of respondents reported feeling unsafe when moving through District 2. The most dangerous times were reported as nighttime and early morning, followed by late afternoon to evening—periods typically associated with a lack of lighting, reduced street activity, and visual barriers. The main factors identified as generators of insecurity included, firstly, the absence of commercial or social activity (mentioned by 27%), followed by poor public lighting (23%), and the presence of physical or visual barriers such as walls, dumpsters, or enclosed facilities (12%). Other recurring issues included deteriorated sidewalks and roads, a lack of urban signage, and the presence of homeless individuals or people with substance abuse problems. While these elements may not be directly linked to crime, they significantly affect subjective perceptions of safety (Figure 5).

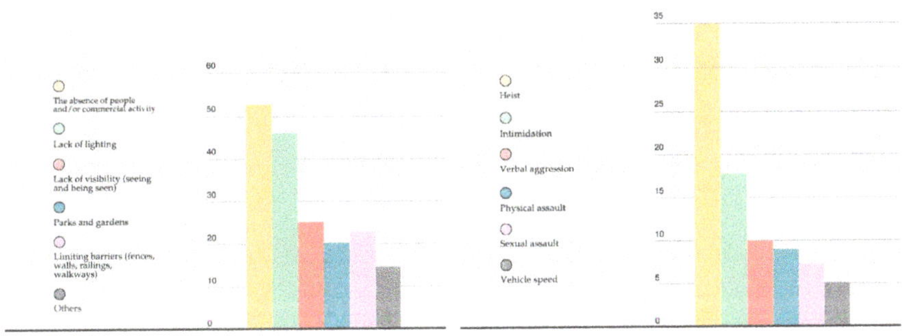

(**a**) (**b**)

Figure 5. (**a**) Distribution of perceived causes of insecurity in District 2 (La Paz) and (**b**) types of violence experienced in District 2 (La Paz). The authors.

Regarding perceived violence, the main forms were theft (over 40%), intimidation (19%), verbal aggression, physical aggression, sexual harassment, and fear of crossing certain streets. Notably, 30 individuals reported having experienced robbery, intimidation, or assaults while alone in public spaces.

4.2. Identity-Based Patterns of Insecurity

While the representativeness of certain identity groups within the sample is low—or, in some cases, nonexistent—and therefore not statistically significant, their inclusion nonetheless offers a valuable approximation of broader patterns and emerging trends. The disaggregated analysis of the digital surveys revealed notable differences in the perception of insecurity among various social groups, particularly in relation to gender, sexual orientation, and Indigenous identity. This intersectional approach made it possible to highlight how diverse social identities not only face different risks in public space but also shape and experience insecurity in distinct ways.

When asked whether they generally felt unsafe moving through District 2, participants' responses revealed significant disparities. As shown in Table 2, 71.4% of women declared feeling unsafe, compared to 38.6% of men. The perception of insecurity was particularly acute among Indigenous women (83.3%), and participants who simultaneously identified as female, Indigenous, and LGBTQ+ (100%). Similarly, Indigenous men (80%) reported high insecurity rates, contrasting with lower figures among non-Indigenous men. These findings indicate that experiences of fear are not distributed uniformly but shaped by intersecting social identities that influence one's exposure to risk and access to safe public space.

Table 2. Distribution of reported feelings of insecurity in public space by gender, sexual orientation, and indigenous identity. The authors.

	Feelings of Insecurity	Total of Participants	% of Participants Feeling Insecure
Female	70	98	71.4%
Female + LGTBQ+	4	7	57.1%
Female + Indigenous	5	6	83.3%
Female + LGTBQ+ + Indigenous	1	1	100%
Female + Other/Prefer not to say	3	5	60%
Male	27	70	3.6%
Male + LGTBQ+	1	3	33.3%
Male + Indigenous	4	5	80%
Male + LGTBQ+ + Indigenous	-	-	-
Male + Other/Prefer not to say	1	2	50%

Beyond the general sense of insecurity, the survey explored what spatial and environmental conditions contribute most to the perception of unsafety (Table 3). Among female respondents, the most frequently cited factors were lack of public lighting (62.96%), absence of people or commercial activity (62.96%), and poor visibility (31.48%). These indicators were even more pronounced among women who identified as LGBTQ+ or Indigenous, for whom 100% highlighted the first two factors.

Table 3. Triggers and spatial conditions contributing to perceived insecurity by social group. The authors.

	Female	Female + LGTBQ+ or Other	Female + Indigenous	Male	Male + LGTBQ+ or Other	Male + Indigenous
Lack of public lighting	62.96%	100%	50%	41.18%	0	50%
Poor visibility (see and seeing)	31.48%	0	25%	23.53%	0	0
Absence of people or commercial activity	62.96%	100%	25%	64.71%	100%	0
Parks and gardens	27.78%	0	0	23.58%	0	0
Limiting barriers (fences, walls, walkways, etc.)	18.52%	0	0	11.76%		0
Others	• Lack of road safety for pedestrians • Presence of people with alcohol or drug problems • Existence of alleys • Absence of police presence • Lack of signage			• Presence of establishments selling alcohol • Presence of people with alcoholism or drug addiction problems • Presence of criminals		

Male respondents expressed similar concerns, though with lower frequencies. A total of 64.71% of men noted the absence of social activity as a key issue, while 41.18% cited insufficient lighting. Male Indigenous respondents mirrored the concerns of women regarding lighting and isolation.

These data show that insecurity is closely tied to the absence of social life and the physical discontinuities of the urban environment, confirming the value of a gender-sensitive reading of public space.

In addition to the primary spatial triggers—such as poor lighting, low visibility, or lack of commercial activity—several participants identified additional context-specific factors grouped under 'Other'. For women, these included the absence of pedestrian safety infrastructure, the presence of people with substance abuse issues, poor signage, narrow alleyways, and a lack of visible police presence. These factors reflect not only spatial deficiencies but also a sense of institutional neglect that heightens vulnerability.

Male respondents, in contrast, emphasized the presence of alcohol-selling establishments, public intoxication, and criminal activity, highlighting a perception of risk linked more to potential aggressors than to environmental neglect.

These qualitative inputs reveal that feelings of insecurity are shaped not just by physical space but also by symbolic cues, social abandonment, and institutional absence, all of which vary significantly across social groups.

The survey also examined the types of experiences or threats that contribute most to individual feelings of fear (Table 4). Among women, the most commonly reported concerns were robbery (39.62%), intimidation (20.75%), and sexual harassment (13.21%). These forms of aggression often occurred in poorly lit or unpopulated areas and were compounded by symbolic dimensions of exclusion or vulnerability.

Table 4. Perceived types of violence in public space disaggregated by gender, sexual orientation, and indigenous identity. The authors.

	Female	Female + LGTBQ+ or Other	Female + Indigenous	Male	Male + LGTBQ+ or Other	Male + Indigenous
Physical assault	11.32%	0	0	18.75%	0	0
Sexual harassment	13.21%	0	0	0	0	0
Verbal aggression	5.66%	0	75%	6.25%	0	50%
Crossing the street/run over	9.43%	0	0	0	0	0
Intimidation	20.75%	0	0	18.75%	0	0
Robbery	39.62%	100%	25%	56.25%	100%	100%

Men, in contrast, identified robbery (56.25%) and physical assault (18.75%) as the primary sources of fear. Indigenous respondents of both genders emphasized verbal aggression as a central issue, suggesting a layer of symbolic violence that accompanies their presence in urban space.

These patterns reinforce the idea that perceptions of insecurity stem from both material risks and subjective, embodied experiences, which vary depending on one's social position and identity.

Finally, participants were asked what urban features or policies might improve their sense of safety (Table 5). Women prioritized public lighting (81.63%), traffic-calming infrastructure (76.53%), and cleanliness and maintenance of public spaces (80.61%). Among Indigenous women, support for improvements was unanimous across almost all categories, especially for green areas, pedestrian crossings, and active commercial façades. Notably, both Indigenous women and men expressed a higher valuation of cultural heritage and symbolic urban elements—such as monuments, murals, or historical references—as factors that positively influence their perception of safety, suggesting a strong link between identity, memory, and spatial security.

Table 5. Urban features that could improve perceived safety by social group. The authors.

	Female	Female + LGTBQ+ or Other	Female + Indigenous	Male	Male + LGTBQ+ or Other	Male + Indigenous
Public Lighting	81.63%	70%	75%	77.14%	80%	40%
Traffic lights and traffic calming elements	76.53%	60%	100%	64.29%	60%	60%

Table 5. *Cont.*

	Female	Female + LGTBQ+ or Other	Female + Indigenous	Male	Male + LGTBQ+ or Other	Male + Indigenous
More space on sidewalks and active mobility lanes (bike and skates)	71.43%	40%	75%	64.29%	80%	80%
Zebra crossings nearby and with good visibility	71.43%	50%	100%	60%	60%	60%
Signage and orientation	66.33%	50%	75%	60%	80%	60%
Existence of green areas	53.06%	60%	100%	51.43%	60%	60%
Heritage and symbolic elements	42.86%	50%	75%	35.71%	40%	80%
Limiting barriers (fences, walls, railings, walkways, etc.)	44.90%	60%	50%	48.57%	60%	80%
Cleaning and maintenance of spaces and urban furniture elements	80.61%	80%	100%	71.43%	60%	60%
Existence of shops and activities on the ground floors	54.08%	70%	100%	65.71%	80%	20%
Existence of passive spaces (parking lots, garage entrances, alleys, etc.)	43.88%	60%	25%	54.29%	40%	40%

Men also cited lighting (77.14%) and the presence of ground-floor businesses (65.71%) as safety-enhancing features. Interestingly, respondents identifying as LGBTQ+ often expressed preferences similar to women, particularly regarding visibility and the presence of social activity.

These responses point to a shared demand for more active, accessible, and well-maintained public spaces while also showing how marginalized groups articulate additional priorities based on their everyday experiences and vulnerabilities.

4.3. Spatial Categorization and Participatory Cartography

On a qualitative level, results from collective mapping and participatory workshops revealed how physical conditions intersect with symbolic and emotional dimensions. The maps created during the sessions produced spatial categories grounded in everyday experience:

- 'Dark places': poorly lit, visually fragmented, with minimal activity or surveillance.
- 'Areas to pass through quickly' or 'dead streets': perceived as dangerous despite being central to daily routes.
- 'Dangerous corners': symbolic nodes of tension, often tied to specific memories or recurring incidents.

These descriptions point not only to objective deficiencies but also to a sensitive dimension of urban space, shaped by fear, memories of past incidents, or the circulation of shared narratives.

The participatory process enabled the development of several thematic maps, each focused on a specific aspect of insecurity:

- Physical characteristics map: identified areas with deteriorated infrastructure, low visibility, and environmental or building neglect. Poor sidewalk conditions, lack of lighting, and closed or underused spaces were common issues.
- Environmental conditions map: highlighted elements such as garbage accumulation, foul odors, overgrown vegetation, excessive noise, or visual pollution—seen as indicators of institutional neglect and urban risk.

- Mobility and routes map: revealed daily routes that are avoided, dangerous stairways, problematic intersections, and streets with heavy traffic. These spaces, often unavoidable for pedestrians, became actual 'insecurity bottlenecks.'
- Use and appropriation map: contrasted areas perceived as protected (e.g., sports fields, active plazas, street fairs, etc.) with spaces without clear function or closed off by institutions (e.g., police stations, hospitals, stadiums, etc.), which contribute to urban fragmentation and disconnection.

The collective work demonstrated that insecurity factors do not act in isolation but overlap. Poorly lit areas with visual barriers and no social presence, for example, were systematically identified as particularly problematic. This overlap allowed the creation of a map of unsafe zones based on the density of such factors—a key tool for prioritizing future interventions.

4.4. Synthesis Mapping of Urban Insecurity

The final product of the participatory diagnosis process was a collaboratively produced insecurity map of District 2, which synthesized the diverse findings from digital surveys, collective mapping, interviews, and territorial observations (Figure 6). This spatial representation does not claim objectivity or statistical precision but rather offers a situated and politically informed reading of urban space from the perspective of its inhabitants.

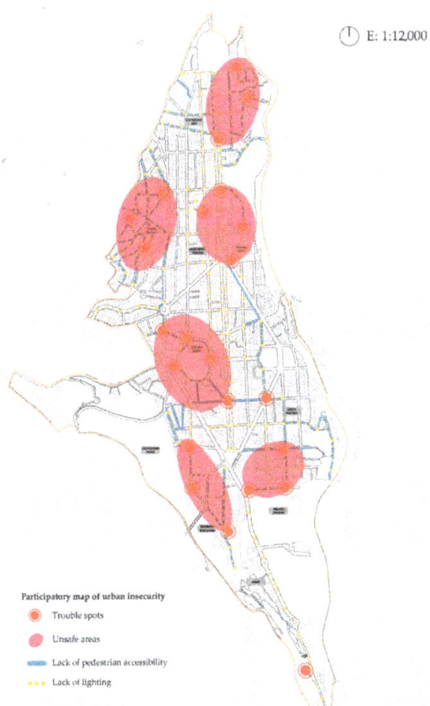

Figure 6. Participatory map of urban insecurity in Municipal District 2 of La Paz. Collective elaboration based on surveys, workshops, and collaborative mapping. 'Map of Gender Insecurity' Project, 2022.

Through the accumulation and layering of factors—both spatial and symbolic—participants were able to identify and collectively validate the areas perceived as most problematic. These included poorly lit streets, dead-end alleys, zones lacking commercial activity or pedestrian presence, and areas characterized by barriers such as walls, fences, or overgrown vegetation. These triggers of fear often overlapped with narratives of past incidents or collective imaginaries transmitted within the neighborhood.

The resulting map visualizes not only 'hotspots' of insecurity but also routes that are avoided, emotionally charged zones, and nodes of opportunity for reclaiming or transforming space. Critical points emerged in areas surrounding public infrastructure (such as stadiums, hospitals, or police buildings) that, despite their institutional presence, were perceived as opaque, inaccessible, or exclusionary.

The map is not intended as an objective or definitive representation of the territory but rather as a tool for social and political reading of urban space, bringing to light everyday violence and spatial inequalities from a citizen perspective. Its value lies in having been built through the mutual recognition of technical and popular knowledge, and in its usefulness as input for gender-sensitive, care-centered, and proximity-based urban planning.

5. Discussion: Urban Space, Gender, and Intersecting Vulnerabilities in Participatory Education

The findings of the participatory diagnosis in District 2 of La Paz confirm what feminist and critical urban studies have long emphasized: The experience of public space is not neutral but deeply shaped by gender, ethnicity, age, and sexual orientation, as well as by structural inequalities and symbolic hierarchies that manifest in spatial configurations and social norms [33–35]. Fear in the urban environment—especially among women, LGBTQ+ individuals, and Indigenous residents—emerges not only from direct experiences of violence but also from a collective perception of risk rooted in both material conditions and cultural constructions of territory.

The disaggregated analysis of survey data revealed significant variations in how different social groups perceive insecurity. While women overall reported higher levels of fear (71.4%), the rate increased substantially among Indigenous women (83.3%) and those identifying with more than one axis of marginalization (e.g., Indigenous + LGBTQ+, 100%). Among men, the general perception of insecurity was lower (38.6%), though Indigenous men also reported elevated levels (80%). These divergences underscore the need for urban analysis that recognizes and makes visible how multiple identities intersect to shape urban experience.

Moreover, the triggers of insecurity varied not only in intensity but also in nature across groups. While women highlighted poor lighting, the absence of commercial activity, and lack of visibility, Indigenous respondents emphasized symbolic aspects such as the absence of cultural markers or community reference points. LGBTQ+ participants, though fewer in number, emphasized social threats—such as discrimination, substance abuse, or exclusion—rather than just spatial factors. This heterogeneity calls for planning approaches that move beyond universalist notions of 'safety' and embrace differentiated and context-sensitive responses.

These insights also reinforce the idea that urban insecurity is not reducible to crime statistics or isolated incidents. As in previous studies [36,37], perceived safety is shaped by a constellation of spatial, emotional, and institutional factors. Physical neglect (e.g., broken sidewalks, garbage accumulation, etc.), lack of lighting and signage, and the presence of opaque or fragmented urban edges all compound the experience of fear. Conversely, areas perceived as safe tend to exhibit continuous activity, good maintenance, proximity to services, and symbolic or cultural visibility—especially valued by Indigenous respondents.

Methodologically, this project built upon the experience of the "Mapa del Miedo" (Fear Map) in Valladolid (Spain), where women collaboratively identified areas perceived as dangerous (https://www10.ava.es/portalva/apps/webappviewer/index.html?id=42

7d25467a764f2d9c0c287d3fc3b314 (accessed on 19 August 2025)). However, the La Paz case expanded this approach by integrating environmental, symbolic, and social dimensions and placing greater emphasis on intergenerational and intercultural dialogue. In this sense, it demonstrated the importance of embedding participatory urbanism in localized, reflexive, and inclusive frameworks.

Crucially, the process in La Paz was not only diagnostic but also pedagogical. Developed within a university cooperation project, it operated simultaneously as a research-action initiative and an educational experience. This dual nature made it possible to test teaching models grounded in territory-based learning, collaborative knowledge production, and situated research methods [38,39]. Students were not passive observers but active co-researchers—facilitating surveys, interpreting data, and engaging with community actors—thus embodying a role of facilitators rather than detached experts [40,41].

This contributes to a broader rethinking of architectural and urbanist education. In recent years, multiple experiences in Latin America and elsewhere have sought to bridge the gap between academic training and urban reality, especially by engaging students in processes of citizen participation, co-design, and social cartography [42,43]. This project aligns with those efforts, suggesting that pedagogical innovation and urban justice are mutually reinforcing. The development of a 'Map of Gender Insecurity' not only yielded relevant planning insights but also fostered a model of expanded education—capable of building critical awareness, civic agency, and collective imagination for safer and more inclusive cities.

6. Conclusions

The 'Map of Gender Insecurity' project in District 2 of La Paz offers valuable contributions to the fields of gender-sensitive urbanism, participatory diagnostics, and socially engaged pedagogy. The experience demonstrated how participatory research methods, combined with feminist and intersectional perspectives, can make visible everyday spatial injustices and contribute to more inclusive urban knowledge production.

One of the most relevant outcomes was the possibility to identify significant differences in how insecurity is perceived and experienced by distinct social groups. Women, Indigenous populations, and LGTBQ+ individuals reported higher levels of fear and emphasized distinct triggers—such as the absence of lighting, pedestrian infrastructure, or social presence. These divergences highlight the necessity of intersectional approaches in urban planning and security policies, which are often built upon neutral or universalized assumptions that overlook the differentiated vulnerabilities of marginalized communities.

From a community perspective, the participatory process fostered meaningful local engagement and empowered residents to critically reflect on their environments. The diagnosis served as both a technical and symbolic tool to articulate claims, identify priorities, and influence municipal planning through its integration into the "Seguras, No Valientes" program. By establishing a platform for dialogue between citizens, public officials, and academic actors, the project contributed to building civic capacities and strengthening the legitimacy of local governance structures.

From an academic standpoint, the initiative functioned as a testbed for pedagogical innovation. It helped bridge the gap between university education and real-world urban challenges by adopting a model that combines critical pedagogy, feminist research, and inter-institutional cooperation. This experience reinforced the value of transdisciplinary teaching that is embedded in territory, socially responsive, and oriented toward public transformation.

However, the study is not without limitations. Firstly, the representativeness of some social groups—especially Indigenous and LGTBQ+ subgroups—was low, limiting the statistical generalization of the findings. Nevertheless, the inclusion of these voices, even in smaller numbers, helped to shape a more nuanced and politically aware understanding of urban insecurity. Secondly, the specific institutional and territorial context of La Paz—particularly the openness of local actors and alignment with broader gender policies—played a key role in enabling the project's implementation. This context may not be replicable elsewhere without substantial adaptations. Another limitation concerns the final insecurity map (Figure 6), which does not disaggregate responses by gender or other identities. Although the digital survey responses were georeferenced, the map was primarily built upon collaborative mapping exercises conducted with community members. This choice privileged symbolic expression and collective interpretation over strictly quantitative spatial analysis. Future applications of the methodology could incorporate layers of geospatial disaggregation or visualization gradation to enrich technical accuracy without losing the participatory essence of the tool.

Despite these constraints, the project offers a replicable methodological model for integrating education, research, and civic action. It contributes to expanding current theoretical frameworks on urban fear, participatory planning, and feminist pedagogy. The combination of spatial analysis, intersectional data disaggregation, and community-based co-production of knowledge can inspire future initiatives seeking to democratize urban design and reshape architectural education.

Ultimately, the project highlights that perception of insecurity in the city is not only a matter of safety but also of visibility, voice, and power. Addressing it requires transforming both the urban environment and the ways we learn to engage with it.

Author Contributions: Conceptualization, S.G.Á.; methodology, S.G.Á. and I.F.; formal analysis, S.G.Á.; investigation, S.G.Á. and I.F.; resources, I.F.; writing—original draft preparation, S.G.Á.; writing—review and editing, S.G.Á. and I.F.; visualization, S.G.Á.; supervision, I.F. All authors have read and agreed to the published version of the manuscript.

Funding: The project that is presented by this research was funded by Municipal Democracy Support Program (PADEM) of Solidar Suiz. This research itself did not receive external funding.

Data Availability Statement: The raw data supporting the conclusions of this article will be made available by the authors upon request.

Acknowledgments: The authors would like to express their sincere gratitude to all the participants and institutions that made the development of this project possible. During the preparation of this manuscript, the author used ChatGPT (GPT-4o, OpenAI, 2025) for the purposes of structuring and organizing content. The model did not generate original ideas or perform data analysis. The authors have reviewed and edited the output and take full responsibility for the content of this publication.

Conflicts of Interest: The authors declare no conflicts of interest.

Notes

1 According to official data from the Gender Equality Observatory for Latin America and the Caribbean, Bolivia ranks fourth among the most dangerous countries in the region for women, with a total of 108 femicides and 46 infanticides, the Department of La Paz reporting the highest incidence of cases [25].

References

1. Congreso Nacional. Ley 1551 de 20 de abril de 1994 de Participación Popular. Bolivia, 1994. Available online: https://siteal.iiep.unesco.org/sites/default/files/sit_accion_files/12.ley_ndeg_1551_1994_1.pdf (accessed on 19 August 2025).
2. Abramo, P. La ciudad com-fusa: Mercado y producción de la estructura urbana en las grandes metrópolis latinoamericanas. *EURE* **2012**, *38*, 35–69. [CrossRef]
3. Valentin, L.; Cruz, L.D. Mecanismo de participación ciudadana en las políticas públicas en américa latina. *Rev. Políticas Públicas* **2017**, *10*, 30–54.

4. Ziccardi, A.; Di Virgilio, M.M. *Ciudades Latinoamericanas: La Cuestión Social Y la Gobernanza Local: Antología Esencial*; CLACSO: Buenos Aires, Argentina, 2020.
5. Arango, S.; Lopez, A. Diseño Urbano Participativo del Espacio Público. *Una Herram. De Apropiación social. Bitácora Urbano Territ.* **2021**, *31*, 13–26. [CrossRef]
6. Barrera, A. Innovación política y participación ciudadana tendencias democráticas en los gobiernos locales. In *Otro Desarrollo Urbano: Ciudad Incluyente, Justicia Social Y Gestión Democrática*; Poggiese, H.A., Cohen Egler, T.T., Ribeiro, A.C.T., Eds.; CLACSO: Buenos Aires, Argentina, 2009; pp. 105–116.
7. Urbanismo Ciudadano en América Latina: Superlibro de Acciones Cívicas Para la Transformación de Las Ciudades. Inter-American Development Bank Publications. Available online: https://publications.iadb.org/es/publications/spanish/viewer/Urbanismo-ciudadano-en-America-Latina-superlibro-de-acciones-civicas-para-la-transformacion-de-las-ciudades.pdf (accessed on 18 June 2025).
8. Alcántara-Ceballos, Y. La producción de conocimiento urbano-arquitectónico. ¿Una colonización cultural? *Rev. Bitácora Urbano Territ.* **2019**, *29*, 185–192. [CrossRef]
9. Valenzuela Montes, L.M.; Demerutis Arenas, J.Á. Editorial: La enseñanza del urbanismo y la planificación en Iberoamérica. *Rev. Bitácora Urbano Territ.* **2022**, *33*, 7–14. [CrossRef]
10. Paisaje, T. Escuchar y transformar la ciudad. In *Urbanismo Colaborativo Y Participación Ciudadana*; Libros de la Catarata: Madrid, Spain, 2019.
11. Borja, J.; Muxí, Z. *El Espacio Público: Ciudad Y Ciudadanía*; Electa: Barcelona, Spain, 2003.
12. Lefebvre, H. *El Derecho a la Ciudad*; Ediciones Península: Barcelona, Spain, 1978.
13. UN-Habitat. *New Urban Agenda*; Habitat II: Quito, Ecuador, 2016.
14. De Sánchez Madariaga, I. *Urbanismo Con Perspectiva de Género*; Instituto Andaluz de la mujer: Sevilla, Spain, 2004.
15. Castañeda, M.A.; Gimelfarb, L.S.; Cordero Robles, M.R.; Valletto, M.B.; Torres Pagnussat, L.; Millicay, S.Y. Urbanismo con perspectiva de género: Características y su desarrollo en América Latina y Argentina. *Estud. Hábitat* **2024**, *22*, e135. [CrossRef]
16. Goessling, K.P. Learning from feminist participatory action research: A framework for responsive and generative research practices with young people. *Action Res.* **2025**, *23*, 48–70. [CrossRef]
17. Falú, A. *Mujeres en la Ciudad: De Violencias Y Derechos*, 1st ed.; Ediciones SUR: Buenos Aires, Argentina, 2009.
18. Pretty, J.; Guijt, I.; Scoones, I.; Thomson, J. *Guía del Capacitador Para el Aprendizaje y Acción Participativa*; A Dirección de Programas de Investigación y Desarrollo (DPID)—Universidad Núr: Santa Cruz, Bolivia, 1997.
19. Freire, P. *Pedagogía Del Oprimido*; Siglo XXI Editores: Mexico City, México, 1970.
20. Giroux, H. *Critical Pedagogy. Handbuch Bildungs-und Erziehungssoziologie*; Springer: Berlin, Germany, 2021. [CrossRef]
21. Escaño, C. Educación hacker: Una pedagogía crítica (inter)creativa para los comunes del conocimiento. In *La Otra Educación. Pedagogías Críticas Para El Siglo XXI*; Aparici, R., Escaño, C., García, D., Coords., Eds.; Universidad Nacional de Educación a Distancia: Madrid, Spain, 2018; pp. 35–43.
22. Puig, J.M.; Martín, X.; Rubio, L. ¿Cómo evaluar proyectos de aprendizaje-servicio? *Voces Educ.* **2017**, *2*, 122–132.
23. Shanti, C.; Gerstenblatt, P.; Frisk, S. Putting the pieces together: Critical service learning and social work education. *Soc. Work Educ.* **2022**, *41*, 1643–1659. [CrossRef]
24. Ayuntamiento de Valladolid. Available online: https://www.valladolid.es/es/actualidad/noticias/abierta-consulta-ciudadana-actualizacion-ampliacion-mapa-mi (accessed on 10 June 2025).
25. Violencia Feminicida en Cifras: América Latina Y El Caribe, No 1. Observatorio de Igualdad de Género de América Latina Y El Caribe (Cepal), 2024. Available online: https://repositorio.cepal.org/server/api/core/bitstreams/69e978aa-ff89-4afb-afbb-e5d39904b9b1/content (accessed on 15 June 2025).
26. De Gobierno Autónomo La, P. *Cartillas Macrodistritales: Macrodistrito Centro*; Secretaría Municipal de Planificación para el Desarrollo: La Paz, Bolivia, 2018. Available online: https://sitservicios.lapaz.bo/cartillas/descargas/CARTILLA_CENTRO.pdf (accessed on 15 June 2025).
27. Ascher, F. *Los Nuevos Principios Del Urbanismo*; Alianza Editorial: Madrid, Spain, 2004.
28. Fals-Borda, O. The application of participatory action-research in Latin America. *Int. Sociol.* **1987**, *2*, 329–347. [CrossRef]
29. Kemmis, S.; McTaggart, R. Participatory Action Research: Communicative Action and the Public Sphere. In *The Sage Handbook of Qualitative Research*, 3rd ed.; Denzin, N.K., Lincoln, Y.S., Eds.; Sage Publications Ltd.: London, UK, 2005; pp. 559–603.
30. Ruiz Hidalgo, D. El aprendizaje basado en proyectos en la educación superior: Docencia y evaluación. In Proceedings of the VIII Jornadas de Doctorandos de la Universidad de Burgos [Recurso Electrónico], 2022, pp. 273–283. Available online: https://libros.ubu.es/servpubu-acceso-abierto/catalog/view/38/22/29 (accessed on 19 August 2025).
31. Bohórquez Rueda, J.A.; Moreno Plazas, J.A. Urbanismo participativo. Una aproximación pedagógica compleja. *Designia* **2021**, *9*, 103–131. [CrossRef]
32. Mancilla, I.; Habergger Lardoyet, S.A. Los proyectos colaborativos Universidad-territorio: Principios y prácticas de educación-investigación- acción necesarias para la transformación social. *Rev. De Educ. A Distancia* **2023**, *23*, 1–32. [CrossRef]

33. Col·lectiu Punt 6. *Urbanismo Feminista. Por Una Transformación Radical de Los Espacios de Vida*; Virus Editorial i Distribuïdora: Barcelona, Spain, 2019.
34. Kern, L. *Ciudad Feminista: La Lucha Por El Espacio en Un Mundo Diseñado Por Hombres*; Bellaterra Edicions: Manresa, Spain, 2021.
35. León-Escribano, C.R. Violencia y género en América Latina. *Pensam. Iberoam.* **2008**, *2*, 71–91.
36. Michaud, A.; Paquin, S. The security of urban women: Practice, research, and partnerships. *Inj. Prev.* **2002**, *8*, iv15–iv16. [CrossRef] [PubMed]
37. UN Women—Headquarters. *Safe Cities and Safe Public Spaces for Women and Girls. Global Flagship Initiative: International Compendium of Practices*; UN Women: New York, NY, USA, 2019. Available online: https://www.unwomen.org/en/digital-library/publications/2019/01/safe-cities-and-safe-public-spaces-international-compendium-of-practices (accessed on 15 May 2025).
38. Caro, M.A.S.; Díaz, M.P.B. Aprendizaje activo en Urbanismo: Aproximación global desde una formación local. In Proceedings of the Jornades Sobre Innovació Docent en Arquitectura Jida'19, Madrid, Spain, 14–15 November 2019; pp. 725–739. Available online: https://upcommons.upc.edu/server/api/core/bitstreams/ab7beac7-cd1d-42e1-99bb-4e1759eb1cb4/content (accessed on 19 August 2025).
39. Hernández, R.C.; López, E.R. Metodologías activas en el urbanismo: De las aulas universitarias a la intervención urbana. In Proceedings of the Jornades Sobre Innovació Docent en Arquitectura Jida'22, Reus, Spain, 17–18 November 2022; pp. 365–367. Available online: https://upcommons.upc.edu/server/api/core/bitstreams/268460d5-9d3f-4a68-a467-6c640d9099f3/content (accessed on 19 August 2025).
40. Alvarado, N.B.; Lobo, E.H.R.; Martínez, A.A.V.; Ochoa, F.M.; Buelvas Ede, J.H.; Méndez, J.C.P. El urbanista actual como agente dinamizador del territorio sostenible. *Cidades. Comunidades E Territ.* **2024**, *48*, 48. Available online: https://journals.openedition.org/cidades/8980 (accessed on 19 August 2025).
41. Ministry of Development. *Agenda Urbana Española*; Publication Center of Ministry of Development: Madrid, Spain, 2019. Available online: https://www.aue.gob.es/ (accessed on 19 June 2025).
42. Valdivia Loro, A. Enseñanza del urbanismo en Latinoamérica. Revisión narrativa en Scopus. *Bitácora Urbano Territ.* **2022**, *33*, 15–26. [CrossRef]
43. Aguilera-Martínez, F.A.; Vargas-Niño, P.A.; Serrano-Cruz, N.I.; Castellanos-Escobar, M.C. Estudio de los imaginarios sociales urbanos desde las prácticas pedagógicas. *Rev. Arquit.* **2015**, *17*, 3–5. [CrossRef]

MDPI AG
Grosspeteranlage 5
4052 Basel
Switzerland
Tel.: +41 61 683 77 34

Land Editorial Office
E-mail: land@mdpi.com
www.mdpi.com/journal/land

www.ingramcontent.com/pod-product-compliance
Lightning Source LLC
Chambersburg PA
CBHW040741250926
51850CB00007BB/1377
* 9 7 8 3 7 2 5 8 7 4 6 6 8 *